COMPARATIVE EMERGENCY MANAGEMENT

Examining Global and Regional Responses to Disasters

COMPARATIVE EMERGENCY MANAGEMENT

Examining Global and Regional Responses to Disasters

DeMond Shondell Miller • Jason David Rivera

CRC Press
Taylor & Francis Group
Boca Raton London New York

CRC Press is an imprint of the
Taylor & Francis Group, an **informa** business

CRC Press
Taylor & Francis Group
6000 Broken Sound Parkway NW, Suite 300
Boca Raton, FL 33487-2742

Printed in the United States of America on acid-free paper
10 9 8 7 6 5 4 3 2 1

International Standard Book Number: 978-1-4398-0491-9 (Hardback)

Library of Congress Cataloging-in-Publication Data

Comparative emergency management : examining global and regional responses to
disasters / edited by DeMond Shondell Miller and Jason David Rivera.
p. cm.
Includes bibliographical references and index.
ISBN 978-1-4398-0491-9
1. Emergency management. 2. Disaster relief. I. Miller, DeMond Shondell, 1973- II. Rivera, Jason David, 1983-

HV551.2.C6345 2011
363.34'8--dc22 2010048890

Visit the Taylor & Francis Web site at
http://www.taylorandfrancis.com

and the CRC Press Web site at
http://www.crcpress.com

Contents

SECTION II AFRICA AND THE MIDDLE EAST

SECTION III EUROPE

Foreword: Insights on the Role of Regional Collaboration

Disaster management has long been implemented and researched in silos and treated as national, internally handled issues. Events such as the Indian Ocean tsunami have started to make people realize that the silo approach is not working. Because hazards and disasters do not respect national borders, we must introduce bold initiatives that reduce barriers to building disaster-resistant regions. This is an increasingly difficult issue when you consider that 85% of natural hazards impact people in developing countries [United Nations International Strategy for Disaster Reduction (UNISDR) and Inter-Parliamentary Union Disaster Risk Reduction 2010]. As a result, the concepts of disaster reduction, risk reduction, and disaster mitigation have become popular topics. In the past ten years, direct disaster damage costs have risen from US$75.5 billion in the 1960s to almost 1 trillion dollars [Munich Re 2002; *Centre for Research on the Epidemiology of Disasters* (CRED) 2009].

Although disaster risk reduction may be the topic *du jour*, we as scholars, practitioners, community leaders, and citizens must strive to work together to be more inclusive in addressing issues ranging from poverty to sustainable development to climate change. This can be accomplished through many collaborative avenues that work to improve construction, education, policy changes, economic development and stability, and social development. However, these collaboratives among disaster researchers and practitioners must also involve stakeholders at all levels of government and community. Regionalism is one of the tools that can be used to make this happen.

The Importance of a Regional Collaborative Approach to Disaster Mitigation and Preparedness

Why is regional collaboration so important? It is so basic and straightforward. However, in reality, it is probably one of the most important questions we can ask. As the world continues to globalize and countries become more interdependent, it is essential that we start looking across and outside our borders (whatever they may be—city, district, state, country) to address the issues of risk.

We see it time and time again—disasters know no borders. The 2004 Indian Ocean tsunami directly impacted 13 countries on two continents and indirectly many, many more [AlertNet 2010]; Hurricane Katrina made a direct hit on four states in the United States but forced people in all 50 states to seek refuge; and the 2010 floods in China have impacted 28 provinces and more than 140 million people [*China Daily* 2010]. These events clearly illustrate the need for regional cooperation.

The key becomes defining "regional." Is it a series of states within a single country? Or a series of countries that border each other? Possibly a series of countries that share a fault line or a tsunami potential? It may be all of these things in a single country, depending on the risk and the hazards. Or does it even have to be defined? And who should define it? I would offer that the concept is self-defined in any way needed to help protect people and their livelihoods. We need to get beyond the confines of definitions and look at the outcomes. These outcomes are what will protect people. If different methods reduce disaster risk and help to protect people (and not put others at risk), then we should consider them.

To make a determination on what is "good" mitigation, there needs to be solid research from which to pull from. The disaster researchers of the world need to unite and start to work together. There needs to be integration across disciplines in the name of good science. Social scientists need to be work with physical and natural scientists to develop research that considers all aspects of an event—the geophysical nature of the hazard and risk as well as how people make decisions and the socioeconomic and cultural influences. This is simple enough but is not being practiced across the board. The Integrated Research on Disaster Risk* (IRDR) program is attempting to accomplish this by bringing together researchers from

* The IRDR program, established in Beijing, China, is cosponsored by the International Council for Science (ICSU), the International Social Sciences Council (ISSC), and the International Strategy for Disaster Reduction (ISDR). It is designed to address the impacts of disasters on regional and global scales and brings together the combined talents of the natural, socioeconomic, health, and engineering sciences from around the world. IRDR will focus on hazards related to geophysical, oceanographic, climate, and weather-triggered events—and even space weather and impact by near-Earth objects. The IRDR has three major research objectives: (1) to address the gaps in knowledge and methods for the effective identification of disaster risks; (2) to better understand just how decisions can contribute to hazards becoming disasters—or reduce their effects; and (3) to develop knowledge-based actions that will reduce risk and curb losses (for more information, see www.irdrinternational.org.

various disciplines to study disaster risk reduction problems from an integrated approach through both research and consultative forums. This allows countries and regions to learn from each other as well as modify successful programs to fit their social and political context.

The collaboration among researchers also allows comprehensive findings to be shared with practitioners who subsequently can implement merited approaches to help protect people. To better facilitate this process, the research community needs to do a better job of transferring their knowledge and findings to the practitioner community in language and formats that are more easily understood. This will allow for good mitigation and preparedness to be implemented, improved upon, and shared. We need to start doing a better work together to improve disaster risk reduction. This means across borders, disciplines, and sectors.

The Role That Regional Partnerships Play in Disaster Mitigation and Preparedness

Regional partnerships play a key role in disaster mitigation and preparedness. A flood can cross a border—why shouldn't the solution to the flood? The recovery and mitigation need to be done in cross-border collaboration as to not increase risk to those downstream or on the other side of the border. Communities are often living and working on both sides of a given border. Because few communities exist in isolation, mitigation efforts need to be designed in ways that do not view communities as such.

The goals must be self-defined. Sitting in my office in Beijing, I cannot tell you what a regional collaboration should look like for the Andean ridge countries of South America. They must define that within the context of their political and social needs. But what we need to be doing is sharing information. LA RED does just that in Latin America and works in conjunction with organizations throughout the region; other regional organizations include Centro de Coordinación para la Prevención de los Desastres Naturales en América Central (CEPREDENAC) and the Comité Andino para la Prevención y Atención de Desastres (CAPRADE). Each region has its own unique issues, concerns, and vulnerabilities to natural hazards; however, where stakeholders have similar goals and common interests, sharing what works and what does not work can become a common practice.

We need to stop arbitrarily trying to set goals and plans for other countries and recognize that knowledge comes from multiple sources. At the end of the day, what works well in Asia may not work in Africa or Europe. We must allow for unique differences while still practicing good mitigative techniques. We cannot "impose" our "best practices" on others. Personally, I do not like the term "best practices." How do we determine what is best? Everything is filtered through our personal cultural, educational, and political background. Who can single out any one practice and say it is better than something else? If it is helping to protect the people and the property from disaster, then *it is* a "best practice." There are many local and regional emergency management strategies, and all have been modified to fit the contexts and culture in which they are used. However, for the most part, we fail to disseminate and promote these successes. Disaster researchers and practitioners need to get better at capturing the successes and promoting them in all arenas.

Moreover, when we use the term *lessons learned*, it begs the question, have we really *learned* any lessons? If we have, then why are we still making the same mistakes? Why do losses from natural disasters continue to rise? Capturing what works (and does not work) as well as a better understanding *why* things are working (or not) is key. Through the FORIN* initiative, the IRDR is trying to do just that. Through various research techniques, scholars and practitioners will be able to trace back the origins of the disaster and the fundamental causes by probing more deeply into the complex and underlying causes of growing disaster losses. Such efforts will require a new commitment than previous research conceptualized, new institutional arrangements, and broader interdisciplinary teams [Burton 2010].

Sustaining Regional Collaborations

Regional collaborations can be sustained over time—if the stakeholders have invested the time to develop positive relationships. To just sign a partnership agreement is not saying much other than "we like how you think" or "that seems like a good idea that we can work with." It becomes a matter of taking steps beyond that to make the regional collaboration work. The agreement needs to be specific in what each organization can do and how they can work together. It is much like a seed a farmer plants. The seed will not grow without water, sunlight, and nurturing. Regional collaborations are the same way. The partnership agreements are nothing more than the seed planted in the ground.

It is important to recognize that many countries have acknowledged these issues and ideas and have begun to address them. Worldwide, more than 60 countries have developed National Platforms for Disaster Risk Reduction. Disaster risk reduction works. The Multi-Hazard Mitigation Council found that, for every dollar spent on mitigation, it saves four dollars next time. This is demonstrated by China's effort to reduce the impacts of flood. They have spent US$3.15 averting losses estimated at US$ 12 billion [Department of International Development (DFID) 2004].

As you read through the chapters looking at emergency management around the world, I encourage you to think about this fact and what could be done in terms of additional mitigation and development if money were better spent. According to the World Bank and the U.S. Geological Survey, if US$40 billion were invested in mitigation and preparedness, the worldwide economic losses from disasters could

* The Forensic Disaster Investigations are one of the research initiatives of the IRDR [see Burton 2010, p. 36].

be reduced by US$280 billion [DFID 2004]. Now think, if the $280 billion saved were reinvested in mitigation, it would amount to a savings of over $1,960 billion worldwide. The economics alone demonstrate that we must shift the focus from response to mitigation. This book provides insights into topics of disaster and hazard management that emphasize regional approaches that continue to be salient. Each section focuses on issues that influence the development of regional collaborations within different geographic regions of the world in order to illustrate the dynamics at play across different communities, cultures, nations, and international relations. I challenge each of you to reflect upon the examples provided throughout this book so that we, as a global community, may move away from a responsive culture of emergency management policies and practices and toward regional-collaborative (however you define *regional*) mitigation strategies and policies that reduce disaster and hazard risk.

Jane E. Rovins, PhD, CEM
Executive Director
Integrated Research on Disaster Risk
Integrated Research on Disaster Risk International Programme Office
Beijing, China

References

AlertNet. 2010. Deadliest tsunami on record. Retrieved from http://www.alertnet.org/db/crisisprofiles/SA_TID.htm?v=at_a_glance (accessed on July 24, 2010).

Burton, I. 2010. Forensic disaster investigation in depth: A new case study model. *Environment* 52:36–41.

Centre for Research on the Epidemiology of Disasters. 2009. *Annual Disaster Statistical Review 2008: The numbers and trends.* Geneva: Centre for Research on the Epidemiology of Disasters.

China Daily. 2010. Floods kill 1,072 in China this year. Retrieved from http://www.chinadaily.com.cn/china/2010floods/2010-08/04/content_11100122.htm (accessed September 10, 2010).

Department of International Development. 2004. *Disaster risk reduction: A development concern (A scoping study on links between disaster risk reduction, poverty and development).* London: ODI.

Disaster Risk Reduction, 2010. *An instrument for achieving the Millennium Development Goals.* Geneva: Inter-Parliamentary Union with United Nations International Strategy for Disaster Reduction.

Multi-Hazard Mitigation Council. 2002. *Natural hazard mitigation saves: An independent study to assess the future savings from mitigation activities.* Washington, DC: Multi-Hazard Mitigation Council.

Munich Re. 2002. *Topics: annual review, natural catastrophes 2002.* Munich: Munich Re United Nations International Strategy for Disaster Reduction and Inter-Parliamentary Union.

Preface

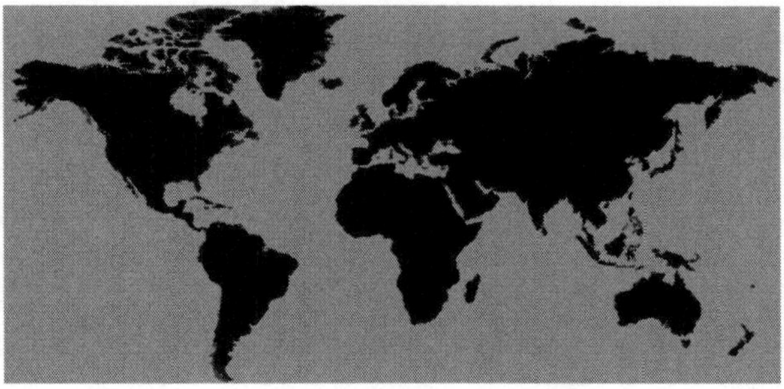

The twenty-first century has witnessed some of the most devastating disasters in recent history. More striking than in the past is the impact that these events have on the global human society as opposed to just local populations. Although local communities tend to suffer the brunt and initial consequences of disaster, many disasters today have spillover effects that are detrimental to regional government structures, nations, and even a host of nations that inhabit a geographical region. Researchers have acknowledged the fact that disasters, natural as well as technological, are occurring with greater frequency and magnitude throughout the world. In addition to these types of disaster, the world continues to be plagued by human-induced tragedies such as political and social violence, which are in some cases just as devastating to the natural and urban environment as severe natural phenomena. When we observe some of the most recent devastating events to impact entire regions, such as the Deepwater Horizon BP oil spill in the Gulf of Mexico, the Indian Ocean tsunami, Hurricane Katrina, the recent 7.0 magnitude earthquake in Haiti, and flooding in West Africa the notion of developing better emergency management policies, procedures, and cooperation becomes all the more relevant. In the aftermath of these events, this book serves as a call to action. It is a call for citizens of the twenty-first century to recognize and act to reduce regional infrastructure vulnerability while building secure interdependent networks sustained by trust among regional stakeholders evidenced by informal and formal agreements to work to resolve problems.

Although the costs of the these events are initially measured in lives taken, or the number of people missing, there are other less tangible impacts that have the ability to result in subsequent disasters for individual communities, subnational regions, nations, and even international relations. Changes in economy, international or internally displaced persons, political destabilization, violence, and a whole host of other issues affect the manner in which populations and societies recover from disaster, but also the success of that recovery. At the point where these external pressures begin to impact the recovery of societies, it is sometimes too late and difficult to make impactful policy changes that will have short-term advantageous results for disaster-affected populations. For this reason, changes must be made at the other end of the disaster management spectrum, during the disaster management, mitigation, and response-planning stages of emergency management. By placing more time and effort into emergency management planning, and focusing on mechanisms that can streamline and standardize emergency mitigation and response across political subdivisions, many of these subsequent disaster impacts can be avoided, thereby increasing the potential of local and regional societies to recover from disasters.

This volume contains the work of researchers investigating ways in which societies experiencing regional environmental threats have been forced to find new ways of regionally coping with vulnerabilities to disaster events, and have entered into new ways of developing emergency management policies at the subnational and international levels. We, along with the contributors, offer this research as an opportunity for thinking creatively in hopes that these lessons are integrated into new development projects in an equitable manner that not only is advantageous for specific geographic populations, but for all human society. Furthermore, we bring forth this work as a way to foster dialogue that will serve as a catalyst for the reduction of social vulnerability and build local and regional capacities to withstand environmental assaults. In doing so, we believe this book will contribute to the establishment of a subfield of regional disaster interdisciplinary study to combine sociology, public policy, economics, disaster studies, history, business, emergency management, critical infrastructure, tourism, and peace studies to address, among other things, social, financial, and physical vulnerabilities, risks, organizational resilience, war, and ethnic conflict as contributing externalities to regional security.

We express our sincere appreciation to all the contributors for their tireless work and continued research in the disaster research field. It is through this forum, *Comparative Emergency Management: Examining Global and Regional Responses to*

Disasters, that we share unique, independent yet related case studies that serve to inform, make important recommendations, and empower societies to become more prepared for the challenges in humanity's future.

Jason D. Rivera
DeMond S. Miller

The Editors

DeMond Shondell Miller is a professor of sociology and environmental studies and director of the Liberal Arts and Sciences Institute for Research and Community Service at Rowan University, Glassboro, New Jersey. He has worked as principal investigator to facilitate research projects involving natural and technological disasters, environmental issues, and community satisfaction. His primary area of specialization is environmental sociology (disaster studies and the study of the social construction of place), community development and community organizing, and social impact assessment. Dr. Miller has presented and published several professional papers; recent examples of such work can be found in *Space and Culture: An International Journal of Social Spaces, Journal of Black Studies*, *The Journal of Public Management and Social Policy*, *Sociological Spectrum*, and *The International Journal of Culture, Tourism and Hospitality Research*. Recently, he has contributed to several edited volumes including *Through the Eye of Katrina: Social Justice in the United States* and *The Sociology of Katrina: Perspectives on a Modern Catastrophe*; he is the coauthor of *Hurricane Katrina and the Redefinition of Landscape* with Jason D. Rivera (2007) and coeditor of *African American and Community Engagement in Higher Education: Community Service, Service Learning and Community-Based Research* with S. Evans, C. Taylor, and M. Dunlap (2008) and a coeditor of *How Ethnically Marginalized Americans Cope with Catastrophic Disasters: Studies in Suffering and Resiliency* (2010) and *Community Disaster Recovery and Resiliency: Exploring Global Opportunities and Challenges* (2010), with Jason D. Rivera. He is currently engaged in research on international environmental policy, coastal and maritime sustainable tourism, and the ongoing social impacts of climate change in the Mediterranean, Hurricane Katrina and the Deepwater Horizon Oil Spill.

Jason David Rivera is a research associate in the William J. Hughes Center for Public Policy at The Richard Stockton College of New Jersey. His research focuses on social vulnerability to natural and manmade disasters with an emphasis on minority experiences. Additionally, his research highlights institutional structures that have historically perpetuated social vulnerability within minority and low-income communities. His research findings have been incorporated into policy recommendations that make mitigation, response, and recovery more efficient and effective. Examples of his work can be found in the *Journal of Black Studies*, *Journal of Applied Security Research: Prevention and Response in Asset Protection*, *Terrorism and Violence*, the *Journal of Public Management and Social Policy*, *Sociological Spectrum*, *The Journal for the Study of Radicalism*, *Space and Culture*, *The Sociology of Katrina: Perspectives on a Modern Catastrophe*, *Through the Eye of the Storm: Social Justice in the United States*, and *Dangers in the Incommensurability of Globalization: Socio-Political Volatilities*. He is a coauthor of *Hurricane Katrina and the Redefinition of Landscape*, with DeMond S. Miller (2007) and a coeditor of *How Ethnically Marginalized Americans Cope with Catastrophic Disasters: Studies in Suffering and Resiliency* (2010) and *Community Disaster Recovery and Resiliency: Exploring Global Opportunities and Challenges* (2010), with DeMond S. Miller.

Contributors

Filip Aggestam
Institute of Forest, Environmental and Natural Resource Policy
University of Natural Resources and Life Sciences
Vienna, Austria

Alex Altshuler
Faculty of Health Sciences
Ben-Gurion University of the Negev
Beer-Sheva, Israel

Rémy Bossu
European-Mediterranean Seismological Center
Arpajon, France

Beryl Cheal
Disaster Training International
Seattle, Washington

Sotiris Chtouris
Department of Sociology
University of the Aegean
Mytilene, Greece

Virginia I. Clerveaux
Department of Civil and Environmental Engineering
Gunma University
Gunma, Japan

Yunfeng Deng
China Academy of Safety Science and Technology
Beijing, China

Ioannis M. Dokas
Cork Constraint Computation Center
University College Cork
Cork, Ireland

Scott Drimie
Johannesburg, South Africa

John Feehan
Cork Constraint Computation Center
University College Cork
Cork, Ireland

Franclin Foping
Cork Constraint Computation Center
University College Cork
Cork, Ireland

Stephen C. Fortier
School of Engineering and Applied Sciences
George Washington University
Washington, D.C.

Sébastien Gilles
European-Mediterranean Seismological Center
Arpajon, France

Sheryl L. Hendriks
Food Security
University of Pretoria
Pretoria, South Africa

Stephanie Hodge
Long Island City, New York

Mark Hutter
Department of Sociology
Rowan University
Glassboro, New Jersey

Syed Imran
Cork Constraint Computation Center
University College Cork
Cork, Ireland

Chuansheng Jiang
China Academy of Safety Science and Technology
Beijing, China

Benjamin W. Kelly
School of Engineering
University of Guelph
Guelph, Ontario, Canada

Sultan Khan
School of Sociology and Social Studies
University of KwaZulu-Natal
Durban, South Africa

Patrick Kilby
Applied Anthropology and Participatory Development (MAAPD) Program
Australian National University
Canberra, Australia

Gilles Mazet-Roux
European-Mediterranean Seismological Center
Arpajon, France

DeMond S. Miller
Liberal Arts and Sciences Institute for Research and Community Service
Rowan University
Glassboro, New Jersey

Anatoly Oleksiyenko
Division of Policy, Administration and Social Science
Faculty of Education
The University of Hong Kong
Hong Kong, China

Philip Duku Osei
Faculty of Social Sciences
University of the West Indies
Mona, Jamaica

J. Steven Picou
Department of Sociology
The University of South Alabama
Mobile, Alabama

Jason D. Rivera
William J. Hughes Center for Public Policy
The Richard Stockton College of New Jersey
Pomona, New Jersey

Fréderic Roussel
European-Mediterranean Seismological Centre
Arpajon, France

Jane E. Rovins
Integrated Research on Disaster Risk (IRDR)
IRDR International Programme Office
Beijing, China

Balfour Spence
Applied Disaster and Emergency Studies (ADES) Department
Brandon University
Brandon, Manitoba, Canada

Flora Tzelepoglou
Department of Sociology
University of the Aegean
Mytilene, Greece

Kathleen O. Vito
School of Health Sciences
The Richard Stockton College of New Jersey
Pomona, New Jersey

Kim Williamson
Australian National University
Canberra, Australia

Acknowledgments

Collaborating on this volume has been challenging, intellectually stimulating, and enjoyable. We are grateful to all of the staff at Taylor and Francis, namely Mr. Mark Listewnik and Ms. Jessica Vakili, Mr. Richard Tressider and Ms. Amor Nanas. Moreover, we thank the contributors, insight who shared their collective research and professional experiences.

We also thank Mr. John J. Reiser, Mr. Christopher Gonzalez, and Mrs. Marianne McCulley for their contributions to this collaborative effort.

We extend our appreciation and gratitude to Mrs. Cynthia Bush for all of her editorial assistance in making this book a reality.

Finally, we wish to express our special appreciation to Mrs. Andrea Rivera for the encouragement, editorial guidance, and continued support throughout the years to see this and other projects come to fruition.

Tragedy Has Brought Us Together: Responding to New and Emerging Regional Catastrophes

DeMond S. Miller and Jason D. Rivera

> Disasters are always more than physical catastrophes with social consequences; they are also the outcome of a complex and long-running interplay of power and geography that works to shield some people from the whims of nature or the negligence of dangerous industries while exposing others to the unwanted fallout of these acts of nature and man [Kroll-Smith 2001, p. 175].

Tragedy on a Regional Scale

The cataclysmic nature of Cyclone Nargis, the floods in Pakistan, the Indian Ocean tsunami, and Hurricane Katrina has redefined a "new normal" in regions and captured the attention of the entire global community. As disasters are becoming defining moments in modern societies, regional crises are becoming a more common threat to the stability of nations with the destabilization of entire regions hanging in the balance. The first decade of the new millennium closes and with it a heightened sense of awareness and understanding of how natural and built environments, ill-crafted social policies, poverty, and other social vulnerability factors in disaster-prone areas exacerbate the cumulative impact of crisis. In fact, the new normal [Pinkowski 2008], as we define this concept, can be viewed as a 1) complex series of rational choices, 2) a willingness to accept individual burdens of sheltering disaster risks, 3) and resources needed to shelter those risks within a socially-constructed structure of risk and vulnerability to disaster avoidance. This exploration seems increasingly necessary because disasters, as sociologically conceptualized, are increasing in the United States and elsewhere due to increased development of coastal or riverbank cities [Achenbach 2008], the parallel increase in population density in disaster-prone areas [Picou and Marshall 2007], and politically influenced decisions about building shortcuts in disaster-prone areas [Johnson 2008], and as society knowingly or unknowingly shoulders more of the burdens of disaster risks. Regional vulnerabilities from environmental degradation, climate change, human intervention, and security remain pressing issues for all of us and should be undertaken to reduce community risks in all phases: prevention, mitigation, preparedness, response, and recovery. Disaster-resilient regions first start with disaster-resilient communities that are empowered so that community members can cope with the adverse effects of multiple hazards, including earthquakes, wildfires, tsunamis, hurricanes, terrorist attacks, and other such catastrophes, that occur and send shock waves throughout a region.

Without adequate planning and collaborations, there is little, if anything, governments within a region can do to prevent the loss of life and property with the rapid onset of a disaster. New paradigms and strategies are needed to address the complex problems of jurisdictional boundaries, disparities among neighboring countries, and how social, cultural, and political factors influence how communities perceive and address hazards, risks, and vulnerabilities. Branscomb and Michel-Kerjan [2006] maintained that "[h]ow societies learn to work collectively to achieve the goal of safety and security and to sustain it in the long term is thus of prime importance" [p. 395]. A collective approach among stakeholders would require, at a minimum, mutual understanding and trust to make collaborations work [Gonzalez 2010]. An example of attempts to build mutual understanding and trust through regional collaboration by addressing longstanding national hostilities between neighbors became more apparent during the 2010 Haitian earthquake. The countries of Haiti and the Dominican Republic share the island of Hispaniola and they share a common set of vulnerabilities to risks; however, the Dominican Republic did not suffer extensive damage as did Haiti. Nevertheless, the Dominican Republic put aside decades of hostility and responded swiftly and generously to the crisis in Haiti [Margesson and Taft-Morales 2010]. Although the two countries have a long history of hostility toward one another, Presidents Preval and Fernandez have been working in recent years to bridge cultural, political, and social gaps toward having a more cooperative relationship; this has been reflected in the Dominican response of sending relief supplies and personnel. Furthermore, the Dominican Republic facilitated aid delivery through use of its airports, roads, and ports; it has stopped repatriation of undocumented Haitians and opened its border to injured Haitians, thousands of whom have been treated in both public and private hospitals [Margesson and Taft-Morales 2010].

The collaboration and actual work of responding and recovering from disasters is performed by government agencies, private sector companies, nonprofit organizations, and citizens [Stewart, Kolluru, and Smith 2009]. With proper planning, key collaborations among stakeholders can facilitate a rapid response in the midst of disasters. Because private corporations have extensive supply and network chains and local distribution systems already in place, a more rapid assessment of the extent and magnitude of the situation can be gathered. Recent examples of regional catastrophes of global significance include the 2004 Indian Ocean earthquake and tsunami,* Hurricane Katrina in 2005,† the 2010 Haitian earthquake,‡ and the Deepwater Horizon BP oil spill.§ Numerous other examples exist to illustrate the increasing role regional crises play in our lives. Each of the above disasters represents an unprecedented national, regional, and global response.

Because we live in an interconnected world of complex networks with linked economic, social, communication, and food distribution systems, any disruption in the local or regional infrastructure quickly affects others; when some are vulnerable, all are vulnerable in that the welfare and lives of significant portions of the population are in harm's way.¶ As these chapters suggest, opportunities for initial public–private collaborations can result in a more timely response at the local, state, and national levels to make the long and arduous process of restoring communities less difficult following a tragedy. Strategic approaches to enhancing institutional capacity (forged by partnerships among stakeholders) to implement

* The 2004 Indian Ocean earthquake and tsunami occurred on December 26, 2004 and represents the largest natural disaster in modern history. Caused by a 9.0 earthquake under the seabed of Sumatra, Indonesia, it claimed the lives of more than 227,000 people in 14 countries with economic losses estimated to be over 10 billion (Robinson and Jarvie 2008; World Bank 2005).

† Hurricane Katrina made its first landfall along the Gulf Coast region of the United States in the American state of Louisiana. Katrina's wind, storm surge, and subsequent flooding of New Orleans and other places along the Gulf Coast became known as the most costly disaster with over 1,836 dead and 135 missing (LA Department of Health and Hospitals, 2006).

‡ The largest earthquake ever recorded in Haiti devastated parts of the country, including the capital, on January 12, 2010. The quake, centered approximately 15 miles southwest of Port-au-Prince, had a magnitude of 7.0. A series of strong aftershocks have followed (U.S. Geological Survey 2010). The Haitian government estimated that 230,000 people died with another 1,000,000 homeless (Associated Press 2010).

§ The Deepwater Horizon oil spill, or BP Oil Spill, the largest accidental marine oil spill in the history of the petroleum industry is the result of the April 20, 2010 Deepwater Horizon drilling rig explosion releasing approximately 4.9 million barrels (780×10^3 m^3) of crude oil. It was estimated that 53,000 barrels per day (8400 m^3/d) were escaping from the well just before it was capped on July 15, 2010 (Robertson and Krauss 2010; Telegraph 2010; USA Today 2010).

¶ At the time of writing, Pakistan experienced the worse flooding, which began in July 2010 after record heavy monsoon rains. The Khyber Pakhtunkhwa province of Pakistan was worst affected. More than 1,600 people were killed, thousands were rendered homeless, and more than 20 million people are affected. See http://www.2010pakistanfloods.com/

planning and prevention strategies ultimately maximizes a community's ability to withstand mass emergencies. These partnerships serve as vital links to an efficient primary initial response, subsequent recovery efforts, and the initial stages of building a more resilient community that positions resources for current and future disasters.

Understanding the Region and the Importance of a Regional Approach

In an important lecture, Paasi [2000] called on everyone to rethink the concept of region and its significance. With the increased interactions and flow of information, capital, people, and information across geopolitical lines, Paasi maintained that we need to review our current conceptualizations of regions and regionalism. While "[c]oncomitantly the 1990s have witnessed a resurgence of regions and boundaries on the one hand in academic discourse and in concrete research, on the other hand, in everyday life struggles of various social groupings . . . [a]n important challenge in much of current research has been to reflect the emerging forms of governance in a situation where globalization and the apparent re-scaling of state governance are fusing our traditional thinking on spatial scales and categories" [Paasi 2000, p. 1]. Furthermore, Paasi [2000] maintained that with the evolution of the traditional notion of a region within a global context as a social construct and that "[s]pace and spatial patterns are not independent of social, cultural and natural processes but, as it has been strongly emphasized, space is not a causal power which would as such determine social processes, rather social (and cultural) and spatial are constituents and outcomes of each other" [p. 2].

To this end, this volume underscores the problems society faces when addressing disasters of regional significance. Also, through this volume, the authors maintain that the challenge is not to curb or lessen the need for local approaches to increase resiliency, but rather to seize the opportunities it provides. Furthermore, we recognize a variety of inequalities and note that "social vulnerability is partially the product of social inequalities. . . . however, it also includes place inequalities—those characteristics of communities and the built environment, such as the level of urbanization, growth rates, and economic vitality, that contribute to [i.e., constitute] the social vulnerability of places" [Cutter, Boruff, and Shirley 2003, p. 243; see also Iversen and Armstrong 2008] and placing an emphasis on the "need for strengthening community resilience, building public–private partnerships, reaching out to marginalized community residents and their trusted institutions, and developing consensus-based coordinating mechanisms at the interorganizational, community, and intergovernmental levels" [Tierney et al. 2006, p. 76; see also Iversen and Armstrong 2008]. For these reasons, we argue that there is a pressing need for engaging in regional collaborations built on local networks and multinational disaster concerns.

For instance, the sheer magnitude of the Indian Ocean tsunami involved several countries, all grappling with similar issues. The transboundary, multijurisdictional issues had to have an international cooperative focus that was grounded in not only local realities, but also regional interests that could restore stability through formal intergovernmental relations. Regional collaboration where two or more governments can take several forms and respond to several types of problems by experiences, strategies, and knowledge used in addressing similar challenges. It is important to share what strategies work in a way that produces evidence-based learning so that similar problems can be addressed within different regions that face similar challenges.

An example of a region-to-region collaboration is the United State's Gulf Coast region seeking the expertise of other areas around the world where coastal cities are below sea level, such as The Netherlands, after Hurricane Katrina. Other regions of the world have interregional strategies to cope with natural hazards by addressing gaps in institutional frameworks, risk identification, knowledge management, governance, and emergency response. To this end, the *African Regional Strategy for Disaster Risk Reduction (*ARSDRR*)** seeks to:

1) Increase political commitment to disaster risk reduction
2) Improve the identification and assessment of disaster risks
3) Enhance knowledge management of disaster risk reduction
4) Increase public awareness of disaster risk education
5) Improve governance of disaster risk reduction institutions
6) Integrate disaster risk reduction into emergency response management

The *African Regional Strategy for Disaster Risk Reduction* [2004] builds on existing risk reduction institutions and programs available in African countries and aims to reduce disaster risks in the development and application of policies, strategies, and practices that minimize vulnerabilities and prevent, mitigate, or prepare for the adverse impacts of hazards within the broad context of sustainable development. To do so, the *African Regional Strategy for Disaster Risk Reduction* also sets forth strategic directions to achieve the five objectives. For example, the strategic directions to increase political commitment to disaster risk reduction include strengthening lobbying and advocacy for political commitment, responsibility, and accountability, while increasing institutional frameworks for risk reduction. It also calls for the African continent and member countries within regions of the continent to recognize hazards and vulnerability factors as dynamic and a need for the improvement

* The Strategy was adopted by African Ministers at the 10th Meeting of the African Ministerial Conference on the Environment (AMCEN), June 26–30, 2004, and submitted to the AU Assembly Summit, where the Strategy was positively received by Heads of State at the 3rd Ordinary Session of the Assembly in Addis Ababa, Ethiopia, July 6–8, 2004, with a call to develop a Program of Action for its implementation.

in the quality of information about disaster risks. While improving the identification, assessment, and monitoring of hazards, vulnerabilities, and capacities, the ARSDRR seeks to formalize greater integration, coordination, and exchange of information among stakeholders, advance efforts to increase the degree to which academic institutions and research centers enhance knowledge management for disaster risk reduction are also being made.

International diplomacy (interregional environmental diplomacy) used to enhance regional resilience through governance involves institutionalizing risk reduction from the local level to the regional level in such a way that emphasizes decentralized power structures and more coordinated partnerships that involve adapted local-level risk management strategies based on local knowledge, evidence-based strategies, and best practices so that large bureaucracies do not hinder aid when it is needed most. For these situations, the development of a national platform for response and disaster management may not always be most beneficial. Rather, an increased international cooperation and coordination strategy may work best when regions suffer from the same catastrophic events. In this way, nations or states from around a region can harmonize policies, organizational agreements, and partnerships and increase public participation in the planning and implementation of risk reduction interventions.

Finally, when an unavoidable catastrophe strikes, regional emergency responses must be coordinated to limit the impacts of the disaster. To this end, emergency responders and regional leaders must be able to coordinate efforts in a manner that resources are not duplicated, and public–private partnerships can be forced to strengthen contingency planning and other preparedness measures in emergency management. For example, during the 2010 Haitian earthquake, Latin American countries responded to Haiti's crisis with immediate provision of emergency supplies, personnel, and pledges of financial and other assistance for its long-term recovery. Members of the Organization of American States (OAS) pledged humanitarian, financial, and other support to Haiti [Margesson and Taft-Morales 2010]. The OAS Group of Friends of Haiti met on January 14th to coordinate search and rescue efforts, prompt donations, and discuss ways to promote recovery. The 15-member Caribbean Community (CARICOM), of which Haiti is also a member, mobilized its disaster emergency response system to assist Haiti, and several members have sent emergency supplies or promised financial assistance. Additionally, the Caribbean Disaster Emergency Management Agency assembled a response team to assess conditions in Haiti as well [Caribbean Media Corporation 2010; also see Margesson and Taft-Morales 2010].

The Region as a Shared Risk and an Embedded Future

Presently we find evidence of the risks shouldered by the embedded social structures that exist within regions. In this work, each chapter may approach the conceptualization of region differently; however, they all stress the common interest in finding ways to reduce environmental risks and vulnerabilities to human-induced, natural, and technological disasters. With the number of vulnerabilities that exist (i.e., drought, global climate change, rapid spread of disease, shifts in the balance of threats where the poor shoulder the greater amount of environmental hazards, population growth, the rise of megacities and fast urbanization, scarcity of drinkable water, declines in food production, and acceleration of toxic warfare) and their policy implications, we need to recast our arguments in the terms of security—not only at the local and national levels, but also at the regional level. For example, with the onset of global climate change and intense urbanization, cities worldwide face several water management problems that include shortages, contamination, and security issues in response to the uncertain access to safe potable water.

The role of disaster awareness in building more resilient societies and communities through education and knowledge transfer has been repeatedly highlighted, particularly in relation to the preparedness, mitigation, emergency response, and recovery stages of the disaster management cycle [Nielsen and Lidstone 1998]. The importance of hazard awareness promotion through the provision of and access to disaster information and knowledge is increasingly being recognized as a critical strategy for the mitigation of the social, economic, and environmental impacts of disasters and, by extension, the enhancement of regional sustainable development (see Chapter 3). What is at stake, we argue, is the very existence of global environmental, social, cultural, and economic security. The case studies in the book all address unique ways to enhance local and regional efforts to reduce vulnerabilities. It is through this forum, *Comparative Emergency Management: Examining Global and Regional Responses to Disasters*, that we share unique interdependent, yet related, case studies that serve to inform, make important recommendations, and empower societies to become more prepared for future challenges.

Structure of the Work

This book has five thematic sections rooted in regional responses to natural, technological, and human-induced catastrophes that highlight the role of regional collaborations. Each chapter is structured so that the primary focus centers on a region by using a case study and identifies what impact that catastrophe had on that region. Moreover, attention to the stakeholders that are brought together from diverse cultures and political and economic systems to collaboratively address common problems are used to illustrate best practices in addressing regional threats through sustainable partnerships that last over time.

Although each chapter represents an independent case study, the authors discuss collaborations within or across country borders. Overall, each chapter focuses on proactive approaches to building long-term structures that handle catastrophes as they occur. By highlighting regional concerns with a specific case study, the authors make a case for the importance of the collaboration across regions that ultimately strengthen and enhance the resiliency of those in harms way. Section I, The Americas, addresses issues to help countries increase their preparedness and mitigation strategies. This section focuses on the collaborations within and among countries in the Caribbean, North America, Central America, and South America in combating regional threats including floods and hurricanes, landslides, and earthquakes. Chapter 1 by Kathleen Vito describes select disasters in specific regions around the United States. Vito maintains that local businesses, educational institutions, civic groups, law enforcement, and public health groups play a specific role that can impact response and recovery. Focusing on internal organizations, creating alliances, and taking on additional responsibilities while relying less on government services will have more long-term benefits. Planning for future emergencies and disasters is ongoing and will need to be a continuous and coordinated process for public health to succeed in its mission to protect and improve the health of our communities [Morrow 2007].

Chapter 2 by Philip Osei, maintains that fashioning modern emergency management systems for the region and individual islands requires extraordinary acumen in communication and collaboration and a wealth of resources. Effective emergency management is premised on the comprehensive integration of emergency plans at all levels of state and nonstate actor involvement. Network governance is important at all phases of the disaster risk reduction process: mitigation, preparedness, response, and recovery [Osei 2007]. To cope with the complexity of natural phenomena, institutional arrangements, and related management systems should incorporate different actors from different areas of society. Thus, an emphasis on the establishment of multiactor structures, consisting of both public and private actors, is a common denominator [Osei 2007].

In Chapter 3, Virginia Clerveaux and Balfour Spence delves into the relationship between disaster awareness and loss reduction in relation to disaster risk management planning. They describe how economic factors can impede the promotion of proper awareness within differing communities in the Caribbean. Globalization has brought forth new challenges. It is up to all involved sectors to create and incorporate up-to-date techniques that better communicate risk reduction information to all affected communities. Introducing a game that can be amusing while at the same time provide useful information to children in disaster-prone areas may ultimately help communities overcome many challenges in disaster recovery.

In Chapter 4, Hutter and Miller argue that, in the aftermath of Hurricane Katrina, tourist and convention planners, city leaders, cultural institutions, and private businesses (such as tour guides) vied for opportunities to coordinate, control, and produce an image of the city that ultimately emphasized the cultural revitalization of New Orleans, which served to develop rules and mechanisms for policies governing the use of culture as the cornerstone of both the tourism industry and economic development of the New Orleans region. These key stakeholders in the tourism industry ultimately govern how places are branded and rebranded and how a destination's symbolic economy becomes the primary driver for rebuilding after a disaster.

Chapter 5 by Benjamin Kelly describes the attempt to address sanitation and environmental change in Ontario, Canada. The Learning Alliance is a group of engineers and scientists that is committed to the development and maintenance of ecologically sustainable technologies and strategies that will reduce risks associated with water management. Although there are plans that are designed specifically to address freshwater conservation, this group has been recruiting other engineers and scientists, businesses, and governments, and pooling knowledge to improve conservation efforts.

In Section II, Africa and the Middle East, issues of security and protection from harm are detailed. The Middle East and North African region (MENA) is prone to various types of disasters caused by natural hazards, such as earthquakes, floods, drought, and landslides. Disasters impose increasingly heavy human and property losses on countries in this region, which hampers their socioeconomic development. Governments and other actors in the MENA region have been developing various strategies targeting disaster preparedness and response; however, disaster risk reduction linked to national development processes is not yet integrated in the planning and policy paradigms of the region. This section seeks to understand how comprehensive and integrated approaches to disaster management and risk reduction are essential in building resilience to disasters, thus reducing human and property losses.

In Chapter 6, Filip Aggestam and Stephanie Hodge discuss the role of stakeholder participation in relation to transboundary water governance projects. The management of a transboundary water body is a daunting task. Despite great leaps forward in our understanding of how to manage limited resources, current water governance constitutes an increasingly complex web that integrates technical and scientific knowledge, legal requirements, socioeconomic aspects, and multi-stakeholder participation [Baggett et al. 2006]. With proper strategies in place, decision-makers may understand and decide on objectives, policy development, implementation, and participation strategies within the context of community-based improvement of resilience and risk reduction.

In Chapter 7, Sheryl Hendriks and Scott Drimie focus on the food crisis in Africa. Since 1998, Africa has experienced an average of 20 food emergencies a year [Economic and Social Council 2005]. A majority of the world's hungry live on the African continent, yet national and regional leaders have failed to progress toward an ultimate goal of decreasing the number of hungry people by half in the year 2015. Many of the decisions were ill-advised and focused on long-term goals rather than on the current crisis impacting the regional relationships among neighboring communities. The way communities and regions in Africa reacted gives insight into the cultural, economic, and social structure of nations and regions on the continent.

In Chapter 8, Sultan Khan describes the actions of faith-based organizations and their response in disaster situations. Faith-based organizations play a major role in the response effort of many humanitarian disasters that occur. Khan focuses on a study of South African Muslim diaspora. Most of the Muslim populations, due to their faith, practice forms of social giving during human disasters. Their motivation for response, the assistance methods used, the deployment of workers after a disaster, and the level of interaction between Muslims and other aid are thoroughly discussed in this chapter.

In Chapter 9, Alex Altshuler maintains that interorganizational cooperation in the field of preparedness for war-caused disaster between local emergency managers and various stakeholders is important. He examines a case study of the Jewish and the Arab sector in Israel. The relationship between different types of stakeholders (i.e., national emergency organizations and governmental ministries, disaster relief and volunteer organizations, trade unions, local governments, health management organizations, community centers, philanthropists, residents and professional consultants) is part of the study. Interorganizational cooperation is a key factor in predicting the level of emergency preparedness in both the Arab and the Jewish sectors in Israel. In addition to cooperation on the local level, the authors propose to establish formal regional frameworks for continuous cooperation in which representatives of both the Arab and the Jewish local authorities will be able to cooperate in a more effective way while assisting each other reciprocally.

In Chapter 10, Beryl Cheal discusses the importance that schools play in rebuilding communities that have been affected by disaster. There are now more storms, as well as more human-induced disasters that result in catastrophes. After a disaster, it is crucial that people do all that they can to bring forth the restoration of their community. Schools play a substantial role in this process by aiding in the development of a preparedness plan that is usable for a specific area. Along with this, schools can be used to aid students and their family members in recovering from the devastating affects of the disaster. Community Emergency Response Service (CERT) is a program offered in many communities around the United States that can be used to come up with a strategy [Rich and Kelman 2007]. A similar program called Teen School Emergency Response Training (SERT) has been developed for the youth, primarily between the ages of 15 to 18 [Rich and Kelman 2007]. These programs help communities and students alike in many disaster-related aspects. Even though students are some of the most vulnerable in times of disasters, they can also be of great importance and can even help save lives. Stressing the fact that planning ahead is key will ultimately teach children that knowing what to do, as well as when to do it, is important. Schools play a significant role in assuring regional stability in the aftermath of a disaster.

Section III, Europe, is composed of three chapters. Due to the proximity of the many nations that comprise Europe, security from natural, human-induced, and technological disasters remains an international concern. Since the expansion of the European Union (EU), catastrophic events resulting in loss of life and property remains a priority. In Chapter 11, Rémy Bossu, Sébastien Gilles, Gilles Mazet-Roux, and Fréderic Roussel discuss an innovative way to engage the public in the response efforts to earthquakes. The European-Mediterranean Seismological Centre (EMSC) works alongside the public to reduce the hardships associated with earthquakes. The EMSC allows for public involvement by giving affected citizens an opportunity to share their observations of a disaster. The public can share this important information with other witnesses and the scientific community via the EMSC Web site. Allowing citizens an active role in earthquake response makes the notion of risk more tangible and intuitive, develops a culture of risk, and may increase public receptivity to prevention messages [Bossu el al. 2009].

In Chapter 12, Sotiris Chtouris and Flora Tzelepoglou examine the devastating forest fires that occurred in Greece in 2007. This disaster proved that there was not enough done in terms of preparation. Risk prevention, mainly through information and mobilization of local residents, emerged as a key factor if we want to confront natural hazards in an effective way, mitigating environmental damage and community problems. The unexpected Greek forest fires of 2007 left the affected communities with the daunting task of coordinating service groups that would work together across the region with hopes of overcoming the disaster at hand. Research data confirm that environmental risk perceptions and risk management are crucial social issues not only in Greece, but also in Southern Europe as well because they involve low levels of social participation and social capital [Putnam 1993]. On the other hand, highly developed social networks based on family (nuclear and extended) ties preserve their cohesion [Chtouris 2004], but they fail to function as cooperation networks when emergencies break out and risk management is called for. Further issues include the lack of proper cooperation among all major parties involved. This lack of planning may prove to be detrimental as it did in the Greek forest fires in 2007.

In Chapter 13, Ioannis M. Dokas, John Feehan, Stephen C. Fortier, Franclin Foping, and Syed Imran review the importance of risk reduction in a case study of the water systems of Ireland. The authors maintain that drinking water systems in several regions of Ireland are vulnerable to natural disasters. The lack of action to improve these systems may serve to increase the nation's population susceptibility to a manmade disaster. As a result, governments, state agencies, international organizations, and other bodies are developing disaster risk reduction and proactive risk management strategies. Drinking water systems are clearly crucial; the security of these supplies continue to face challenges which can severely impact communities around Ireland.

Section IV focuses on Asian nations. As a region, Asia is diverse in area, population, geography, natural resources, cultural legacy, colonial experiences, stages of development, and systems of government. With the coastal urban populations of Malaysia, Indonesia, Philippines, China, and Thailand increasing, the pressure on urban infrastructure and vulnerability to hazards increases as the urban poor settle in fragile environments. In Chapter 14, Patrick Kilby and Kim Williamson examine community-oriented disaster response and the lessons learned from the humanitarian response to the Asian tsunami in India and the Jogyakarta earthquake in Indonesia. Disaster management includes the concept of disaster risk reduction. This idea is important for many communities that are forced to endure catastrophic disasters, but this concept sits behind the disaster response effort due to the complexity and evolution of the global environment. Although it is important that vulnerable areas such as Asia and Indonesia are well prepared, the response of these communities after a disaster is crucial. Having a strong disaster response strategy can drastically reduce the number of lives lost and repair damaged infrastructure more quickly, speeding a community's return to everyday life. Disaster response thus emerges as an essential and ongoing component of disaster risk management. The development of government and public response plans is not an easy task to accomplish, but it plays a large role in returning to normalcy.

In Chapter 15, Chuangsheng Jiang and Yunfeng Deng discuss an incident that occurred in China at the Lujia well. On December 23, 2003, the Lujia well suffered a blowout accident. Residents of the affected areas had to be evacuated to assure their safety. A total of four communities were affected. The destruction left 243 people dead and many others injured. An analysis was administered to make sure that incidents like this do not occur in the future. This preparedness plan may assure that lives as well as money are saved. A series of professional standards based on blowout risk prevention and control was developed [China Academy of Safety Science and Technology 2008]. This was the first occurrence that safety science institutions in mainland China had performed systematic technical field conditions after a technological disaster. This action was a step in the right direction in relation to safer workplaces. Before the blowout, 96.4% of villagers knew nothing or very little of the potential hazards of the well. After the blowout, 80% of villagers still remained unaware of the dangers that lie among them [Yuan and Pu 2008]. This number needs to be substantially lower to be prepared for another disaster.

The final section, Global Challenges and Next Steps for the Twenty-First Century, features three components: Chapter 16 by Anatoly Oleksiyenko, our conclusion, and an afterword by J. Steven Picou. In Chapter 16, Oleksiyenko explains the need for more coordination and training for responders to health-related regional and global crises. Global health is the concern of many organizations, including national governments, international agencies, private foundations, pharmaceutical companies, and nongovernmental and academic institutions, among other relevant actors. Problems such as pandemics and disaster medical care bring forth the need for more effective and strategic planning across many stakeholders in the healthcare sector. However, this is not always easily accomplished due to various political, economic, and cultural factors. The issue lies in creating an infrastructure that involves all of the complex organizations in the best way possible. Oleksiyenko asserts that medical schools play a pivotal role within these partnerships of promoting quality global health during times of disaster.

Finally, J. Steven Picou proposes an insightful approach to an effective response that relates to both regional cooperation and individual empowerment. When considering potential climate change catastrophes yet to come, Picou admonishes us to reframe Beck's concept of individualization when viewing emerging regional partnerships, networks, cooperatives, and other collaborative systems for responding to disasters as part of a regional response to help us understand the structural potential of people for responding to catastrophe in the twenty-first century.

These chapters represent a wealth of information that can be used for regional collaboration, which can serve as a catalyst for meaningful social change that impacts lives by decreasing vulnerabilities to regional environmental threats. Long-lasting social change is the most influential way in which emergency management and disaster prevention can be instituted across international boundaries. Only through further discussion and expansion of the research presented in this and similar volumes can we go from analysis to policy formation, to policy application, and to life-saving action.

References

Achenbach, J. 2008. Man to blame for Iowa flooding? *The Washington Post*. Retrieved June 19, 2008 from www.msnbc.com/id/25254541.

Associated Press. 2010. Haiti raises earthquake toll to 230,000. Retrieved August 16, 2010 from http://www.washingtonpost.com/wp-dyn/content/article/2010/02/09/AR2010020904447.html.

Baggett, S., Jeffrey, P., and Jefferson, B. 2006. Risk perception in participatory planning for water use. *Desalination* 187:149–158.

Bossu, R., Gilles, M. R., and Gilles, S. 2009. The importance of in-situ observations for rapid loss estimates in the Euro-Med region. EGU General Assembly 2009. *Geophysical Research Abstracts* Vol. 11.

Branscomb, L. W., and Michel-Kerjan, E. O. 2006. Public-private collaboration on a national and international scale. In *Seeds of disaster, roots of response: How private action can reduce public vulnerability*, eds. P. E. Auerswald, L. M. Branscomb, T. M. La Porte, and E. O. Michel-Kerjan, 395–403. New York: Cambridge University Press.

Caribbean Media Corporation. 2010. Caribbean heads of state to visit Haiti to assess quake damage. *BBC Monitoring Americas*, January 14, 2010.

Chtouris, S. 2004. *Rational symbolic networks*. Athens: Nesos Editions.

Cutter, S. L., Boruff, B. J., and Shirley, W. L. 2003. Social vulnerability to environmental hazards. *Social Science Quarterly* 84:242–261.

Economic and Social Council. 2005. President's summary at the special event Food Crises in Africa. *ECOSOC Chamber Thursday*, October 27, 2005. New York: UN Economic and Social Council.

Gonzalez, C. 2010. Opportunities and challenges for private-public partnerships in the 21st century. In *Community disaster recovery and resiliency: Exploring global opportunities and challenges*, eds. D. S. Miller and J. D. Rivera, 391–394. Boca Raton, FL: Auerbach/Taylor & Francis.

Iversen, R. R., and Armstrong, A. L. 2008. Hurricane Katrina and New Orleans: What might a sociological embeddedness perspective offer disaster research and planning? *Analyses of Social Issues and Public Policy* 8:183–209.

Johnson, T. 2008. Quake puts focus on poorly built schools. *The Philadelphia Inquirer*, May 17, 2008, p. A2.

Kroll-Smith, S. 2001. Review of *The Northbridge earthquake: vulnerability and disaster* by R. Bolin with L. Stanford. *Contemporary Sociology* 30:175–176.

Louisiana Department of Health and Hospitals. 2006. Reports of missing and deceased. Retrieved August 16, 2010 from http://www.dhh.louisiana.gov/offices/page.asp?ID=192&Detail=5248.

Margesson, R., and Taft-Morales, M. 2010. *Haiti earthquake: Crisis and response.* Congressional Research Service report # R41023. Washington, DC: Congressional Research Service.

Morrow, C. 2007. Public health emergency preparedness and the evolution of public health in the 21st century. *Journal of Public Health Management Practice* 13:439–440.

Nielsen, S., and Lidstone, J. 1998. Public education and disaster management: Is there any guiding theory? *The Australian Journal of Emergency Management* 13:14–19.

Osei, P. D. 2007. Policy responses, institutional networks management and post Hurricane Ivan reconstruction in Jamaica. *Disaster Prevention and Management* 16:217–234.

Paasi, A. 2000. Re-constructing regions and regional identity. Retrieved February 18, 2011 from http://socgeo.ruhosting.nl/colloquium/Paasi1.pdf.

Picou, J. S., and Marshal, B. K. 2007. Katrina as paradigm shift: Reflections on disaster research in the twenty-first century. In *The sociology of Katrina: Perspectives on a modern catastrophe*, eds. D. L. Brunsma, D. Overfelt, and S. J. Picou, 1–20. Lanham, MD: Rowman & Littlefield.

Pinkowski, J. 2008. The new normal. In *The homeland security handbook*, ed. J. Pinkowski, 3–24. Boca Raton, FL: CRC Press/Taylor & Francis.

Putnam, R. 1993. *Making democracy work: Civic traditions in modern Italy.* Princeton, NJ: Princeton University Press.

Rich, H., and Kelman, I. 2007. Teens learn to save lives—and do. *Disaster and Social Crisis Research Network Electronic Newsletter*, 32:6–7.

Robertson, C., and Krauss, C. 2010. Gulf spill is the largest of its kind, scientists say. *New York Times,* August 3, 2010.

Robinson, L., and Jarvie, J. 2008. Post-disaster community tourism recovery: The tsunami and Arugam Bay, Sri Lanka. *Disasters* 32:631–645.

Stewart, G., Kolluru, R., and Smith, M. 2009. Leveraging public-private partnerships to improve community resilience in times of disaster. *International Journal of Physical Distribution & Logistics Management* 39:343–364.

Telegraph. 2010. BP leak the world's worst accidental oil spill. Retrieved August 15, 2010 from http://www.telegraph.co.uk/finance/newsbysector/energy/oilandgas/7924009/BP-leak-the-worlds-worst-accidental-oil-spill.html.

Tierney, K. 2006. The red pill. Retrieved May 17, 2008 from http://understandingkatrina.ssrc.org/Tierney/printable.html.

United Nations International Strategy for Disaster Reduction. *African regional strategy for disaster risk reduction.* 2004. Retrieved August 16, 2010 from http://www.unisdr.org/africa/af-hfa/docs/africa-regional-strategy.pdf.

USA Today Staff. 2010. Obama, in Gulf, pledges to push on stopping leak. Retrieved May 27, 2010 from http://www.usatoday.com/news/nation/2010-05-27-oil-spill-news_N.htm?csp=34news.

U.S. Geological Survey. 2010. The largest earthquake ever recorded was the 9.5 magnitude 1960 Chile earthquake. Retrieved January 13, 2010 from http://earthquake.usgs.gov/earthquakes/recenteqsww/Quakes/us2010rja6.php and http://earthquake.usgs.gov/earthquakes/world/historical_mag.php.

World Bank. 2005. Sri Lanka: Preliminary damage and needs assessment Joint Asia Development Bank. Retrieved August 16, 2010 from http://www.adb.org/Documents/Reports/Tsunami/sri-lanka-tsunami-assessment.pdf.

Yuan, L., and Pu, C. 2008. Investigation of sufferers' mental health after the gas blowout accident on December 23. *Chinese Journal of Health Psychology* 12:321–324.

THE AMERICAS

I

Chapter 1

Forging Partnerships within Geopolitical Regions of the United States for Mutual Aid and Mass Prophylaxis Intervention

Kathleen O. Vito

Contents

Introduction

The purpose of this chapter is to describe selected disasters in regions of the United States and the impact of these disasters on the development of disaster preparedness, response plans, and organizations from a historical context. From the growth of American Red Cross disaster services and the utilization of the Salvation Army in Galveston in 1900 to new federal programs after 9/11 such as the Medical Reserve Corps and Emergency Services Advanced Registration of Volunteer Health Professionals, hazard management has focused more on response and recovery. After Hurricane Katrina, both the public and private sectors in the United States shifted focus to risk reduction and building resiliency. The realization that businesses, educational institutions, civic groups, faith-based organizations, law enforcement, public health, and health-care delivery systems all have a stake and a role in disaster preparedness and response has stimulated new partnerships and models of intervention with national, state, and local government. With less reliance on federal and state response services, local regions will better be able to meet the immediate needs of the population during a disaster. Newer models of service delivery such as community-run places of distribution increase the ability of a local region to build community resistance to chaos by boosting the surge capacity of local resources. The development of new models creates changes in the ways organizations usually function in the community, as well as in regulations and policies.

Policy Formulation Process for Disaster Preparedness and Response

The process of policy formulation for all hazards in the United States has been one of reaction and response to events that have disturbed our sense of control. Emergency management and disaster response initially were spun off of local and regional efforts to improve living conditions after disasters. The policy was the product of a set of political compromises made at the Constitutional Convention in 1787. Our government was designed to be a system of dual authority between the federal government and the states. Over the centuries, federalism has evolved with periodic shifts in authority between the state and federal governments over which should be in charge of all types of concerns, including the health and safety of citizens. In reality, when it comes to natural and manmade disasters, the analogy has been described by some as rowboat federalism because all levels of government find themselves in the "same boat, tossed by the same waves, and dependent on each other's paddles" [Hoover 2005]. Because of the trend of policy formulation as a reaction to an event over the past two centuries in the United States, policy change and the development of programs and agencies has focused on recovery and not prevention.

All disasters begin at the local level where identification of the disaster and initial response has historically been. Interstate mutual aid compacts from state to state without the consent of Congress were specifically outlawed by the U.S. Constitution. Not until 1996, with the passage of the Emergency Management Assistance Compact, were the federal barriers to states sharing resources in times of disasters lifted [Hughes 2004]. It was only after 9/11 that a new federal policy advocated mutual aid agreements between local governing bodies and their states as part of the National Strategy for Homeland Security. Regulations developed in response to new federal policies have continued to rely on the local government level as the first responder and decision-maker about whether local resources can handle the situation.

Historical Foundations of Emergency Response in the United States

The Tribal Society of Native Americans developed as a result of mutual need of aid to survive and thrive. Communities formed as villages, with emerging roles as leaders, planners, hunters, healers, gatherers, builders, clothing-makers, cooks, and caregivers. Native Americans in the northeast are known for their tendency to live in multifamily dwellings (long houses) to share shelter, heat, food, and caregiver responsibilities in the harsh winter environment. In less harsh geographic areas, the family unit would occupy a single dwelling but in close proximity to others. In populations where the custom was to make seasonal moves to follow the movement of game, the entire community moved as one, with each member dependent on the others to fulfill their role. Mutual aid and assistance was the way of living. Tribes developed mutual aid with neighboring tribes, but some were notorious for making war against each other. Even in war, the victorious tribe would assume responsibility for the women and children left homeless and defenseless and absorb them into their tribe. In early colonial settlements of European immigrants, the Native American mutual aid and assistance was key to the survival of new settlements. Although the settlements at Plymouth and Jamestown survived and thrived because of the mutual aid and assistance, the settlement at Roanoke failed. While the true story of why Roanoke failed and what happened to the English men and women that settled there will never be known, there has been some conjecture that Eleanor Dare, the first white infant born in the New World, was spared and was raised by the Croatian Indians as one of them.

Colonial bucket brigades and sanitation regulations were first established in the New World as mutual aid and assistance. Voluntary cooperation and assistance could mean the difference between life and death for the colonists if there were food shortages, major illnesses, or other disasters. No one family could fight a house or barn fire by themselves. The first volunteer firefighting company is attributed to Benjamin Franklin in 1736. The early volunteer fire companies were based on the premise that sooner or later everyone would need the help themselves that they gave to others. When a devastating fire in New Hampshire destroyed much of an entire town, the Congressional Act of 1803 was passed [*Congressional Digest* 2005] to establish a mechanism for the federal government to provide disaster relief funds. Policy set by the original Fire Disaster Relief Act of 1803 set a precedent for more than 100 individual acts of ad hoc legislation. Not every disaster in the United States received federal assistance. For example, in 1887, President Cleveland vetoed emergency funds for victims of a drought in Texas, stating that there was no basis in the Constitution for the federal government to grant relief funds for events not related to public service [Barnett 2000]. Local officials often would decline to request federal aid, concerned that the community would be stigmatized. During this time, many communities banned together to build basic shelters known as pest houses to house people in isolation rather than approach the government for assistance. The pest houses were staffed by volunteers and relied on donations of firewood, clothing, and food from the community.

Some of the earliest established local boards of health in the United States are attributed to Massachusetts in 1850. Local laws were established to set standards and supervisory functions of sanitation for the health of the public [Tulchinsky 2009] or public hygiene, as it was termed in the centennial discourse presented to the International Medical Congress held in Philadelphia in 1876 [Bowditch 1877]. Early in the history of public health, the goal was preventative medicine in an effort to prevent the spread of communicable disease. The use of isolation and quarantine was the planned response if prevention failed. Public health officials worked in isolation from other agencies that might have had a vested interest in preparedness, response, and recovery to disasters from other hazards. The yellow fever epidemic in Philadelphia was an emergency event that brought public health together with new partners from the community for supplies, food, tents, clothing, health-care workers, and other needed services. It was in this type of emergency that the immediate need for human resources and supplies were more important than possible fiscal assistance from the federal government if another act of Congress could be passed.

The U.S. Civil War was the beginning of military physicians and volunteer nurses in roles related to emergency management. The military model of management of large numbers of wounded and people with diseases became the model of care that is with us today for mass casualty events involving private citizens. The military system is a structured and graded response system: identification of ground zero, triage, and rescue of the injured that proceeds rearward to an acute care site, and, last, retrieval of the dead [Dara 2005]. The American Red Cross was chartered in the United States in 1881 by an act of Congress to provide formal disaster services in times of peace because one of the outcomes of volunteer nurses on Civil War battlefronts was increased rates of survival of the wounded and ill [Gebbie 2006].

In 1899, the British-based Salvation Army was chartered in the United States. The role the Salvation Army was to play in disasters and emergencies was similar to that of the already established American Red Cross. Both organizations responded to the disaster in Galveston in 1900. Both organizations sent relief workers and volunteers from around the country to assist with recovery. The contingent from the American Red Cross included Clara Barton herself. In 1909, the American Red Cross added professional nursing as a specific service.

The American Red Cross was called to action from 1918 to 1919 for nursing services. The flu pandemic that started with servicemen returning from Europe in World War I broke out in the general public. The pandemic quickly overwhelmed health-care systems and mortuary services. Volunteer nurses from the American Red Cross and other organizations cared for large numbers of patients placed on cots and makeshift beds in large ballrooms of hotels, school gymnasiums, and tents. Nursing care was supportive to help people recover if they could and comfort those who would die. At times, there were so many bodies to be buried at once that they piled up in makeshift morgues and were buried in mass graves.

Federal public health and health-care systems policy toward disaster relief remained piecemeal by event until the 1930s, when Roosevelt's New Deal helped shape federal assistance, including the Reconstruction Finance Program for rebuilding after a disaster. Although assistance was still disaster-specific, federal medical personal, supplies, and transport were made available on request from a state. No longer was it necessary to enact a separate act of Congress.

Before, during, and after World War II, the United States suffered a major polio epidemic. Isolation of those who were ill occurred in hospitals across the country. Makeshift hospital wards to house hundreds of patients in iron lung chambers were developed in places such as gymnasiums and ball rooms. Quarantine of healthy family members of polio patients was common. With local systems of neighborhood volunteers, families received supplies of fresh food dropped off on their doorstep. Communication with their friends and neighbors was limited to interactions through a closed door or window [Krawslecki 2009].

After World War II, in anticipation of air raids and potential nuclear attacks, the United States developed sophisticated systems of civil defense in the 1950s and 1960s. The era of civil defense is considered to be the birth of modern emergency management [Coppola 2007] and is known as the Cold War era. The Cold War prompted the federal government to fund programs of preparedness under the auspices of the Defense Department's Civil Preparedness Agency. This was a time of great uncertainty. Early warning systems for the public were designed and tested on a regular basis. People were encouraged to build backyard bomb shelters and stockpile food and water in case of a nuclear attack. Public buildings that housed shelters and supplies were marked with Civil Defense logos for easy identification by the public. School children practiced drills for sheltering in place by tucking under their desktops. Although not public knowledge until later, the federal government built an underground facility for Congress and key government offices on the grounds of the Greenbrier Resort in West Virginia that was equipped with food, supplies, and the latest in communication equipment. The plan was for the President to be housed in another facility in case there was need to evacuate Washington, DC. Because most of the plans for continued operation of the federal government in case of a direct attack were secret, there was not much opportunity for mutual alliance development. During the Cold War, the new Centers for Disease Control developed as a concept of epidemic intelligence surveillance in case of biological welfare. While the military was focused on nuclear attacks, federal health officials promoted the idea to the public that they had more to fear from water and food that were sabotaged [Fee 2001].

One of the least known efforts of mutual alliance at the local level occurred during the early use of polio vaccines. From 1953 to 1955, more than an estimated 200,000 volunteers helped run mass prophylaxis clinics as part of clinical trials to test the new polio vaccine by Salk by assisting families with children through clinic lines, giving out lollipops and Polio Pioneer buttons, and staffing the reception, recovery areas and clinic stations. The clinic design and logistics of volunteer mobilization were repeated in communities across the country in 1962 when the oral immunization developed by Sabin was approved for use. The federally sponsored project employed volunteers from all walks of life to handle crowds, set up lines for distribution, do advance publicity, and assist nurses, pharmacists, and physicians in the distribution of the Sabin vaccine. Additional nurses were recruited from the ranks of retired nurses, hospital-based nurses, school nurses, and nursing students to help public health nurses place two drops of vaccine on a sugar cube for each person coming through the clinic. Civic groups from churches, service clubs, scouts, and parent–teacher associations all assisted. The event was nicknamed Sabin Sundays because many were held on Sundays in church and public school buildings [Shields 1963]. It is not known how many Sabin Sunday clinics were held across the country because the event was not well documented in the literature. The young daughter of a public health nurse in 1962, I was a volunteer at several Sabin Sundays in Schenectady, New York. Not aware of the significance of the event at the time, in doing research for this chapter I realized that the days spent putting sugar cubes into medicine cups for the nurses at these public clinics were part of a national program.

During the 1960s, a series of severe hurricanes struck the costal United States. The federal response was to enact the Federal Disaster Assistance Administration with the Department of Housing and Urban Development. The Departments of Health and Civil Defense, among others, continued to be separate entities. One particular event in 1969, Hurricane Camille, prompted the American Red Cross to develop and offer specific disaster training courses for nurses. The Disaster Relief Act of 1974 was an attempt to address the lack of coordination in disaster relief operations from the federal level [Koenig 2007] because the Departments of Housing and Urban Development, Defense, and Health each had various components of emergency response and disaster relief within their jurisdictions.

In the early spring of 1976, the military health system notified President Ford of a new strain of flu that was present in the United States [Lakoff 2008]. Many service men at Fort Dix in New Jersey were ill and one had died. The epidemiological investigation indicated that this virus, named swine flu, was spreading from human to human and had already escaped into the civilian population in the area. Federal and state officials worked together in an unprecedented alliance to develop an immunization and a plan to vaccinate the entire population before the fall flu immunization season. Mass vaccination of the public occurred for the first time in the United States since Sabin Sundays. Volunteers from the Red Cross, public health, and health-care delivery systems were trained to use a new vaccination device nicknamed the swine flu gun. This automated device was developed by the military and allowed injections to be given to multiple people without the need for preparation of individual doses for injection in syringes. It was only later determined that the swine flu guns carried a risk of potentially infecting persons receiving the vaccine from persons in front of them in the vaccination line. The swine flu program reached more than 40 million Americans before it was disbanded in mid-December of that year due to several deaths and illnesses of Guillain–Barré Syndrome related to the flu vaccine. The swine flu epidemic did not develop as was predicted. Lessons learned included a plan to stockpile vaccine, to develop logistical plans for mass immunization, and to vaccine the general public once it was apparent that an epidemic was developing.

The Three Mile Island nuclear power plant accident in 1979 illustrated the problems with piecemeal emergency preparedness. At the time of the accident, more than 100 federal agencies were involved with some type of disaster relief. The Federal Emergency Management Agency (FEMA) was established by President Carter to absorb several other agencies at the federal level. It was not until 1993 that the leadership of FEMA was under a director with emergency management experience. After the end of the Cold War, civil defense funding was directed to disaster relief, recovery and mitigation programs within FEMA. With more attention to an all-hazards approach, FEMA was able to coordinate the federal response for real events and also to provide programs to educate state and local officials for emergency response, provide equipment for local and state responses, and fund efforts by local communities to teach their residents how to survive a disaster.

Although the American Red Cross continued to provide disaster relief services and training programs for volunteers, it was not until 1991 after Hurricane Hugo and the earthquake in San Francisco that the American Red Cross decided to add a mental health component for the disaster workers and survivors. The president of the American Red Cross was Elizabeth Dole who stated, "We recognize that a crucial aspect of disaster relief, beyond providing food and shelter, is helping victims and survivors cope with their losses" [Heaps 2009, p. 1]. A precedent was set for more formal relationships among community partners by the signing of an official statement of understanding between the American Red Cross and the American Psychological Association to develop a disaster response network of local psychologists willing to volunteer their services in a disaster. Formal disaster mental health services from volunteers through the American Psychological Association proved to be vital later in the disasters of 9/11 and Hurricane Katrina. The American Red Cross estimates that the type and percentage of mental health services provided by their volunteers in recent disasters has included 30% education, 39% direct stress management intervention, 25% problem-solving, and the remaining 15% in advocacy for additional services and referrals.

In 2006, the Post-Katrina Emergency Reform Act was aimed at reorganizing FEMA and granting the agency more authority to remedy gaps in emergency preparedness and response that became apparent in the poorly coordinated preparedness and response to the hurricane. However, the legislation also put FEMA under the direction of the Department of Homeland Security. In a way, this seems circular in development. FEMA grew out of the civil defense program when the Cold War ended, only to return to a program established to fight a war on terrorism.

Reaction to disasters and emergencies has created a patchwork system of preparedness and response policy that Birkland [2007] has termed a "model of event-related policy change." The limitations of this model are many but include problems with the nature of the focusing event for each policy or new program. The focusing event can be anything from a crisis to a catastrophic event. Disasters can be sudden or progressive. Birkland [2007] has suggested that the focus on homeland security following 9/11 had a "corrosive influence on the nation's preparedness for natural disasters." Models of disaster preparedness that do not emphasize formal partnerships for mutual aid and alliances are vulnerable for failure, as the breakdown of any interdependent system creates the possibility of cascading failure [Gerber 2007].

Mutual Alliances in Contemporary Emergency Management and Response

What began as a program in 1970 of the California Fire Service to improve coordination between local, state, and federal agencies when fighting large-scale forest fires with volunteers from out of state, has become the widely used incident command system for civilian response to disasters and emergencies. Under the National Incident Management System, all formal groups of health-care, public health, government, fire, police, emergency management, and emergency medical services are trained to respond to emergencies and disasters within a common incident command system. A common reference to the interagency and interdisciplinary response is to refer to all partners as first responders. A first responder agency is a local agency that is expected to respond first to a disaster or emergency and assess the scope of the disaster and the need for additional responders outside the local area. First responder agencies now plan, practice, drill, and exercise together to forge alliances with each other and become better acquainted with the skills and capacities of each other.

The terrorist attack on the mainland of the United States in September of 2001 was unheralded in the history of emergency, disaster preparedness and response. FEMA was the lead agency for response, but nothing in previous emergency planning had prepared the agency for such a situation. The Salvation Army was the first relief agency at Ground Zero, setting up mobile canteens within a half-hour of the first plane crash. The Salvation Army continued in service at Ground Zero for 11 months, serving meals to the rescue workers, providing rest and hygiene areas, work boots, gas masks, eye drops, lip balm, and personal hygiene supplies as needed for the round-the-clock operations of rescue and recovery of human remains. One of the lessons learned from the 9/11 disaster response and recovery was that large numbers of well-meaning individuals showed up onsite to help. These responders were nicknamed the SUVs, for spontaneous uninvited volunteers.

The federal government created programs for disaster medical response teams and mobile federal medical stations staffed by U.S. public health service health professionals for deployment when local responders require assistance. First, fulltime public health service staff members are used to their maximum surge capacity. Next, reserve levels of personnel are deployed by the government if the disaster is large or the response and recovery must be sustained over a long period of time. The need for pre-identification, training, and coordination of volunteers for times of disaster gave rise to the federal development of Citizen Corps, Medical Reserve Corps, and Emergency Services Advanced Registry of Volunteer Health Professionals as structures and programs for community volunteers and health professionals. An important aspect of the training in the programs is for personnel not to deploy themselves to a disaster area, but rather wait until contacted to serve. These programs were designed to create local and national mutual aid alliances of disaster workers and health professionals and received additional momentum when Congress approved the Pandemic and All-Hazards Preparedness Act of 2006 and created the Office of the Assistant Secretary for Preparedness and Response (ASPR) for natural and manmade disasters. The purpose of the new federal agency is to integrate institutional response plans, knowledge of different partners, and techniques of intervention into a modern context that raises awareness of public health as a matter of national security. Key to this piece of legislation is a plan to develop an integrated system of public health preparedness that extends the system to mandated mutual alliances at the federal, state, and local levels. For example, hospital preparedness grants were moved from the Health Resources Services Agency to ASPR and the grant recipients must demonstrate effective and working partnerships among themselves, their local health department, and the federal Emergency System Advanced Registration for Volunteer Health Professionals. Since the start up of ASPR, the focus of federal grants for bioterrorism has shifted to pandemic flu, all hazards, and support for the strengthening of local networks and alliances for critical local public health infrastructure. A key element is that all partners must be compliant with the use of the incident command structure.

The shootings of students and faculty by a lone gunman at Virginia Tech University in April of 2007 illustrated the need for many diverse organizations to have a common incident command structure as a framework for organized response. Once the shootings were over, the immense work to find all the victims, triage and transport them, keep track of where they had been taken, notify the families, and provide services to the traumatized student body, faculty, and staff was initially not well organized. The hospitals, university, public health, and law enforcement all had individual disaster plans in place, but the response and recovery needed to be under a unified command to be functionally effective as quickly as possible. Once an emergency operations center was established by the state health department, there was marked improvement in efficiency for both services and communication within and outside ground zero. A reception center was established for the parents of slain and injured students. The media was set up in a separate area away from the main campus with frequent briefings by a designated public information officer. The after-action analysis of the incident determined that most university staff were not familiar with the incident command structure even though it is mandated for all state employees in Virginia. State university employees were somehow exempted from this requirement [Virginia Tech Review Panel 2007].

Coordination of Diverse Local Stakeholders by Public Health

The current national response plan states that local jurisdictions must be able to respond and begin recovery without outside help from the federal government for 72 hours [Lindell 2007]. Use of the public health system as the local lead agency for emergency preparedness and response is a relatively new role for health departments. No longer working in silos of communicable disease and environmental hazards, the expectations for local public health workforces to respond has been greatly expanded, due in part to federal funding from the Bioterrorism Act of 2002 [Ziskin 2007]. Grants to states from the Department of Health and Human Services include money for public health initiatives in the event of another pandemic. Health promotion and education of the community on topics such as handwashing, prevention of flu and stressing that emergency preparedness begins with personal family emergency preparedness are grant requirements, as are mutual aid alliances, local and regional drills, and training.

Future of Mutual Alliances in Emergency Preparedness and Response

Planning for future emergencies and disasters is ongoing and must be a continuous and coordinated process for public health to succeed in its mission to protect and improve the health of our communities [Morrow 2007]. New partners will emerge over time and former partnerships and services may be restructured. The goal is to get all the response partners to the table. Strategies to strengthen the public health system with community partners will improve community preparedness. A resilient community will not be as reliant on state and federal resources. Partnerships at the local level facilitate identification of local needs and gaps in current services. Mutual alliances will boost public health and emergency response infrastructures and provide community partners with mechanisms for information-sharing and coordination. For example, in New Jersey, the State Department of Health and Senior Services has grouped programs dedicated to disaster preparedness within one section of the health department under the management of a senior assistant commissioner. The state level oversees county level operations for public health emergency preparedness and response called the Local Information Network and Communication System (LINCS) [Ziskin 2007]. Each public health discipline in a LINCS agency participates in statewide working meetings on projects that build mutual alliances between counties. The LINCS agency, which represents the county health department, meets quarterly with local community partners to forge alliances and working relationships. The quarterly meetings include the following suggested partners (see Table 1.1).

Multiple professional organizations besides public health organizations have individually and collectively established task forces, set standards for core

Table 1.1 Examples of Community Partners with NJ LINCS Agencies

Local township and city health departments
Health-care systems including home care services
County emergency management
Local municipal emergency planning committees
Police and police departments
County sheriff and prosecutor offices
Community-based disaster groups
Health-care outreach coalitions
Retired and senior citizen programs
Schools and universities
Emergency medical services
Area military organizations such as the National Guard or Coast Guard
Local businesses
Critical infrastructures such as power plants
Others

competencies for health professionals and have developed educational programs to prepare the health-care workforce for emergency preparedness and response [Lyznicki 2007]. Specialty organizations have advanced the science of effective disaster response through the promotion of disaster research and consensus-building in their respective area of health care. For example, the American Academy of Pediatrics has issued expert guidance articles on how to treat and respond to children who are victims of terrorism, disasters, and other public health emergencies [Markenson 2006].

In anticipation of a pandemic flu, area facilities and emergency rooms will be overwhelmed. One partnership plan for expanding surge capacity care is to establish acute flu care centers affiliated with hospitals. People will be directed to flu assessment and reception sites for diagnosis and treatment instead of emergency departments. If possible, people may be sent home to recover but they will need home care services such as health assessments for complications, medication, and food deliveries. Volunteers from the Salvation Army, Red Cross, and Medical Reserve Corps will be called in to help home care agencies much in the way that Meals on Wheels programs work now. In-patient care provided at a flu care center will be less acute than for critically ill patients who require ventilators, but hospital staff will need supplemental staffing by volunteers from other communities to handle the burden of temporary additional bed capacity of the health-care delivery system.

Another need for partnership building is in the development of shelters for people with special needs or with pets. In an emergency, the Red Cross opens, supplies, and staffs shelters. They cannot take care of people who need assistance with medical procedures or activities of daily living. They cannot take in pets unless the animal is a service animal. Mutual alliances with community Red Cross Chapters have developed innovative plans for special needs shelters to coexist onsite with regular shelters. The Red Cross will provide the critical infrastructure support such as cots, heat, lighting, and food. Staffing for the special needs sections will come from volunteers such as the Medical Reserve Corps, health professional students, and community emergency response teams of the Citizens' Corp. Innovative planning for shelters for pets and livestock is another area for mutual alliances between volunteers to care for the animals and physical space for the animals. Colleges of veterinary science and community animal response teams are two organizations already forging alliances with public fairgrounds and other suitable spaces for large-scale animal shelters. Another innovative alliance is for a community partner to provide a separate shelter for animals near a Red Cross shelter, so that pet owners can care for their pets housed in a separate building.

Planning for public quarantine and isolation is another area for community partnerships. Plans are in place to invoke laws forbidding public gatherings if necessary to break the cycle of human-to-human infection of novel viruses. Partnerships with schools and colleges before a disaster will allow for a transition to online learning if schools have to be closed for several weeks. Partnering with local businesses as part of the planning will help managers evaluate essential projects for potential as work from home options. Some grocery store chains have begun to develop processes for phone or internet shopping orders and home deliveries. In the past, local pharmacies maintained home delivery of prescriptions. Outreach for mutual alliances should include the businesses in the local area to make them aware of possible needs for this shift in service to home-based and delivery. Encouragement to develop plans based on anticipated needs will be more effective if initiated prior to an emergency.

Several models exist for mass prophylaxis of the population for situations such as anthrax exposure or a disease outbreak requiring vaccination. Models used in the various top officials (TOPOFF) exercises used variations of pull models where people are expected to come to special public clinics called places of distribution. The public health agencies responsible for places of distribution management will become overwhelmed quickly in a real event because there are limited human resources to staff all of the places of distribution needed for a jurisdiction in a 72-hour timeframe. Push places of distribution are another model where community partners help distribute medication for prophylaxis. One idea that has been exercised in large municipalities has been the use of the postal service. Fixed facilities or closed places of distribution using key personnel from schools, churches, civic groups, and businesses to distribute mass prophylaxis to their student body, staff, members, or employees is another model that has been developed. In these models, the medications and supplies will be provided by the public health agency to the fixed facility. Push and closed places of distribution are examples of mutual aid and alliances at a formal level with signed memos of understanding.

No matter what designs of places of distribution are used, they will require the use of extra health professionals if the prophylaxis is an injectable substance. With that in mind, an innovative project to train surge capacity vaccinators to administer injections with traditional syringes, needleless injection systems, and the technique of small pox immunization (which requires a scratch technique no longer taught in nursing schools) was developed by the New Jersey Department of Health and Senior Services LINCS public health nurses. The 2-hour free vaccinator training with a practice session is targeted at nursing students, school nurses, and emergency medical technicians. The training provides the volunteer nurse with 2 hours of approved continuing education credit toward registered nurses' relicensure in New Jersey. A half-hour, "just-in-time" refresher training program was also developed to be offered as needed.

Mutual alliances are not always planned or formal. After 9/11, an estimated half million people left Manhattan in an impromptu water evacuation [Clarke 2006] of barges, fishing boats, pleasure boats, and ferries. No one was in charge of the operation, nor was it part of a preconceived plan. No one evacuating had ever practiced for this in a drill. There were many stories of personal heroism of strangers helping each other to evacuate the World Trade Center complex that day. Ripley's [2008] tales of ordinary people caught in disasters highlights the human drive for survival and the humanity to come to the aid of others. People are more likely to act in a prosocial or therapeutic way if they have been given honest information about the event, understand the significance of the threat, and can see what action would be in their own best interest [Wilson 2007], even if they were initially unprepared for this.

The concept of silver linings from disaster [Clarke 2006, p. 129] is an important lesson as one attempts to study past disasters as rehearsal for current and future disasters because they are a normal part of life. Disasters are phenomena with patterns that result from complex causes. For example, the patterns of destruction from hurricanes help drive new designs and technologies in rebuilding. Disasters can be great educators and opportunities to develop community partnerships. Mutual alliances forged from tabletop drills and exercises at the local and regional levels were activated for response to the novel flu pandemic in 2009 to 2010. Storage facilities for medical supplies were provided by private warehouses that had engaged in partnerships with local health departments as part of the Strategic National Stockpile initiative from the federal government. Proper cold storage with security and back-up systems for temperature monitoring were provided by community hospital pharmacies when vaccine storage needs in local health departments exceeded their capacity. Public schools, churches, and business locations were used as clinic locations first for CDC-specified targeted priority groups. These locations were later made available for more clinics for the general public once vaccine was available for nonpriority groups. Needs for additional staffing for public health clinics were met by volunteers from public school staff, the Medical Reserve Corps, community emergency response teams, private industry, and university and college faculty and students in the health sciences. Coordinating this major public health response within days of receipt of delivery of vaccine would not have been possible without previous partnership and alliance work. Vaccines were delivered at the local level in lots of less than 3000 doses at one time, making the coordination of clinic operations in several locations a constant daily process for more than 6 months. Although it seems that the 2009 to 2010 pandemic did not result in widespread illnesses and deaths compared to previous pandemics, the experience provided real-time opportunities to refine previous plans into workable response strategies for the future.

Conclusion

Today's emergency and disaster preparedness and response policies and programs are shaped from earlier times in the history of the United States from responses to natural disasters, epidemics, pandemics, civil defense, and homeland security. The federal government continues to exert influence but not control over much of the emergency response capability and preparedness in the United States. Lessons learned from previous successes and failures strongly point out the failure of ad hoc decisions and crisis management. Mutual alliance-building is nearly impossible once the event occurs. Policy formulation needs to be systematic, deliberate, and evidence-based. The United States cannot continue to formulate reactionary and event-driven policy.

In these more modern times of all-hazards preparedness and response, a new order of mitigation, preparedness, response, and recovery is in order. The optimum time to establish partnerships with all the stakeholders and responders is not during a disaster. Multiple local partnerships are the key to successful disaster preparedness and response because communities must be able to initiate local response and recovery. Local communities are expected to sustain the local response and recovery for 72 hours. Partnerships for local and regional preparedness and response will build resilience for any type of disaster as the ability to provide what is needed, where it is needed, and when it is needed is inherent if the community itself can be its own first responder. Drills and exercises for all types of hazards or disasters must be designed and played out using mutual alliances.

References

Barnet, B. 2000. US government natural disaster assistance: Historical analysis and a proposal for the future. *Disasters* 23:139–155.

Birkland, T. 2007. *Lessons of disaster: Policy change after catastrophic events.* Washington, DC: Georgetown University Press.

Bowditch, H. [1877] 1972. *Public hygiene in America.* New York: Arno Press.

Clarke, L. 2006. *Worst cases.* Chicago: University of Chicago Press.

Coppola, D. 2007. *Introduction to international disaster management.* New York: Elsevier.

Dara, S., R. Ashton, C. Farmer, and P. Carlton. 2005. Worldwide disaster medical response: A historical perspective. *Critical Care Medicine* 33:52–56.

Fee, E., and T. Brown. 2001. Preemptive biopreparedness: Can we learn anything from history. *American Journal of Public Health* 91:721–726.

Gebbie, K., and K. Qureshi. 2006. A historical challenge: Nurses and emergencies. *Online Journal of Issues in Nursing* 11. Available from http://nursingworld.org/ojin (accessed April 24, 2009).

Gerber, B. 2007. Disaster management in the United States: Examining key political and policy challenges. *Policy Studies Journal* 35:227–238.

Heaps, R. 2009. The American Red Cross and Psychology's Disaster Response Network: A historical perspective. Retrieved from http://www.apa.org/practice/drnhistory.html (accessed April 14, 2009).

History of federal disaster mitigation: Evolution of the Federal Emergency Management Agency. 2005. *Congressional Digest* 84:258–288.

Hoover, M. 2005. Rowboat federalism: The politics of U.S. disaster relief. [online publication]. *Monthly Review Foundation.* Available from http://mrzine.monthlyreview.org/hoover281105.html (accessed April 19, 2009).

Hughes, A. 2004. Homeland security: Interstate mutual aid. *Spectrum: Journal of State Government* 77:16–17.

Koenig, H. 2007. Emergency response system in the United States. *Southern Medical Journal* 100:924–925.

Krawslecki, J. Personal communication. February 16, 2009.
Lakoff, A. 2008. From population to vital system. In *Biosecurity interventions*. A. Lakoff and S. Collier (Eds.). New York: Columbia University Press.
Lindell, M., C. Prater, and R. Perry. 2007. *Introduction to emergency management*. Hoboken, New Jersey: John Wiley & Sons.
Lyznicki, J., I. Subbarao, G. Benjamin, and J. James. 2007. Developing a consensus framework for an effective and efficient disaster response health system: A national call to action. *Disaster Medicine and Public Health Preparedness* 1 (Suppl 1):S51–S54.
Markenson, D., S. Reynolds, and the Committee on Pediatric Emergency Medicine. 2006. The pediatrician and disaster preparedness. *Pediatrics* 117:340–362.
Mileti, D. 1999. *Disasters by design: A reassessment of natural hazards in the United States*. Washington, DC: Joseph Henry Press.
Morrow, C. 2007. Public health emergency preparedness and the evolution of public health in the 21st century. *Journal of Public Health Management Practice*. 13:439–440.
Ripley, A. 2008. *The unthinkable: Who survives when disasters strike and why*. New York: Crown Publishers.
Shields, H. 1963. Sabin Sundays. *American Journal of Nursing* 63:109–111.
Tulchinsky, T., and E. Varavikova. 2009. *The new public health*. 2nd ed. New York: Elsevier.
Virginia Tech Review Panel. 2007. Mass shootings at Virginia Tech April 16, 2007. Retrieved from http://www.governor.virginia.gov/TempContent/techPanelReport-docs/FullReport.pdf (accessed February 10, 2010).
Wilson, S., B. Temple, M. Milliron, C.Vazquez, M. Packard, and R. Bruce. 2007. The lack of disaster preparedness by the public and its affect on communities. *Internet Journal of Rescue and Disaster Medicine* 7. Available from http://web.ebscohost.comezproxy.stockton.edu:2048 (accessed April 19, 2009).
Ziskin, L., and D. Harris. 2007. State health policy for terrorism preparedness. *American Journal of Public Health* 97:1583–1588.

Chapter 2

Network Governance in Emergency Management in the Caribbean

Philip Duku Osei

Contents

Introduction

This chapter is intended as an exploratory essay on network governance in emergency management in the Caribbean. The Caribbean is an archipelago situated east of the North and South American land masses. By its nature and location, it faces a host of seismic activity and vagaries of weather conditions, including a hurricane season that extends from June to November every year. To give a sense of the magnitude of the hazards experienced in the region, an illustration from Jamaica will be instructive. Jamaica has experienced an increase in the frequency of natural events associated with weather, including tropical depressions, tropical storms, hurricanes, droughts, and landslides over the past 35 years. The impact of these hazards has caused major social dislocation and serious economic costs.

Between 1998 and 2004, the country experienced 10 major weather-related disaster events with significant economic implications. In 1998, the country had three events, the cost of which was approximately J$200,000,000. In 2001, Hurricane Michelle resulted in damage valued at 0.8% of the gross domestic product and cost the country 2.8% of government revenue and grants. In 2002, the damage was valued at 0.7% of the gross domestic product. In 2004, Hurricane Ivan resulted in damage of J$35 billion, approximately 8% of gross domestic product. The magnitude of the costs associated with natural events seems to be escalating. In Haiti alone, the human costs associated with weather conditions always assume tragic proportions. Coupled with this are manmade hazards that are ever present because of the potential of maritime accidents posed by the shipment of hazardous material including oil and nuclear material. Historically and geopolitically, the Caribbean region has had variegated colonial and postcolonial experiences, having been conquered and ruled previously by Spanish, British, French, and Dutch imperial powers. By nature, as small island developing states (SIDS) sharing the Caribbean Sea to the west and the Atlantic Ocean to the east and the legacy of differing governance institutions bequeathed by the colonialists and perpetuated in the postcolonial era, fashioning modern emergency management systems for the region and individual islands required extraordinary acumen in communication, collaboration, and a wealth of resources.

With this extraordinary geographical and governance complexity, praxis, and political realities, these characteristics of the region must be fully appreciated in conceptualizing the nature of institutions, governance styles, and financing of emergency management in the Caribbean. Another fascinating feature of this complexity that must also be borne in mind in our conceptualization of emergency management in the Caribbean region concerns the identity and constitution of the region. There are continental Caribbean countries, such as Guyana, Suriname, and Belize, that are closely related to the Anglophone Caribbean regional bloc—the Caribbean Community (CARICOM)—and who, by virtue of common history and efforts at integration for purposes of trade, have functioned as a region whose common destiny is bound together (Figure 2.1). These identities have implications for the outcomes of collaborative endeavors and individual country action for risk reduction and emergency management. From my observation of country experiences and regional collaboration in dealing with emergencies, it seems network governance and management are the most appropriate concepts that can be employed if one is to fully understand the complexity of the issue. This choice of theoretical framework is justified because of the presence of independent country

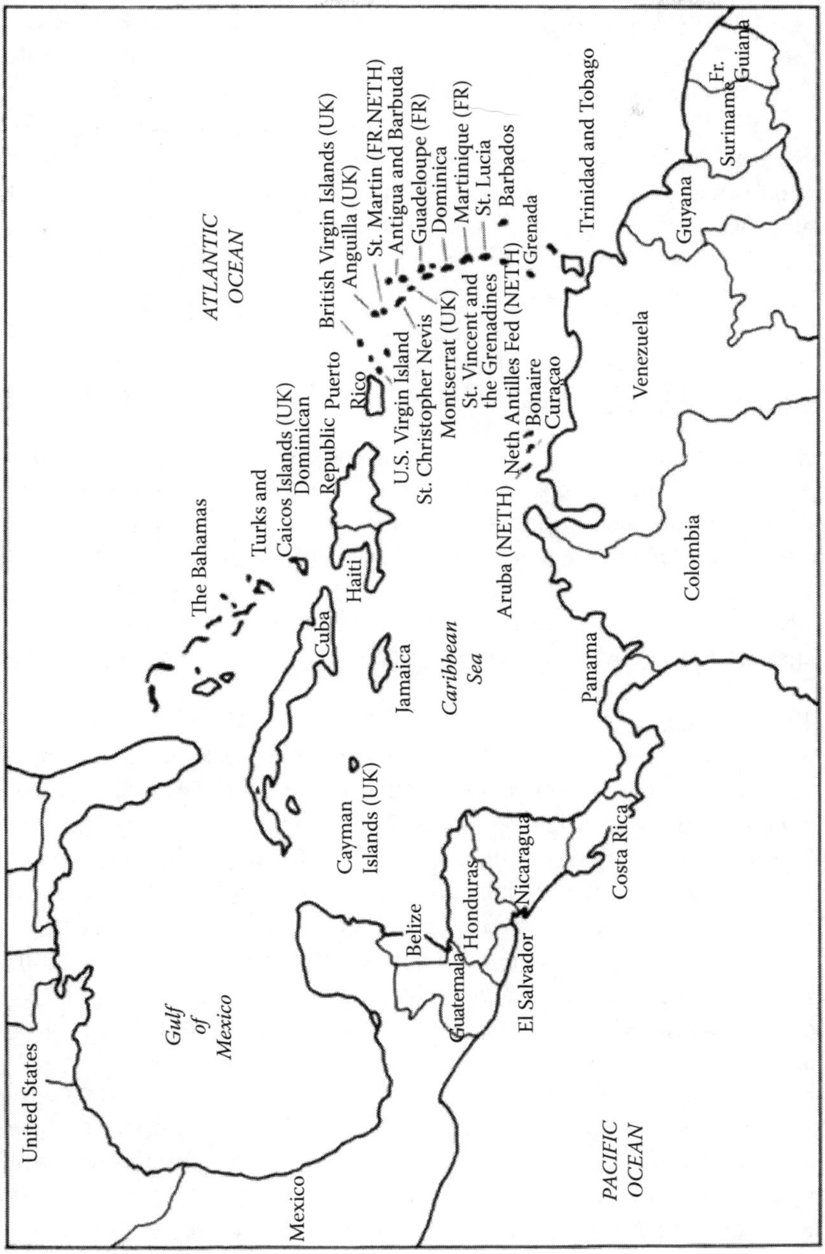

Figure 2.1 Map of Latin America and the Caribbean.

institutional arrangements, as well as the existence of strong political inclinations to have regional frameworks for resolving and addressing disastrous events that might overwhelm individual countries.

From the foregoing, emergency management in the Caribbean can be conceptualized as a four-tiered arrangement of institutions and organizations consisting of multilateral, regional, national, and subnational policy frameworks and the financing and delivery systems. Emergency management, so conceptualized, is a public good and has been delivered and improved as such over time and space. The main purpose of this chapter is to unravel the complex relationships and collaborative arrangements for maintaining the system and responding to emergencies when they occur. Therefore, our examination will focus on system maintenance, system sustenance, system efficacy, and response outcomes.

Theoretical Framework

Network governance connotes increased efficiency, specifically the integration of processes for the measurement and improvement of organizational efficiency and end-user satisfaction. It emphasizes data-sharing and collaboration between internal teams and external partners. Effective emergency management is premised on the comprehensive integration of emergency plans at all levels of state and nonstate actor involvement. Network governance is important at all phases of the disaster risk reduction process: mitigation, preparedness, response, and recovery.

Additionally, managing a portfolio of provider networks requires a form of public management different from the hierarchical model to which governments and citizens are accustomed. Partnerships will not protect the government from not owning up to its responsibility of providing the public with quality services that are justly delivered. As the challenges posed by the global environment become more complicated, governance models as one-size-fits-all solutions are quickly becoming outmoded. Government agencies are becoming more important as generators of public value rather than direct service providers within the web of multiorganizational, multigovernmental, and multisectoral relationships that increasingly characterize modern government [Goldsmith and Eggers 2004, p. 8].

To cope with the complexity of natural phenomena, institutional arrangements and related management systems should incorporate different actors from different areas of society. Thus, an emphasis on the establishment of multiactor structures, consisting of both public and private actors, is a common denominator. The concepts of comanagement and governance are also useful. They incorporate an implicit assumption about the establishment of social networks based on different logic than political–administrative hierarchy. Briefly, these networks can be described as social structures made up of nodes (actors) that are connected via a multitude of links (e.g., in the form of information flows, exchanges of goods, legal relations) [Carlsson and Sandstrom 2006].

After consolidating the findings from reviews of seven umbrella or network organizations with the aim of identifying effective network governance structures and practices, Edgar [2002, p. 3] posited the following:

- Network organizations that demonstrate credibility through sound governance, organizational stability, and the capacity to bring informed expertise or opinion to policy discussions are more likely to exert effective influence on public policy development than those without such credibility. This is more difficult in networks that have a disparate membership and deal with multiple issues.
- Rules for decision-making need to be clear. Variations in cultural approaches to decision-making need to be accommodated in a way that balances respect for traditional values with the need for efficient use of available time.
- Networks that receive core funding from a government source generally have their relationship with government defined by a contribution agreement or some other form of contractual agreement that defines the relationship and expectations. This type of interdependent relationship encourages a collegial and collaborative approach to policy formulation and delivery.
- Policy formulation appears to work best when there is a flexible approach to participation of minority groups taking part in all discussions, taking part in discussions related to a specific topic, or simply keeping interested parties informed of developments so they can engage in the policy formulation process as they see it important to their own interests.
- Effective advocacy also requires an ongoing positive relationship with government, including regular, informal communication with government to stay in the loop and the ability to anticipate new issues and bring them to the table.
- Most effective policy networks use an annual general meeting or annual conference to identify priorities for policy work and begin the policy development process. Resolution sessions at such meetings, combined with a degree of flexibility for the national governing body to initiate policy work between Annual General Meetings (AGMs), strengthen the organization's effectiveness.
- Entrepreneurial strategies to diversify revenues and escape "single source revenue dependence" [p. 3] are key to developing essential organizational flexibility and independence. Grassroots fundraising has been an important element in the strategy of successful networks.

Goldsmith and Eggers [2004] highlighted that networks can serve a range of impromptu purposes, such as creating a marketplace of new ideas inside a bureaucracy or producing the maximum possible public value, greater than the sum of what each lone player could accomplish without collaboration. There is a new thrust toward network governance because of the advantages offered by the model in addressing the myriad challenges faced by governments. The movement to networks includes some of the following advantages:

- Specialization.
- Innovation.
- Speed.
- Flexibility.
- Increased reach.
- Networks force companies and bureaucracies to concentrate on their core mission and exploit the expertise of best-of-breed providers [Goldsmith and Eggers 2004, p. 29].
- Networks foster learning and continuous improvement by providing more timely access to a broader knowledge base than is possible within a single organization [Goldsmith and Eggers 2004, p. 31].
- Networks are more nimble and flexible than hierarchies. They enable the government to bypass stultifying procedures that slow personnel movement, and they leverage both private sector and governmental funds. Networks permit more rapid and usually less politically painful downsizing and upsizing than hierarchies.
- Networks enable governments to increase productivity when new demands conflict with hiring limits.
- Networking helps governments evolve from one-size-fits-all service providers to a one-stop portal for myriad providers [Goldsmith and Eggers 2004, p. 33].
- A network is less limited by rigid government programmatic restrictions and therefore can also fill in funding gaps with private dollars, which generally have fewer restrictions than government funds. The flexibility of the private provider allows a speedy, customized response to clients whose needs are continually changing [Goldsmith and Eggers 2004].
- The network can remove jurisdiction-imposed scale limitations; network partners can provide answers, take risks, or reduce the marginal costs by attracting other government customers [Goldsmith and Eggers 2004, p. 34].
- Networks can help governments expand their reach by enabling them to engage the nonprofit sector's innovative spirit and creativity in efforts to solve major social problems.
- Networks allow innovative government officials to discharge the government's role in solving social problems by supporting not supplanting functioning elements of civil society [Goldsmith and Eggers 2004, p. 37].

Critique of the Network Approach

However, the network model is not considered a panacea for all organizational problems. In fact, it comes with several challenges. The following are some of the weaknesses of the approach:

- A program's success or failure is dependent on whether the network manager masters the challenges of governing by network: aligning goals, providing oversight, averting communications meltdown, coordinating multiple partners, managing the tension between competition and collaboration, and overcoming data deficits and capacity shortages.
- Government networks, however, tend to form to deliver the type of service whose outcomes are sometimes unclear, are difficult to measure, and may take years to be realized.
- The complexity that necessitates a networked solution often makes the accountability process difficult.
- Varied criteria for success makes for messy network management.
- The missions of the organizations within the network will not always align well.
- Goal congruence also occurs when government activates a network but also competes against parts of the network.
- Governments often mistake public–private partnerships and outsourcing as a way to offload the headaches of managing a service and therefore fail to exercise adequate oversight. This then leads to cost overruns and service failures, among other things.
- Network government typically involves coordination between multiple levels of government, nonprofit organizations, and for-profit companies. Each has its own constituencies and, when complexity is high and responsibility unclear, coordination problems can undermine the network. Poor performance by any one organization or the breakdown of the relationships between any two organizations can imperil the performance of the whole.
- The dearth of accurate data may cause networked service delivery to fail just as it does more straightforward outsourcing.
- Good network procurement capacity requires the participation of individuals with broad experience and the ability to see how varied configurations produce different outcomes and how different partners produce different results.
- Uncertainty regarding what will occur when a contract expires fosters aversion to risk-taking and wariness among partners about one another.
- If the benefit of the new network is flexibility and performance, then the government contract manager may exercise the right to terminate or substantially alter the contract, thus aggravating anxieties regarding relationship stability.

The Emergency Management Architecture in the Caribbean

This section examines the emergency management architecture and how it functions in the Caribbean. This is done by dealing with institutional frameworks, structures, and processes designed by the governments of the region. This approach brings to life the organic elements of cooperation and collaboration that also qualifies these emergency service arrangements as a holistic networked system.

Institutions

Structurally, there are four tiers of institutions that directly engage in emergency management in the region. The first tier consists of multilateral institutions that have a regional focus. At this level, actors engage in diplomacy, organizational activities, and arrangements for technical skill-sharing, financing, and support for creating and establishing information platforms and information societies. The second tier is occupied by regional institutions, which in the Caribbean was concretized by the establishment of the Caribbean Disaster Emergency Response Agency (CDERA). The CDERA is a regional attempt at cooperation in disaster and emergency management and response, and it is as strong as the authority imbued in it by the Heads of Government Council of the Caribbean Community. The third tier is composed of national (central) institutions that provide overarching policies, frameworks, and plans for emergency management planning, mitigation, and response. There are 14 national disaster and emergency planning institutions in the Caribbean Community. The non-Anglophone countries have their own systems and institutions, and the British and French Dependencies and Departments normally have systems and institutions that derive their authority from the mainland or metropole.

However, national level plans and policies will not be effectively implemented without the fourth tier of the governance structure, which is composed of local authorities, local communities, and nongovernmental organizations and individuals who are integrally involved in building resilient communities and better response machineries.

The multilateral level institutional framework for disaster management includes the International Strategy for Disaster Reduction (ISDR), also referred to as the Geneva Mandate on Disaster Reduction. This framework encourages networking and information-sharing. The ISDR has been mandated by the UN General Assembly to continue international cooperation to reduce the impacts of El Nino and other climatic variables (CARDIN Directory of Disaster Agencies 2001). The ISDR is the unit for Latin America and the Caribbean and is headquartered in San Jose, Costa Rica.

There are regional institutions for knowledge generation and sharing in relation to disaster management, including the Regional Disaster Information Centre for Latin America and the Caribbean (CRID). This agency exists to promote a culture of prevention in Latin American and Caribbean countries through the compilation and dissemination of disaster-related information and the promotion of cooperative efforts to improve risk management in the region. It started as the Disaster Documentation Centre (DDC), established in PAHO in 1993, and saw some growth in its activities in 1994 with the establishment of the International

Decade for Natural Disaster Reduction (IDNDR). The DDC's activities are underpinned by three main objectives: compilation, analyses of existing documentation, and supportive policy development and research in prevention and disaster-related management supplied by organizations and agencies across the LAC regions.

Stiching Rampenbestrijding en Brandweerzog Nederlandse Antillen-Strirana (National Disaster and Fire Prevention Organisation for the Netherlands Antilles) was funded in August 1997. It is a disaster prevention project for the Netherlands Antilles that is supported and driven by the Dutch Ministry of Internal Affairs in continental Europe. However, the coordinator of the project is based in Curaçao in the Netherlands Antilles (CARDIN Directory of Disaster Agencies 2001).

There is a mélange of other institutions and organizations that are networked around central agencies according to themes, including information-sharing, technical skills assistance, and relief procurement and management after major events. An analysis of missions and frameworks of these multiorganizations highlights the specialties among them:

1. Havana Initiative 2001, which established a collaborative framework for capacity-building as its core activity in the areas of risk management, adaptation, and mobilization of resources.
2. UNEP-CEP deals with global environmental issues and programming.
3. University-based centers and research institutional networks, which deal with meteorological and seismic studies and risk reduction.
4. On-the-ground, project-based approaches, which are normally under the purview of communities and local authorities.

Cooperation has also been forged through high-powered meetings and conferences. For example, there was a Regional Strategy for Disaster Information meeting hosted by the Costa Rica National Emergency Commission in February of 1997 (CARDIN Directory of Disaster Agencies 2001). In 2006, the governments of the Caribbean region, the World Bank, and other multilaterals created the Caribbean Catastrophe Risk Insurance Facility (CCRIF) as an insurance system and a buffer to which regional countries could refer for assistance when severe emergencies occur such as hurricanes and earthquakes.

Existing Collaborative Arrangements

Within the past two decades, there has been an increasing recognition and commitment to network governance approaches to emergency management in the Caribbean. In the Caribbean, an agenda to develop and maintain an institutional framework and capacity for advancing disaster preparedness has emerged. With programs such as the Caribbean Disaster Mitigation Project (CDMP), dating from the mid-1980s, the Organization of American States (OAS) and the United States Agency for International Development (USAID) were two of the first international organizations to focus on disaster mitigation within the region [Bisek et al. 2001]. This was closely followed by calls for the integration of natural and environmental disaster policies into national development planning processes at the 1994 SIDS conference in Barbados when the Programme of Action for the Sustainable Development of Small Island Developing States was adopted [Bisek et al. 2001]. The OAS was also a key implementing agency for the Caribbean Planning for Adaptation to Climate Change (CPACC) project, an initiative funded by the Global Environmental Facility and executed through the UWI Centre for Environment and Development. In May 1997, at the Caribbean/United States Summit, several countries agreed that the region is vulnerable to various forms of natural disasters and made commitments to strengthen the capacity of countries within the region to protect themselves from disasters, to treat investment in planning, preparedness, and mitigation initiatives as developmental priorities, and to decrease the need for emergency response resources in the future. The Bridgetown Declaration came out of this summit.

In 1998, the Emergency Recovery and Disaster Management Program was approved by the World Bank to help the eastern Caribbean countries create a disaster management strategy that was comprehensive. Similarly, the Organization of Eastern Caribbean States (OECS) countries were given assistance in preparedness, mitigation, and postdisaster rehabilitation through the new Strategy and Operational Guidelines for Natural Disaster Management, developed by the Caribbean Development Bank in 1998 [Bisek et al. 2001]. Institutions for managing risks and emergencies were developed in a piecemeal fashion. In October 1999, the mandate of the CDERA was expanded to include the principles of comprehensive disaster management (CDM). This was done within the framework of the 21st Century Disaster Management Action Plan for the Caribbean, which came out of the Third Caribbean Conference on Natural Hazards [Bisek et al. 2001]. This was later followed by commitments to establish a disaster mitigation facility for the Caribbean—a facility that would strengthen the Caribbean Development Bank's capacity to implement effective strategies and help borrowing member countries adopt disaster mitigation policies and practices—with the signing of an agreement between USAID and the Bank in 2000. Instituted support for the development of the facility was given by USAID/OFDA. Additionally, a serious component of the post-Hurricane-George and post-Hurricane-Lenny reconstruction program of USAID included mitigation.

The OAS partnered with USAID/OFDA to execute the Caribbean Disaster Mitigation Project (CDMP) from 1993 to 1999. The activities were in six major streams: community-based preparedness, hazard assessment and mapping, hazard-resistant building practices, vulnerability and risk audits for lifeline facilities, promotion of hazard mitigation within the property insurance industry, and incorporation of hazard mitigation into postdisaster recovery. The outputs from these activities provided a significant body of technical products, which were made available for use within each territory. In the same year, 1999 members and associate members of the ACS signed the Agreement for Regional Cooperation on Natural Disasters. The grouping identified natural disasters as a priority in its first-phase work program and formed the Special Group on Natural Disasters to promote cooperation among the members and interaction with relevant regional and international organizations. Similarly, three projects to address early warning systems, floods and earthquakes, and training programs, education, awareness, and strengthening of Disaster Emergency Management Organizations (DEMOs) emanated from the discussions of a meeting in October of 2000. The projects were intended to achieve three objectives: strengthen the response capabilities of the disaster relief systems of ACS members; utilize existing disaster information centers to create new information related to the activities of the priority areas identified and; develop technical cooperation in emergency relief among members as a mean to strengthen the disaster management organizations. The ACS has also spearheaded a study on wind and earthquake codes for the Greater Caribbean Basin. This study was to be undertaken by the University of the West Indies in collaboration with the Universities of Costa Rica and Pavia (Italy).

The implementation of the Comprehensive Approach for Disaster Management in the Caribbean project in 2000 sought to develop a regional CDM strategy, strengthen CDERA to implement CDM, and build support for CDM at the national level; this was supported by USAID and the United Nations Development Programme (UNDP). During the 1990s, several milestones were achieved out of the UN International Decade for Disaster Reduction, bringing to the forefront a realization among governments, donors, and other key players the need for a comprehensive disaster management strategy.

The program and institutional foundations of emergency management in the Caribbean include community-based, national, regional, international, and nongovernmental organizations. At the regional level, there are four disaster management networks: Meteo Frances, CDERA, Anglo-Spanish networks, and Netherlands Antilles.

The 73-year-old Meteo Frances serves the French Departments of Martinique, Guadeloupe, and French Guiana. Regional direction for the Caribbean is headquartered at Fort de France, Martinique. The center has direct links with the main operational center of Meteo France that is located in Toulouse, France. Each center provides services including weather forecasting, consultancy, and risk assessment. Meteo France was chosen as the manager of the first step in the Caribbean Radar Network Project (CARDIN Directory of Disaster Agencies 2001).

The CDERA is an interregional supportive network of independent emergency units throughout the Caribbean region. The participating member states and agencies of the CDERA include Anguilla's coordinator, Antigua and Barbuda's National Office of Disaster Services, the Central Emergency Relief Organisation in Barbados, the Disaster Management Unit in the Bahamas, National Emergency Management Organisation in Belize, British Virgin Islands' Department of Disaster Management, Dominica's Office of Disaster Management, the National Disaster Management Agency in Grenada, Guyana's Civil Defense Commission, Jamaica's

Office of Disaster Preparedness and Emergency Management, Montserrat's Disaster Management Coordination Agency, the National Emergency Management Agency in Saint Kitts and Nevis, Saint Lucia's National Emergency Management Organisation, Saint Vincent and the Grenadines' National Emergency Management Organisation, and the National Emergency Management Agency in Trinidad and Tobago.

CDERA is the central disaster management organization within the Caribbean region, established by agreement of its member states in 1991 for the mobilization and coordination of emergency disaster relief for member states, mitigating immediate consequences of disaster, promoting a sustainable disaster response capability among members, and securing, coordinating, and channelling reliable and comprehensive information on disasters. Separate and apart from building disaster management capabilities within the region, the organization's main mandate is to coordinate the regional response. The specialized agency grew out of the Pan Caribbean Disaster Preparedness and Prevention Project and relates to CARICOM through the Council for Trade and Economic Development. Despite the agency's dynamic, highly variable operating environment, with a diverse contingent of partners responding rapidly to disasters when they occur, CDERA is constrained by the availability of funding, as are most of its partner governments in the region.

The Pan-American Health Organisation (PAHO) is one of CDERA's key partners, instrumental in developing the Humanitarian Supply Management System (SUMA), a system for the management and distribution of relief supplies. As highlighted by Bisek et al. [2001], PAHO has been engaged in mitigation activities with respect to health and sanitation infrastructure and the preparedness and response mechanisms. The Caribbean People's Development Agency is a regional development-focused nongovernmental organization based in St. Vincent and the Grenadines with a formal agreement with CDERA focused on local community preparedness [Bisek et al., 2001]. Chaired by the UNDP, the Eastern Caribbean Donors Group supports CDERA in coordinating disaster response assistance. The UNDP has also contributed to emergency management within the region by funding the Disaster Emergency Response and Management System Project, which began in 1996 and ended in 2000 and had a major role to play in the initiation of the Comprehensive Disaster Management Strategy.

Furthermore, there is the Institutional Support and Capacity Building for Disaster Management project. Funded by the European Development Fund, this project seeks to strengthen CDERA's capacity as the regional driver of the Comprehensive Disaster Management Strategy, a stakeholder-framed mechanism for prioritizing outcomes and framing cooperation. CDERA is implementing the Institutional Support and Capacity Building for Disaster Management in the Caribbean Project (No. 9 ACP RCA 9). The project's country action plan is designed to respond to the expressed need of countries for direct support in developing and implementing a national CDM Program and is organized along five result areas:

1. Increased effectiveness of CDERA and improved regional coordination.
2. Development of disaster management legislation and regulations in support of CDM.
3. CDM promoted through national disaster awareness campaigns.
4. Increased use of information communication and technologies (ICTs).
5. Enhanced regional capacity for undertaking disaster management and related teaching and research.

It is anticipated that the project will result in increased technical assistance for participating states under a technical assistance protocol arrangement, improved capacity of a CDERA coordinating unit to provide technical assistance in programming design and implementation, and improved horizontal cooperation for support of best practice documentation and experience sharing (http://www.cdera.org/projects/cap/index.php).

The Anglo-Spanish networks refer to the cooperative arrangements between the anglophone Caribbean countries and networks and their Latin American counterparts. This network represents the Latin language bloc. The Netherlands Antilles bloc represents the networks that converge around the Dutch-speaking Caribbean, including countries such as Curaçao and Suriname.

Network Governance of Emergency in Action

This section discusses the management, funding, and working relations at the four multiorganizational levels. The international development partners or donors deal directly with the regional emergency agency and with national governments. CDERA mediates and leverages resources and technical know-how for regional capacity-building and knowledge-sharing. It could be gleaned from Figure 2.2 that there are several networked activities in the functioning of CDERA that look intricate, even viewing governance at this level alone. Adaptive management provides partners with flexibility in the goals and the methods they select to achieve them. The ability of the network is enabled by the central coordination role that CDERA is expected to provide and the efficacy with which it provides it (Figure 2.3). However, even though most of the activities of financing and knowledge-sharing tend to be voluntary, they are simultaneously compelling because failure in this field could have dire consequences. Similarly, progress can only be measured by continuous feedback and evaluation during the implementation of discrete programs instead of by a checklist of performance targets at a program's conclusion. Incremental capacity development has marked this area of human endeavor as authoritative decision-making; the responsibility to effect changes on the ground resides with the governments of the regional countries.

The layers below the regional institutions (the national institutions and the local-level decentralized institutions) also add a certain level of complexity. As argued by

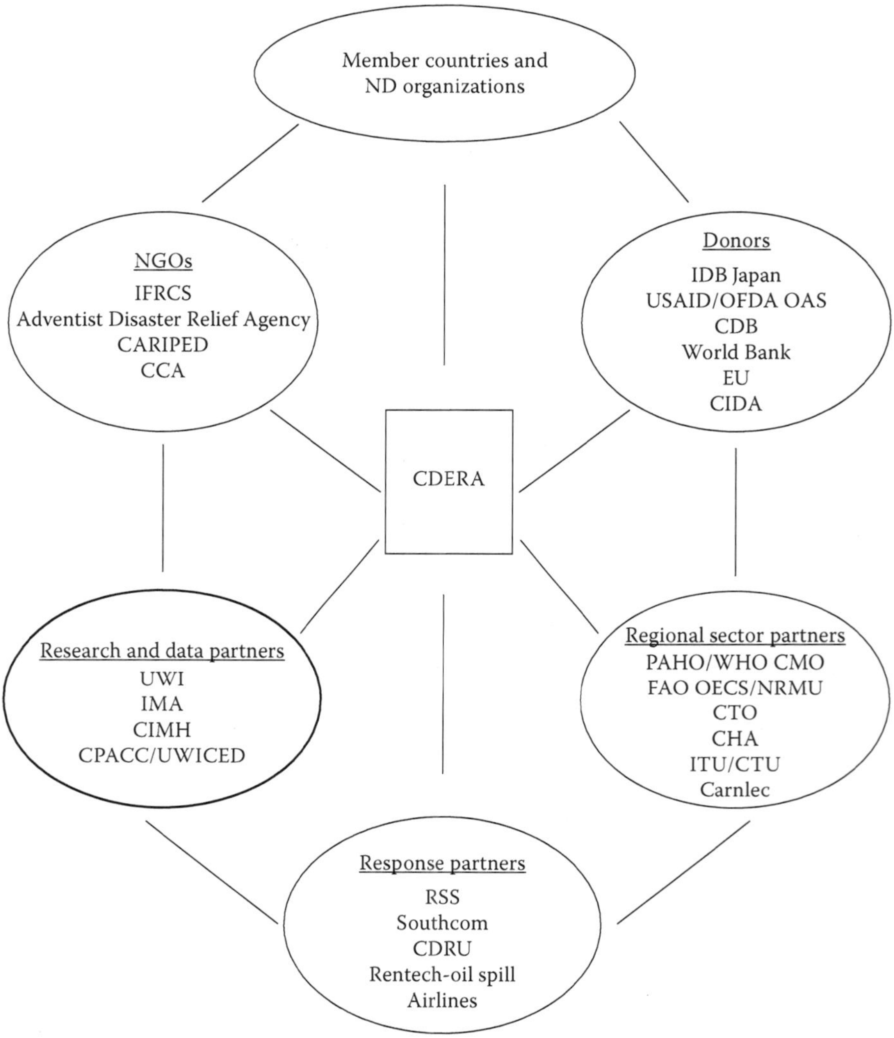

Figure 2.2 CDERA's strategic linkages. (Reproduced from Jones, E. B., et al., *Comprehensive Disaster Management in the Caribbean: A Baseline Study?*, www .cdera.org. With permission.)

Wolfgang and Elbie [2004], using governance of networks means the successful implementation of social institutions, guaranteeing rational decision-making for self-regulation of participants' cooperation and that, as far as network membership is voluntary, the most important task of these institutions is the balance of different interests by appropriate communication management and by generally accepted rules leading to common agreements and joint action. The most difficult task for governance of networks is the process of balancing different interests and

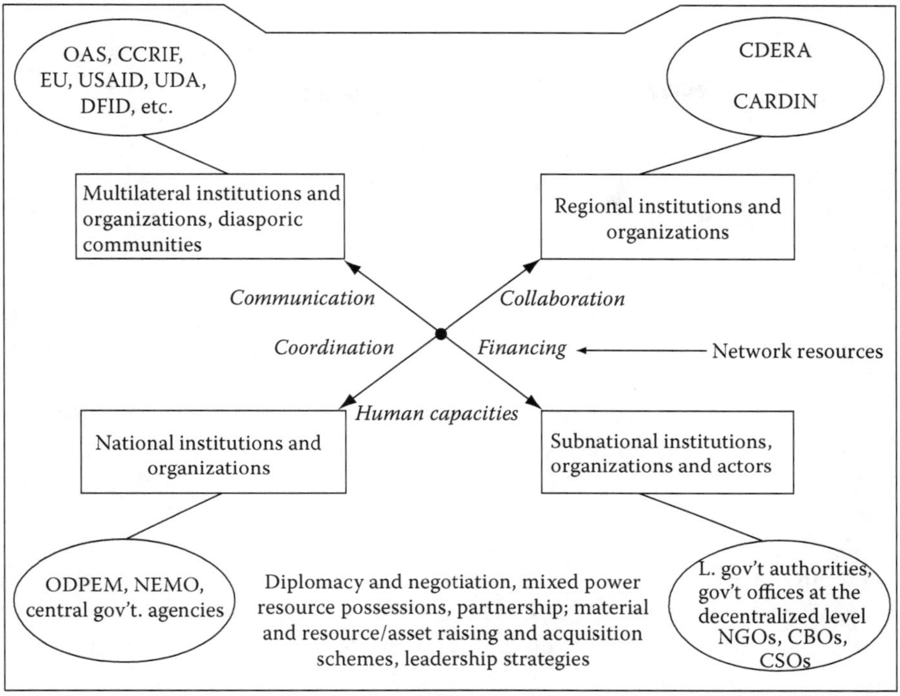

Figure 2.3 Network governance of emergency management.

aggregating them to a communally shared position. Success is reckoned in terms of the networks' ability in the steering of contents and achieving common goals by the local partners [Wolfgang and Elbie 2004].

Policies on disaster management and response are normally set by cabinet and national disaster committees, which are normally led by prime ministers. Over the years, Caribbean governments have learned and drawn lessons from international sources and come to the realization that the management of hazard now consists of more than the event stage itself (response, rehabilitation, and reconstruction), but also includes the pre-event phase (preparedness, risk identification, risk assessment, and risk reduction or mitigation). It has also been recognized that the resources to manage hazards do not necessarily reside in the public service of the countries. As such, national disaster management mechanisms are developed using collaborative and network approaches that include government, nongovernmental, and international organizations and institutions operating within each country with specific responsibility for the management of hazards.

The subnational level has varied institutions, which include local authorities and communities. In those islands with formal local authorities, the functions of disaster management are assumed by the relevant local authority. It must be emphasized that the lack of formal local authorities does not mean that the function is not performed. Indeed, the functions are performed by alternative institutions. The parish or local government level acts in similar ways to the national level, in which bylaws are promulgated by the local authority and coordinated by a district or parish disaster coordinator. The district or parish coordinator is expected to organize the local territorial units and equip them with the requisite knowledge. As such, policy-level decisions and activities related to public awareness, prevention, and response are guided by the local governance system. Research shows that the subnational level tends to be under-resourced, yet the capacity is expected to exist for use at the national level in the context of major natural events [Osei 2007].

It was not possible to conduct primary research in all the CARICOM countries for purposes of writing this chapter, but the author's experience in Jamaica, Saint Lucia, and Saint Kitts and Nevis indicates that the islands have varying capacity for emergency management at the community level. The author's knowledge of local governance systems in these countries suggests that countries with vibrant community development programs will tend to have organizational acumen for responding to emergencies.

Among the three countries named above, Jamaica has the most well-developed community development institutions at both national and local levels. At the community level, it was realized in Jamaica that to create a comprehensive emergency management design, public action in this thematic area requires community acceptance and initiative. Therefore, the Office of Disaster Preparedness and Emergency Management developed a community program that sought to provide communities with the ability to manage local interventions in preparedness and response. These community-level capacity developments get an added boost in times of emergency, when multilateral-level knowledge and know-how are shared with local authorities and communities by agencies such as the United Nations Economic Commission for Latin America and the Caribbean, which assists countries in postdisaster assessment, as happened in Jamaica after Hurricane Ivan in 2004.

Financing Emergency Management

Financing for emergency management and response comes from several sources and is multilayered. Traditionally, financing has been country-driven via the budget processes of each country. Most CARICOM countries are heavily indebted middle-income countries whose economic growth was spurred by the preferential treatment received in trade with metropolitan countries. The benefits of preferential trade are dwindling quickly with the collapse of the sugar and banana markets and the replacement of the existing regime with an economic partnership agreement, which enjoins the Caribbean countries to move toward competitiveness and maturity. The expiration of preferential treatment has been long in coming and, with the drying up of national budgetary resources, some of the countries attempted the establishment of some form of national disaster funds. Unfortunately, inadequate resources were reserved for these funds and every emergency has dictated an invitation by governments to the international community for help.

During emergencies caused by the main hazards (hurricanes, floods, volcanoes, and earthquakes), one element of funding that has acted as a sure buffer has been remittances from the Caribbean Diaspora in North America and Europe. Caribbean citizens abroad use the instrumentalities of their countries' overseas missions to garner support and raise funds to help in relief, reconstruction, and development. Citizens groups donate medical equipment, water-purifying equipment, and other supplies either to the national government or to specific localities in these countries. The preponderant method of reaching the needs of local people in the Caribbean who have suffered catastrophe has been through the direct transfers to families and relatives, and this is partly because of the diasporic communities' distrust of government. Indeed, political cleavages are sharpened after an emergency, when governments and opposition do not trust their own public response agencies to distribute relief supplies equitably. A good illustration of this point could be seen in prime ministerial decisions to set up seemingly neutral agencies in the aftermath of Hurricanes Gilbert in 1988 and Hurricane Ivan in 2004 in Jamaica.

The Caribbean lies directly in the path of hurricanes that originate in the Atlantic Ocean and the region experiences a hurricane season for 6 months of the year. The losses sustained by the countries in any eventful year are enormous. For example, Hurricane Ivan caused losses amounting to 8% of the economy of Jamaica in 2004. Similarly, a World Bank brief [2006] noted that in 2004 four hurricanes—Charley, Frances, Ivan, and Jeanne—wreaked havoc on several small islands in the region, causing combined losses of over USD$4 billion. Furthermore, the aggregate economic losses incurred by the small island states of the Caribbean Basin as a result of storms during the period from 1979 to 2005 is estimated at USD$16.6 billion (in current value), or USD$613 million annually. However, though less frequent, other natural events could be equally destructive, such as the volcanic eruption in Montserrat in 1995.

The economic impact of hurricanes is noted to be highly variable. The World Bank's analysis of the economic impact of storms in the Caribbean is instructive, as could be viewed in the following quote.

> During a 27-year period (1979–2005), 13 years are classified as 'loss free', meaning no significant economic damage was incurred by any Caribbean country; eight years saw a single storm causing losses to a small island state in the region; and finally six years during which significant damage was caused by multiple storms. In rare occurrences, storms can also cause damage to multiple countries. Hurricane Ivan, a Category 3 hurricane, had a measurable impact in eight different countries . . . Apart from storm frequency, the intensity of the storm is a determining factor influencing economic losses. The record shows that storm losses to private property, state infrastructure and other state property can be considerable. When Hurricane Ivan struck Grenada

> in 2004, the loss was calculated at US $800 million, about two times the country's GDP—of which government losses accounted for about 30 percent [World Bank 2006].

As a result of the weak economic resilience of the Caribbean states, the CARICOM heads of government requested the World Bank help them with insuring catastrophe risk during the 2004 season. The governments of the region working in collaboration with the Bank and other donors developed a short-term solution to the tight liquidity issues associated with major natural events in which substantial losses are incurred. This was in the form of the Caribbean Catastrophe Risk Insurance Facility (CCRIF) in 2006. The following are main features of the CCRIF [World Bank 2006]:

- The CCRIF allows CARICOM governments to purchase insurance coverage to finance immediate postdisaster recovery needs.
- The facility acts as a risk aggregator. The CCRIF allows participating countries to pool their country-specific risks into one, better-diversified portfolio. This diversification was expected to result in a substantial reduction in premium cost of 45% to 50%.
- Claims payments will depend on parametric triggers. Index-based (or parametric) insurance instruments pay claims based on the occurrence of a predefined event rather than an assessment of actual losses on the ground. This measurement, made remotely by an independent agency, allows for transparent, low settlement costs and quick-disbursing contracts.
- The facility was created with initial funding from donors. Initial funding was needed to allow the facility to cover start-up costs, retain some of the risk, and access the reinsurance markets where it is most efficient.
- The facility transfers the risks it cannot retain to the international financial markets. This will be done through reinsurance or through other financial coverage instruments (for example, catastrophe bonds). The accumulation of reserves over time should lessen the facility's dependence on outside risk transfer and smooth the catastrophe reinsurance pricing cycle.
- The facility is expected to maintain financial protection to survive 1-in-1000 year events. Should the total insured losses exceed its claims-paying capacity, payouts will be prorated based on the total amount of expected claims compared to the remaining available funds.
- The CCRIF was established as an independent legal entity. It was conceptualized to be created as an insurance captive managed by a specialized firm under the supervision of a board of directors composed of representatives from the donors and participating countries. This board will be supported by the technical advice of a facility supervisor.
- Insured countries will pay an annual premium commensurate with their own specific risk exposure. Parametric insurance products are priced for

each country based on their individual risk profile. Annual premiums typically vary from USD$200,000 to USD$4 million for coverage ranging from USD$10 to USD$50 million.

In the past two years, the CCRIF paid out approximately USD$500,000 each to Dominica and St. Lucia after an earthquake occurred in 2007 and USD$6.3 million to the Turks and Caicos Islands after Hurricane Ike. The CCRIF continues to work on developing new products, such as an extreme rainfall coverage, that will benefit countries throughout the region (www.ccrif.org).

Conclusion

This chapter reveals that there are many disaster and emergency management institutions operating in the Caribbean, but because the region is an archipelago, a networked approach was the only feasible governance mechanism that could ensure that the heavily indebted middle income economies of the region maximized cooperation and collaboration to mitigate the worst consequences of natural events. We also noted that there are four tiered multiorganizational networks and that the national level institutions were the level where authoritative decisions about hands-on emergency management are made. The multilateral and regional levels are more suited to diplomacy and leveraging resources for capacity-building and raising funds. Our analysis also revealed that the decentralized levels of emergency management tend to be less well resourced; however, it is the level where actual impacts are felt and managed.

Network rules and budgetary decisions must be clear and unambiguous and communicated effectively to the decentralized levels of governance systems to ensure the maintenance of capacity. In that way, local disaster and emergency management structures will be ready and available when they are called on to perform in times of severe events.

Through advocacy and collaboration, the SISC have been able to look for extraregional funding and created a multilateral insurance facility in the form of the CCRIF to prevent single-source revenue dependency. The networked governance approach to emergency management has made the Caribbean resilient. For example, countries that were flattened by hurricanes, such as Grenada, have been able to bounce back and rebuild their economies. The linguistic barriers have been partially circumvented and cooperation between the English-, French-, and Spanish-speaking islands has flourished through information-sharing networks, and such systems and approaches must be built on and improved.

A plethora of organizations have contributed and are still contributing to emergency management within the Caribbean. Many of these organizations have been mentioned in this paper, even though a much longer list has been supplied in the work of Bisek et al. [2001] which can be consulted for additional reference.

Finally, despite the involvement of a mélange of organizations in the disaster risk management process, institutional capacity and cooperation in disaster management within Latin America and the Caribbean are always in need of improvements given the changing global climate, the annual cycle of hurricanes and the emergence of new threats such as the Avian and Swine Flu. The management of these new risks might require an extension of the network.

References

Bisek, P., Jones, E., and Ornstein, C. 2001. A strategy and results framework for comprehensive disaster management in the Caribbean. CDERA: Barbados. Retrieved from www.cdera.org (accessed March 28, 2009).

CDERA Virtual Disaster Library. The Caribbean's clearinghouse on disaster information. Retrieved from www.cdera.org/doccentre/background.php (accessed March 28, 2009).

Edgar, L. 2002. Building policy partnerships: Making network governance work. Ottawa: Institute of Governance.

Goldsmith, S., and Eggers, W. D. 2004. *Governing by network: The new shape of the public sector.* Washington, DC: Brookings Institution.

Jones, E. B., Bisek, P. A., and Ornstein, C. 2001. Comprehensive disaster management in the Caribbean: A baseline study? www.cdera.org (accessed March 30, 2009).

Osei, P. D. 2007. Policy responses, institutional networks management and post-Hurricane Ivan reconstruction in Jamaica. *Disaster Prevention and Management* 16:271–234.

Wolfgang, M., and Elbe, S. 2004. Local network governance: Perspectives and problems for the German rural sector. Saarbrücken, Germany: Centre for Evaluation at Saarland University.

World Bank. 2006. Caribbean Catastrophe Risk Insurance Facility: A solution to the short-term liquidity needs of small island states in the aftermath of natural disasters. Retrieved from http://siteresources.worldbank.org/PROJECTS/Resources/Catastrophicriskinsurancefacility.pdf (accessed March 28, 2009).

Chapter 3

Game Approach to Disaster Loss Reduction in Caribbean Multicultural Societies: The Disaster Awareness Game

Virginia I. Clerveaux and Balfour Spence

Contents

The positive relationship between the level of disaster awareness and related loss reduction has long been acknowledged in disaster risk management planning. However, this relationship can be stymied by several social and economic factors that challenge the effectiveness of awareness promotion. The advent of globalization, with its characteristic freer movement of people and capital and resultant increase of multiculturalism globally, has added a new and challenging dimension to this relationship. International protocol and the moral and ethical suasions of disaster management embody the principle of responsibility and care for all individuals within a given jurisdiction during times of emergency. As the global community becomes increasingly multicultural, disaster-risk managers must seek new, comprehensive techniques for communicating risk information for awareness promotion. This chapter highlights a game approach that has been used as an effective tool in the communication of disaster risk information for loss reduction in the Caribbean region. Accessibility is not merely a function of the location in which the information is reserved, but is also related to its formatting, comprehensiveness, and the extent to which populations' perceptions and cultural nuances are considered in the provision of information.

Risk Communication, Multiculturalism, and Disaster Management

Since the mid-1980s, the focus of risk communication has evolved from concern about how to best inform the public about the technical aspects of risk assessment to a process of early and ongoing communication with all concerned groups. The basic philosophy of this new paradigm is that risk communication should not only handle how to communicate risk to vulnerable groups, but also include an understanding of how various groups in the society perceive and respond to risk information. Central to the relationship between multiculturalism and disaster management is the emergent paradigm of inclusive awareness promotion through effective disaster risk communication, a theme that has become a major focus in comprehensive disaster management.

Awareness Promotion and Risk Communication: Emergent Paradigm in Comprehensive Disaster Management

Although science- and technology-related measures for disaster risk reduction have gained significant momentum in recent years, their impact in reducing losses has been overshadowed by a combination of higher levels of exposure and higher incidences of failed development that contribute to greater social and environmental risks. There is now general consensus that the technological solution alone is insufficient to produce sustainable loss reduction from the impact of disasters and the need for greater emphasis on the social context of risk reduction. Much of the focus of this emergent paradigm is targeted at issues related to the provision of reliable risk information and the effective communication of this information to vulnerable populations. Communication of risk information to vulnerable populations is not only contingent on the generation and availability of such information, but also critically, on the accessibility of this information.

Disaster Management and Multicultural Context of the Caribbean

Multiculturalism in the Caribbean is largely a manifestation of the historical patterns of migration within the region. The nature, direction, and magnitude of migration have always been influenced by trends in global and regional socioeconomic development. For example, the slave trade in the eighteenth and nineteenth centuries represents the first major immigration waves to the Caribbean. Following the emancipation of slaves in the nineteenth century, individuals' intraregional movement dominated the pattern of migration and greater numbers of workers sought employment and better working conditions in more vibrant economies. Intraregional migration of Caribbean nationals continued into the twentieth century. However, unlike earlier migration waves in which the pull factor involved employment opportunities in large-scale agricultural and infrastructural projects, this new wave was centered on an emergent and booming oil industry. The oil boom in the 1970s attracted many migrants from the smaller and less developed islands to work in the oil refineries in the dependencies of the Netherlands and the United States, particularly the United States Virgin Islands and the Netherlands Antilles. Likewise, the booming energy sector in Trinidad and Tobago became a magnet for many seeking employment. The global energy crisis in the 1980s reduced the demand for labor in Trinidad and Tobago and new employment opportunities emerged, especially in a maturing tourism sector. During the 1990s, the Caribbean's growing tourism sector increased the demand for construction and service-related workers that could not be supplied by the domestic labor force on some of the smaller Caribbean islands where this new growth occurred. As a consequence, workers from other islands and neighboring Latin American countries, particularly Columbia and Venezuela, filled these gaps.

The establishment of the Caribbean Single Market and Economy in 2005 within the regional integration movement heralded a new era in the intraregional movement of Caribbean nationals and fostered higher levels of multiculturalism, especially in more economically vibrant pull states. This increased level of intraregional multiculturalism is evident in the fact that the absolute number of Caribbean nationals residing in other countries in the region has steadily increased during the past two decades [UNDP 2002]. Although the absolute migrant stock is comparatively small in the Caribbean, it is markedly high in terms of the percentage of total population. On average, approximately 3% of the population of Caribbean countries can be considered migrants. However, this proportion varies considerably from country to country, with the more economically vibrant countries recording the highest proportion. Areas with higher migrant proportions include the British Virgin Islands, the Cayman Islands, Anguilla, the Netherlands Antilles, Aruba, and the United States Virgin Islands. Irrespective of the period and destination of migration, migrants tend to originate primarily from the Organization of Eastern Caribbean States (OECS), Jamaica, Guyana, Suriname, and Haiti.

Deteriorating economic and social conditions, high unemployment (particularly for younger, more mobile people), and little hope for improvements in the foreseeable future constitute the main push factors. Diversity in Caribbean culture and the intraregional migration trends of the region have placed multiculturalism as a central theme in the disaster management discourse of the region. Such multicultural population environments call for new communication and disaster management strategies that allow access to risk information. Therefore, multicultural environments offer a conceptual platform within which organizations, agencies, and nongovernmental organizations can refine their strategies for coping with the heterogeneity of the social environment. Paradoxically, many disaster management decision-makers and informants remain entrenched in functionalist rather than structuralist views of the nature of society [Burrel and Morgan 1979]. Consequently, many disaster managers lack an awareness of the basic factors that must be embedded in an effective disaster management plan in a multicultural context.

Risk Communication Approaches and Caribbean Multiculturalism

The technological era has brought together diverse cultures with unprecedented speed, increasing the need for and challenge of communication across cultures [Banks 2000]. Despite this newly emerging cultural landscape and consequent challenges, disaster-related information and communication continue on the assumption of a homogeneous cultural landscape. To date, there have been no comprehensive, theory-driven, or systematic treatments of multicultural communication in public relations. Although a continued integration process within CARICOM has led to the incorporation of non–English-speaking countries, such as Haiti and Suriname, efforts to address the communication of risk in countries with high levels of multiculturalism are largely lacking from the recently developed Comprehensive Disaster Management Strategy. This chapter addresses this gap in the comprehensive disaster management framework of the region. These gaps in Caribbean disaster risk communication can be summarized in the following manner:

1. Weak knowledge management
2. Traditional mediums of risk communication
3. Role of the media
4. Insufficient capacity-building for risk communication
5. Inadequate inclusiveness in disaster management planning

Weak Knowledge Management

Major shortcomings in Caribbean disaster risk reduction strategies relate to weaknesses in the management of knowledge. Information management and communication, training, and research have posed major challenges to risk managers in the region. As such, there is a paucity of knowledge about important risk management components such as good practices, institutions, and networks of practitioners in disaster risk reduction. In addition, research related to crosscutting issues, such as perceptions and response tendency of migrants and the status and character of indigenous knowledge and its role in disaster risk reduction, are virtually nonexistent in the research agenda.

Traditional Mediums of Risk Communication

Media such as national radio, television, and print has been the traditional approach used by disaster risk managers to promote disaster awareness and risk reduction options to the public. This implies that the availability of such information is limited to the capacity to access and assimilate this information. Migrants are often excluded from information transmitted through these modes, especially where they lack the capacity to access these traditional media and comprehend the information transmitted. Therefore, effective and inclusive risk communication must be cognizant of the communication needs of all people within a jurisdiction and tailor their mode of communication accordingly.

Role of the Media

The role of the media in the communication of risk for overall disaster loss reduction cannot be overstated. The media is a critical instrument for promoting awareness at all levels of society, from political decision-makers to laymen. However, the effectiveness of the media in this regard is a function of the comprehensiveness of the information provided, the language considerations in its presentation, and the extent to which access to the media is available. Some national authorities publish newsletters about disasters but often only overview semiintellectual topics with little relevance to the everyday risk situations of the majority of the people and, in most cases, are not translated into common local languages. These defects limit the use of media in promoting effective public awareness of disaster risk reduction to multicultural and multilanguage groups in the society.

Communication of disaster risk information in multiple languages is an activity that is rarely undertaken in most Caribbean countries and one that holds significant potential for disaster risk reduction. Even in countries where the translation of risk information is undertaken, the process is haphazard. Usually, translation and disaster risk reduction is produced immediately before a threat of hazard (i.e., a hurricane). Because risk perception is something embedded in one's psyche and the provision of risk information is intended to result in behavioral change, the communication of risk information should be done in a manner that would be easily understood by the vulnerable population over an extended period of time. Additionally, there is a need to ensure that the translation of warning information is accurate and reliable.

Insufficient Capacity-Building for Risk Communication

Public education and training are key components in capacity-building for risk reduction. However, this critical component is in its infancy in the disaster management framework of the Caribbean region and the virtual absence of disaster risk reduction in the curriculum of the formal educational system of the region is noticeable. Building this type of capacity is especially relevant in multicultural societies where the formal education system can be used as a conduit to transfer risk information to children of migrant households who would otherwise not have access to this information.

Inadequate Inclusiveness in Disaster Management Planning

If the communication of risk is to meet its intended objective of disaster loss reduction, inclusion of all societal groups must be a focus in the communication strategy. This will avoid marginalization of any group in their access to disaster-related information because people who are marginalized in the early stages are likely to be marginalized later. Therefore, they need to be part of planning from the beginning [Fothergill et al. 1999]. Political and socioeconomic empowerment of marginalized groups enables their participation in policy and decision-making and reduces their vulnerability. Generally, migrants tend to agglomerate instead of integrate with members of the host country; therefore, they are more likely to trust someone they identify with. Being aware of this tendency, disaster managers must identify these migrants or natural group leaders by working with various church groups or other associations that migrants are known to participate in to form a partnership that can be used to provide migrants with risk information or to encourage their participation in policies or community disaster mitigation strategies that will affect their lives, especially during emergency situations.

Disaster Awareness Game: Technique for Awareness Promotion and Risk Communication in Multicultural Environments

The Disaster Awareness Game was developed to address the aforementioned gaps in communication of disaster risk information in the increasingly multicultural landscape of the Caribbean region. It is premised on the concept that in multicultural settings, children can be critical conduits for the transfer of risk information between the formal learning environment and their household and communities. Therefore, the model of the game discussed in this chapter is specifically aimed at younger school children, particularly those between 9 and 11 years old. The concept of this game is that effective communication of risk information to this age group will enhance the overall awareness status of households and communities of these multicultural landscapes. Additionally, promotion of disaster awareness among younger children, irrespective of the multicultural setting, represents future investment and value added in the risk reduction process if the assumption that aware children become aware adults is subscribed to. The vulnerability of children to multiple hazards is discussed in context of the game specific to school children. This vulnerability has been the underpinning stimulus for recent increases in initiatives for the promotion of disaster awareness in schools. Central to these initiatives is the development of techniques for enhancing the disaster knowledge base of young children. The main challenge to the success of these initiatives is ensuring the methodological approach and tools employed in the disaster education of children are not only relevant to their educational capacity but are also interesting enough to sustain their attention. It is in this context that the Disaster Awareness Game was designed to evaluate and promote disaster awareness in children.

Game Approach as a Technique in Disaster Awareness Promotion among Children

The role of disaster awareness in building more resilient societies and communities through education and knowledge transfer has been repeatedly highlighted, particularly in relation to the preparedness, mitigation, emergency response, and recovery stages of the disaster management cycle [Nielsen and Lidstone 1998]. It is in that regard that the Hyogo Framework identifies knowledge of and education on disaster risk reduction as one of the five priorities for action to achieve disaster-resilient communities and nations. Similarly, United Nations International Strategy for Disaster Risk Reduction (UNISDR) has consistently campaigned to make disaster awareness and risk reduction integral to school curricula whether in formal or informal education. The objective is to promote disaster risk education in national school curricula in countries vulnerable to natural hazards.

It is believed that knowledge management and education can assist hazard-prone communities in gaining a better understanding of how to cope with risks and become more resilient. Disaster education is recognized as an essential element in sustainable development because it accelerates the progress of societies toward disaster resilience [UNESCO 2004]. This view is reiterated in the Kobe Report [2005], which states that "education for creating a culture of disaster resilience is an interactive process of mutual learning among people and institutions. It encompasses far more than formal education at schools and universities, and affects all aspects of life through the concerted effort to overcome universal barriers of ignorance, apathy, disciplinary boundaries, and lack of political will present in communities. Education also involves the enhancement and use of indigenous knowledge for protecting people, habitat, livelihoods, and cultural heritage from natural hazards" [p. 2]. The report further postulates that history has taught that inadequate disaster reduction awareness and preparation has repeatedly lead to preventable loss of life and damage in all major natural disasters and that preparation through education is less costly than learning through tragedy.

According to the ISDR Report [2006, 2007], schools are the best venues for inculcating collective values. Recently, there has been a renewal of focus, at both the national and global levels, of the importance of using public education as a key to disaster mitigation. The 1990's Decade for Natural Disaster Reduction emphasized the importance of governments "educating and training their citizens to increase awareness" [Nielsen and Lidstone 1998, p. 14]. It is also widely recognized that an educated public is better able to prepare for, adapt to, and respond to hazards, and that education for disaster reduction is complex but essential to any properly implemented, centrally managed hazard strategy. The implementation of disaster loss reduction programs in schools is increasingly being recognized as one of the key elements in long-term disaster risk reduction strategies. The rationale is that school children can play an important role in real-life future catastrophic events if disaster risk reduction is integrated into school curricula. The importance of disaster awareness in reducing fatalities among children was demonstrated during the Indian Ocean Tsunami. Many lives were saved because a child was able to recall the warning signs of a possible tsunami from school geography lessons. It is in this context that the promotion of disaster education among school children and the subsequent transfer of knowledge to the communities from which these children come can effectively build more disaster-resilient societies, especially in the multicultural landscapes of the Caribbean.

Wisner [2006] echoes this view in his statement that "at all levels, pupils and students from primary schools to postgraduate level can actively study the safety of their own schools and work with teachers and community members to find ways to protect themselves. They can also spread the methods of participatory vulnerability and capacity assessment and hazard mapping to the broader communities surrounding schools and other institutions of education and research of which they are a part" [p. 5]. Similar sentiments are expressed by the ISDR [2006] in the statement that "disaster loss-reduction education for children fosters awareness and better understanding about the immediate environment in which they and their families live and work" [p. 7].

Since children are known to be influential and effective communicators, lessons learned at school will later be transmitted to the home [UNESCO 2004]. The value added of current disaster education intervention in school curricula is that children who get disaster education will have a greater understanding of disasters, the effects of human actions, the consequences of poor environmental management, and the need to promote a new kind of development path that is in greater harmony with nature once they become adults [ISDR 2002]. In other words, education for disaster reduction is an integral part of education for sustainable development because education, knowledge, and awareness are critical to building the capacity for hazard loss reduction [Wisner 2006]. There is an emerging paradigm in which children are not regarded merely as potential victims of hazards and disasters but rather as catalysts for loss reduction. Because children represent the gateway to creating a culture of prevention in society, they can be taught the value of prevention and encouraged to play proactive roles in disaster reduction [Anderson 2000]. It is in that regard that the Hyogo Framework calls for a bottom-up (students, teachers, and communities) and top-down (government, United Nations, international organizations, and nongovernmental organizations) approach to disaster risk reduction. The new paradigm has many challenges, especially in relation to the development of appropriate methodologies and techniques for providing children with information that will build their disaster awareness capacity.

Although the importance of tailoring disaster information and knowledge to the mental capacity of children as a strategy in risk reduction cannot be overstated, the design of children-specific disaster information is one of the greatest challenges to the process. Challenges include the following:

1. Determination of the existing levels of awareness prior to curriculum design as a means of establishing the level of intervention required.
2. Design and development of appropriate tools for evaluating the existing level of awareness and also the knowledge gained from interventions.
3. Ensuring that the techniques employed for imparting disaster knowledge to children are sufficiently interesting and interactive to hold their attention.

The design of methodologies for educating children in disaster loss-reduction issues requires several critical considerations that are highlighted in the following section.

Critical Considerations in Promoting Awareness and Communicating Disaster Risk Information to Children

There is general consensus that the tens of thousands of children who lost their lives in the Indian Ocean Tsunami on December 26, 2004 might have survived had they been equipped with relevant information and skills that would have fostered appropriate responses to reduce their risk. In support of this view, the Asian Disaster Reduction Centre (ADRC) argues that the integration of tsunami disaster education into school curricula is the most effective way of preventing or mitigating a similar tragedy [Suzuki 2006]. This argument echoes views expressed by organizations such as the ISDR [2002] that school education is one of the key elements in long-term disaster risk reduction strategies. Consensus is that the mainstreaming of disaster education in schools must take into account key considerations related to the following:

1. Curriculum development
2. Pedagogy
3. Disaster information tailored to children's mental capacity
4. Measurement of awareness level
5. Children's attention spans
6. Mode of communication
7. Inclusion of children in the decision-making process

Curriculum Development

There is much interest in curriculum and teaching practice as vehicles for promoting disaster-related knowledge among children. However, the incorporation of disaster education into existing school curricula cannot be ad hoc and must be approached within the context of an overall educational system whose strength and functionality is reflected in each constituent curriculum. Interventions for the promotion of disaster education must therefore ensure effective interface with the existing components of the system if the initiatives for risk reduction education are to be realistic and feasible [Wisner 2006].

Pedagogy

Pedagogy refers to the art of effectively imparting knowledge. Infusing disaster education into school curricula requires appropriate pedagogic considerations. Traditional school curricula are largely academic in their orientation; however, disaster education is not only an academic exercise but also a knowledge transfer system that can make the difference between life and death, economic progress and impoverishment, and sustainable development and environmental degradation. It is in this regard that those entrusted with imparting disaster information to children should themselves be adequately and appropriately trained, not only in relation to the content of such knowledge but also in relation to the methodologies of effective communication. If this can be achieved in relation to school curricula, the multiplier effect on the wider society in terms of increased levels of awareness will be astounding [Wisner 2006].

Tailoring of Disaster Information

The tailoring of information toward target groups is a critical component of pedagogic methodology. This is because response to information is a function of a plethora of social and environmental factors that inform their perceptions of risk. As such, the people's cognitive processes in the understanding of disaster risk are as important as social and societal factors [Gutteling and Weigman 1996]. Therefore, disaster information for children must consider issues such as age, level of literacy, local language and dialects, and cultural factors to maximize the detail and wealth of the collective information.

Measurement of Level of Awareness

Rationalization is critical to the success of any intervention. In that regard, the promotion of disaster education in schools must be informed by the existing level of knowledge among children. This necessitates the evaluation of targeted students to establish their existing knowledge in relation to hazards and disasters. Depending on the age group of the targeted children, it is important that the evaluation technique is in sync with their level of literacy and mental capacity. It is widely recognized by educators that games are an effective mode of communicating information to children of all abilities, so game techniques can be used in measuring levels of disaster awareness among children. Several methods can be employed to measure awareness among children and various techniques used in the promotion of awareness. Cognitive pretesting is one of the preferred measurements [Ciochetto and Haley 1995]. Techniques that have been employed are varied. For example, in Jamaica, techniques used include fire and earthquake drills, poster competitions, cultural competitions, songs, dances, skits, exhibitions, and talks [Wisner 2006].

Children's Attention Span

Attention span is an important consideration in the education of young children. The tendency is for a positive relationship between attention span and the level of interactiveness of the teaching technique. It is for this reason that games, simulations, and skits are effective tools for imparting disaster knowledge to children. However, it is important that detail and accuracy of information are not sacrificed for interactiveness. Research on attention spans of various age groups have indicated that tasks that are considered preattentive for adults demand more attention from school-age children. However, children are often more adept in the control of visual attention [Enns 1990, p. 138].

Mode of Communication

Children of the same age can respond differently to techniques employed for their education. Careful consideration must be paid to the mode of communication used in the dissemination of disaster risk information to young children and must cover a range of interactive and visual techniques and, if possible, include hands-on and experiential learning methods [Wisner 2006]. Ideally, a disaster-relevant curriculum would not only impart knowledge related to the natural hazards themselves, but also ensure that various media are used in the dissemination process to allow for the production of understandable and accessible information for all levels of children's mental abilities (e.g., posters, murals, and simple drawings for primary school children).

Inclusion of Children in the Decision-Making Process

The United Nations Convention on the Rights of the Child (CRC) recognizes that a child "is a subject of rights who is able to form and express opinions, to participate in decision-making processes and influence solutions, to intervene as a partner in the process of social change and in the building of democracy" [2009, p. 5]. Most disaster risk reduction measures have excluded the input of children, although it is widely recognized that children are among the most vulnerable groups affected by disasters. In relation to disaster risk reduction, children can play a critical role in informing the contents of risk reduction measures and the techniques for imparting related knowledge. Consultation with children and those who teach and work with them must therefore be integral to the promotion of disaster awareness. The Disaster Awareness Game takes into account the aforementioned considerations.

Disaster Awareness Game

The Disaster Awareness Game combines several tools and techniques designed to achieve the following:

1. Measure levels of hazard and disaster awareness
2. Educate children and adults about hazards and disasters that are relevant to their environment
3. Encourage positive behavior at all stages of the disaster management cycle
4. Dispel myths about hazards and disasters

These tools include presentations on local hazards and the relevant disaster management context, a board game with related question cards, and a score sheet that is used to evaluate levels of awareness prior to and after exposure to the game. The board game also helps inform players of the consequences of poor environmental behavior, especially in relation to hazards.

Objectives of the Disaster Awareness Game Technique

The Disaster Awareness Game is an educational disaster game technique designed to evaluate levels of hazard awareness to determine and prioritize interventions for disaster education. Additionally, the Disaster Awareness Game evaluation process is intended to encourage positive mitigation and response behaviors and to dispel myths about disasters in a fun but intellectual environment. An overarching theme of the Disaster Awareness Game is the relationship between attitudes to disaster information and social vulnerability. Playing the game is intended to assess the level of knowledge of players within the context of the disaster management cycle. In that regard, the game navigates its players through preparedness, prevention, mitigation, emergency response, and recovery and rehabilitation stages. There is a strong emphasis on measures that can be employed to mitigate impact and, by extension, contribute to the sustainability of households, communities, and nations. The weaknesses and gaps in the answers given by players are not interpreted negatively but rather are regarded as opportunities to design appropriate interventions for risk reduction.

Critical Considerations in the Development of the Disaster Awareness Game

Key issues considered in developing the Disaster Awareness Game for school children included design concept, reading and comprehension ability, age range, usability, attention span, and learning considerations.

Design Considerations

The Disaster Awareness Game consists of three levels (basic, intermediate, and advanced) and is adaptable to any level of educational attainment. The board game layout is identical for all levels. The level is determined by the degree of difficulty of the questions contained on the game cards. Currently, the Disaster Awareness Game is in an electronic format that allows users to adapt and format the game for different ability levels by adjusting the degree of challenge of the questions, the types of hazards, and the related questions to reflect local vernacular and hazard experience and exposure. The flexibility of the game allows users to format it to reflect specific areas of disaster management that decision-makers might need to highlight. The adaptation discussed was targeted at Grade 5

Table 3.1 Distribution of Sample by Ethnicity in TCI

Ethnic Origin	*Frequency*	*% Sample*
Jamaica	1	3.03
Dominican Republic	2	6.06
Haiti	8	24.24
Bahamas	1	3.03
Barbados	1	3.03
USA	2	6.06
TCI	18	54.54
Total	**33**	**100**

primary school children and was tested in the Caribbean countries of St. Vincent and the Grenadines (SVG) and the Turks and Caicos Islands (TCI). Unlike the multicultural TCI (Table 3.1), SVG's population is more homogeneous. However, assessment done in SVG would allow comparison on a controlled population. To date, the Disaster Awareness Game is only available in English but translation into Japanese is currently underway.

Reading and Comprehension Ability

The board component of the Disaster Awareness Game process was designed to ensure that players are evaluated on the hazards and related management issues that are relevant to their environment (Figure 3.1). For this reason, the contents of the board component are adjustable and dependent on the players' environment. Once the content of the Disaster Awareness Game board was developed, the next step was to determine the degree of difficulty of related questions. Like the board component, the level of difficulty of the game cards can be easily and electronically adapted to suit the player's environment. In the game design for children, reading and comprehension ability were key considerations.

Consultation with educators of young children preceded the development of the question cards. The preliminary game was intended for Grade 5 (ages 9 to 12 years) primary-level children. The consultation process established the reading and comprehension ability of this age group and the question cards were designed accordingly. Questions on the game cards coincided with the hazard content of the board game, thus covering similar stages of the disaster management cycle. The construction of the questions using a multiple choice format was one of the outcomes of the consultation with educators. Likewise, the limitation of response

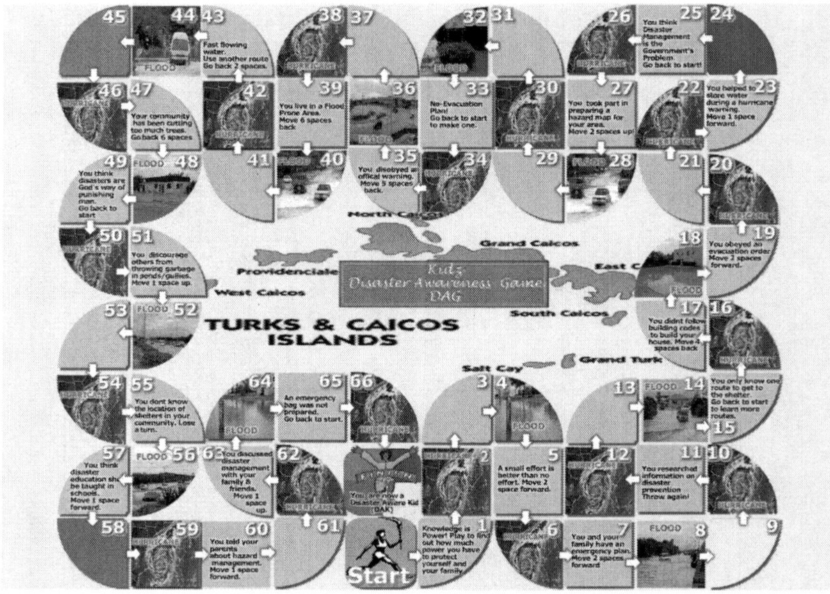

Figure 3.1 Disaster Awareness Game board game.

options to three (Figure 3.2) was also an outcome of this process. Further fine-tuning of questions occurred in relation to simplicity and straightforwardness. Appropriateness of the questions for the targeted students was further assessed by circulating draft questions and response options among a panel consisting of child educators and disaster managers.

Age Range

The age range consideration for playing the game is informed by knowledge of the mental and comprehension ability of different ages. Based on this knowledge, it was decided that the game would be effective in the promotion of disaster awareness among children 9 years and older.

Usability

The primary attraction of the Disaster Awareness Game as an educational tool for disaster awareness and risk reduction is its low resource demand, simple technology, and cost effectiveness. The inputs are low cost and easily available, so the Disaster Awareness Game process can be easily implemented in poor societies where resource availability is a major constraint. Additionally, the digital format of the board game allows facilitators to adapt and modify the game in

DAG Score Sheet

FLOOD				HURRICANE			
No. 1	a b c	No. 26	a b c	No. 1	a b c	No. 26	a b c
2	a b c	27	a b c	2	a b c	27	a b c
3	a b c	28	a b c	3	a b c	28	a b c
4	a b c	29	a b c	4	a b c	29	a b c
5	a b c	30	a b c	5	a b c	30	a b c
6	a b c	31	a b c	6	a b c	31	a b c
7	a b c	32	a b c	7	a b c	32	a b c
8	a b c	33	a b c	8	a b c	33	a b c
9	a b c	34	a b c	9	a b c	34	a b c
10	a b c	35	a b c	10	a b c	35	a b c
11	a b c	36	a b c	11	a b c	36	a b c
12	a b c	37	a b c	12	a b c	37	a b c
13	a b c	38	a b c	13	a b c	38	a b c
14	a b c	39	a b c	14	a b c	39	a b c
15	a b c	40	a b c	15	a b c	40	a b c
16	a b c	41	a b c	16	a b c	41	a b c
17	a b c	42	a b c	17	a b c	42	a b c
18	a b c	43	a b c	18	a b c	43	a b c
19	a b c	44	a b c	19	a b c	44	a b c
20	a b c	45	a b c	20	a b c	45	a b c
21	a b c	46	a b c	21	a b c	46	a b c
22	a b c	47	a b c	22	a b c	47	a b c
23	a b c	48	a b c	23	a b c	48	a b c
24	a b c	49	a b c	24	a b c	49	a b c
25	a b c	50	a b c	25	a b c	50	a b c

Figure 3.2 Disaster Awareness Game score sheet.

accordance with the requirements of the local environment in which it is being played. The game can also be easily modified to focus on single or multiple hazards as well as on specific components of the disaster management cycle. This level of flexibility can be effective in prioritizing the focus of the disaster management education in schools. Also, the board game and question cards can be translated into multiple languages and dialects to ensure more universal relevance.

Attention Span and Learning Considerations

The Disaster Awareness Game is intended to promote disaster education in a relaxed and exciting environment of mental competitiveness and enjoyment. To prevent boredom, length of play, although flexible, was limited to 1 hour. Various means of stimulating interest among children were incorporated into both the board game and the game cards. Colorful pictures, clip art, graphics, and attractive text were used extensively. Because the game is not intended to be a formal examination of player knowledge, the design of the game cards (Figure 3.3) was intended to instill

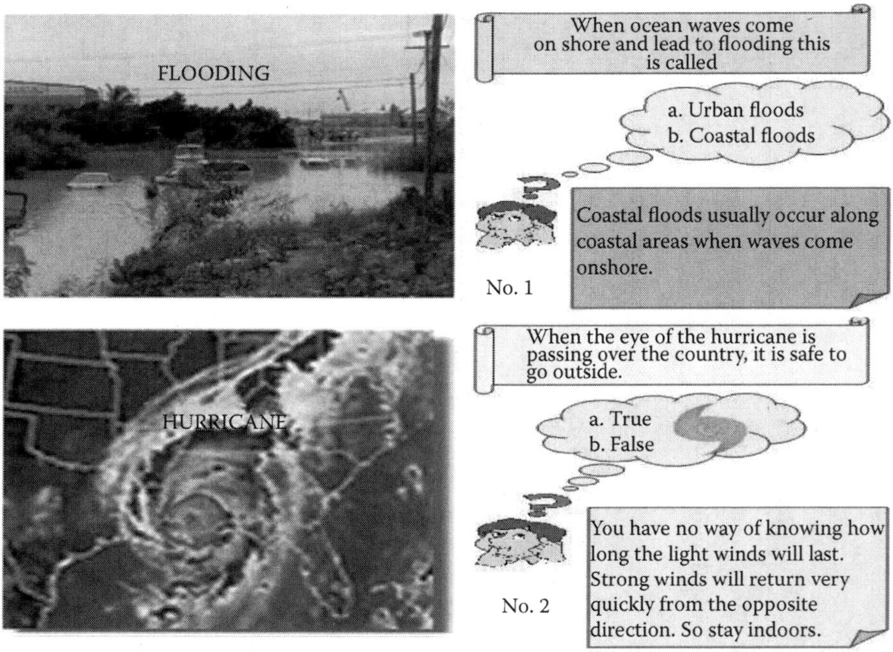

Figure 3.3 Disaster Awareness Game flood and hurricane card sample.

confidence in players by providing the correct response option as well as a brief explanation for the response. Positive disaster management behaviors are encouraged in the game by rewarding players for a correct response by moving one place forward and penalizing players for an incorrect response by moving two places backward. Players can also advance if their game pieces fall on a position that suggests positive disaster management behavior. Likewise, backward movement (penalty) can occur if the game piece falls on a location that indicates detrimental disaster management behavior. This system of reward and penalty helps reinforce the learning of positive disaster management behavior. The pedagogical effectiveness of the Disaster Awareness Game can be influenced by prevailing dispositions of children, especially because these relate to metacognitive skills, reading ability, and general learning attitudes.

Rules and Rationale

To achieve the objectives of the game, it was imperative that the rules of the game be clear, simple, complete, and concise so players could easily grasp them (Figure 3.4).

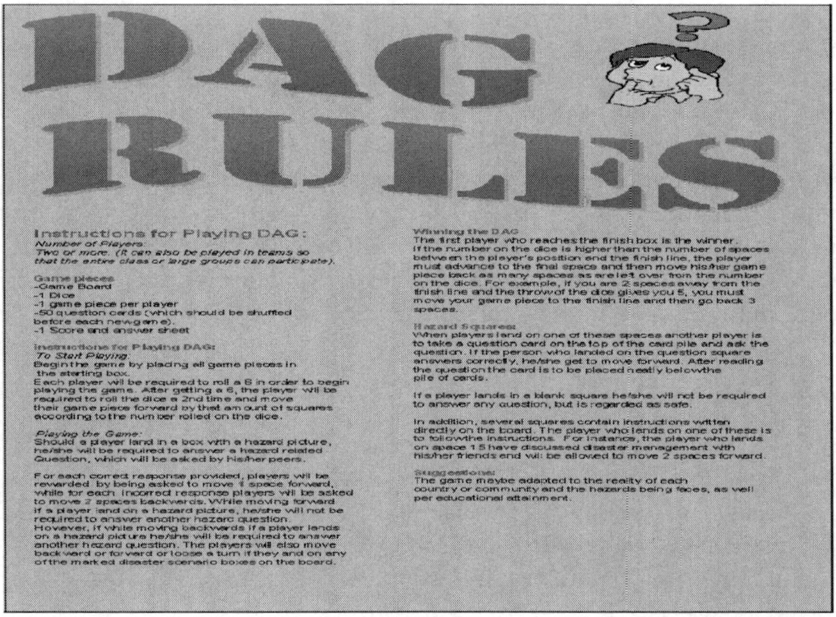

DAG RULES

Instructions for Playing DAG:

Number of Players:
Two or more. (It can also be played in teams so that the entire class or large groups can participate).

Game pieces
-Game Board
-1 Dice
-1 game piece per player
-50 question cards (which should be shuffled before each new game).
-1 Score and answer sheet

Instructions for Playing DAG:

To Start Playing:
Begin the game by placing all game pieces in the starting box.
Each player will be required to roll a 6 in order to begin playing the game. After getting a 6, the player will be required to roll the dice a 2nd time and move their game piece forward by that amount of squares according to the number rolled on the dice.

Playing the Game:
Should a player land in a box with a hazard picture, he/she will be required to answer a hazard related Question, which will be asked by his/her peers.

For each correct response provided, players will be rewarded by being asked to move 1 space forward, while for each incorrect response players will be asked to move 2 spaces backwards. While moving forward if a player land on a hazard picture, he/she will not be required to answer another hazard question.
However, if while moving backwards if a player lands on a hazard picture he/she will be required to answer another hazard question. The players will also move backward or forward or loose a turn if they and on any of the marked disaster scenario boxes on the board.

Winning the DAG
The first player who reaches the finish box is the winner.
If the number on the dice is higher than the number of spaces between the player's position and the finish line, the player must advance to the final space and then move his/her game piece back as many spaces as are left over from the number on the dice. For example, if you are 2 spaces away from the finish line and the throw of the dice gives you 5, you must move your game piece to the finish line and then go back 3 spaces.

Hazard Squares:
When players land on one of these spaces another player is to take a question card on the top of the card pile and ask the question. If the person who landed on the question square answers correctly, he/she get to move forward. After reading the question the card is to be placed neatly below the pile of cards.

If a player lands in a blank square he/she will not be required to answer any question, but is regarded as safe.

In addition, several squares contain instructions written directly on the board. The player who lands on one of these is to follow the instructions. For instance, the player who lands on space 15 have discussed disaster management with his/her friends and will be allowed to move 2 spaces forward.

Suggestions:
The game maybe adapted to the reality of each country or community and the hazards being faces, as well per educational attainment.

Figure 3.4 Disaster Awareness Game instruction sheet.

Effectiveness of the Disaster Awareness Game

Confirmation is a critical component in the development of any evaluation tool. With regard to the Disaster Awareness Game, the confirmation process was intended to identify and eliminate glitches in the process. In SVG, 42 students were evaluated and 33 were evaluated in TCI. For testing of the Disaster Awareness Game, two-day workshops were organized in each of the targeted countries. The goals of the workshops were to evaluate existing levels of disaster awareness within the context of the Disaster Awareness Game and enhance awareness. In SVG, participating students in Grade 5 varied in age from 9 to 12 years (mean: 10.9 years). In TCI, ages also varied from 9 to 12 years (mean: 10.2 years). The evaluation process in the workshop involved a pregame survey of students, an initial game exposure, provision of disaster information using various modes, and postgame assessment.

The involvement of students as well as their teachers and parents as major actors in disaster prevention and emergency preparedness, the fact that workshop activities received the support of both the Ministry of Education and National Disaster Management Organization in each country, and the media coverage and publicity significantly contributed to the smoothness of the evaluation exercise and the quality of data generated. The long-term relevance of the Disaster Awareness Game is related to emerging efforts by Caribbean countries to benchmark the comparative disaster management status among countries as a means of rationalizing and prioritizing disaster management interventions. In that context, the Disaster Awareness Game can be used as a benchmarking tool that allows comparison of levels of disaster awareness among children of similar ages throughout the region and can inform the type and level of disaster intervention required. The communication of disaster information to children was done in stages. First, students' level of hazard awareness was ascertained through a pregame survey and initial exposure to the Disaster Awareness Game and was followed by the provision of disaster information to fill the gaps in students' knowledge of hazard awareness and management. A postgame assessment was performed to determine the effectiveness of the various strategies used in communicating disaster information to children.

Evaluation Methodology

The evaluation methodology was undertaken in four stages:

1. Pregame survey. This stage evaluates the existing levels of disaster awareness among the target population using a questionnaire survey.
2. Initial game exposure. This stage exposes the target population to the Disaster Awareness Game.
3. Provisions of disaster information. This stage provides participants with disaster management information on hazards that are pertinent to their environment.
4. Postgame assessment. This stage evaluates the impact of the game and disaster information on the level of awareness among participants.

Pregame Survey

At the start of the workshop, a questionnaire was administered to participating students and their level of awareness was assessed based on the correctness of their responses. The questionnaires were designed to assess the children's level of awareness, risk perceptions, factual knowledge, and physical preparedness for hazards to which their communities are vulnerable. The questions addressed to students in SVG were related to hurricanes, volcanoes, landslides, mudslides, and flooding. Questions for TCI students were confined to hurricanes and flooding. The questionnaire also assessed the children's prior exposure to specific hazards and to disaster education programs designed to increase awareness, knowledge, and preparedness provided by local disaster management officials, media, or schoolteachers. Ethnicity, culture, and language can be key determinants of variations in the disaster-related behavior of a population and there were significant differences between the two samples in regard to ethnicity. The students evaluated in SVG were homogeneous in their ethnicity (all originated from SVG); ethnicity varied in the TCI. The ethnicity of the target population is relevant to the evaluation process because ethnic background is a major consideration in assessing perceptions,

Figure 3.5 Playing the Disaster Awareness Game in SVG.

attitudes, and behaviors in relation to hazards and disasters. Despite the ethnic homogeneity of the SVG sample, the diversity of locations from which students originated resulted in a variety of hazard experiences, making the evaluation process as varied as the TCI sample, where hazard experiences largely related to countries of origin.

Initial Game Exposure

At this stage of the evaluation exercise, students learned the rules of the game and played the game (Figure 3.5 and 3.6) without any additional information provided to them. The idea behind this exercise was to complement the measurement exercise undertaken in the pregame survey. Where initial game exposure provided a

Figure 3.6 Playing the Disaster Awareness Game in TCI.

Figure 3.7 Provision of disaster information to students in SVG using PowerPoint.

reliability check for the pregame survey, a scorecard was kept to record correct responses. The scoreboard could also be used as a database for determining levels of awareness at this stage. At the end of the exercise, game scores of participating students were compared with the results of the pregame survey to establish reliability.

Provision of Disaster Management Information

Discussions occurred with students who were participating in the two-day Disaster Awareness Game workshop. These discussions were facilitated by simple PowerPoint presentations, videos, and other interactive learning strategies (Figure 3.7 and 3.8).

Postgame Assessment

The posttest assessment consisted of a second exposure to the Disaster Awareness Game and evaluation of the student's performance compared to their previous

Figure 3.8 Provision of disaster information to students in TCI using videos.

exposure and the disaster management information provided. The results of the postgame assessment, discussed below, focus on the influence of the Disaster Awareness Game exercise on awareness of local hazards, risk perception and preparedness, and mitigation.

Pregame Analysis

The pregame survey analysis is discussed in terms of children's natural hazard awareness, their risk perceptions, and their household level of preparedness.

Children's Hazard Awareness

The level of hazard awareness among children is influenced by the extent of their hazard and disaster education as well as their level of exposure, especially within the context of their household. During the pretest exercise, the children's level of awareness was assessed by asking them to identify hazards that have or that might impact their communities. Table 3.2 summarizes the results of this exercise.

In all cases, children in SVG demonstrated a higher level of pretest awareness of hazards common to both environments than children from the TCI. The zero score for children in the TCI in relation to landslides and debris flow and volcanic eruption could be anticipated because these hazards are not part of the TCI hazard profile. The overall higher level of awareness in SVG is likely related to the more

Table 3.2 Children's Perceptions of Likely Hazard Impacts on Their Communities

Hazard	*Turks and Caicos Islands (%)*	*St. Vincent and the Grenadines (%)*
Hurricane	72.7	100
Flood	42.4	93.9
Tropical storm	33.3	78.8
Windstorm	18.2	54.5
Earthquake	15.2	81.8
Drought	9.1	45.5
Wildfire	6.1	12.1
Tsunami	3.0	33.3
Landslide/debris flow	0	93.9
Volcanic eruption	0	97.0

frequent occurrence of these hazards when compared with the TCI. Drought was the exception for hazard awareness; despite extremely low rainfall levels in the TCI, children demonstrated a low level of awareness of the potential impacts of meteorological and physiological drought. The likely explanation is that despite low rainfall levels, the TCI domestic water supply is fairly reliable due to extensive rainwater harvesting and desalinization of seawater. Unlike SVG, where agriculture is the main livelihood activity and drought is most noticeable, agriculture as a form of livelihood is almost nonexistent in the TCI. The importance of hazard education in promoting levels of awareness and risk reduction is underscored. The level of hazard education among sampled children was assessed in terms of exposure and access to disaster information related to two hazards (floods and hurricanes) that are common to the TCI and SVG profiles. Table 3.3 summarizes the results of this pretest exercise.

For both groups, the pretest survey revealed that television programs were the primary source of flood and hurricane hazard information. These programs are mostly seasonal and are broadcast during the Atlantic hurricane season, when impacts from hurricanes and related flooding are most likely. Schools also significantly factor into the disaster education of children, although SVG reported schools as the least important source of hazard-related information in regard to hurricanes. This finding is probably due to extensive information provided by media and parents, which reduces emphasis of disaster information in schools. It is noteworthy that the programs that provide hazard and disaster information to children are not always intentionally educational and can include disaster-themed movies such as *The Day After Tomorrow* [2004], *Volcano* [1997], and *The Perfect Storm* [2000], which are all Hollywood blockbuster movies that are broadcast on television.

Table 3.3 Comparative Sources of Hazard/Disaster Information: Turks and Caicos Islands and St. Vincent and the Grenadines[a]

	Flood			*Hurricane*		
	Information Source					
Sample Source	*Schools*	*Parents or Guardians*	*Television Programs*	*Schools*	*Parents or Guardians*	*Television Programs*
Turks and Caicos Islands (%)	51.5	51.5	93.9%	72.7	63.6	93.9
St. Vincent and the Grenadines (%)	78.8	66.7	100	54.5	81.8	100

[a] Sources of information are not mutually exclusive.

Although the level of hazard education among sampled children might seem relatively high, the extensiveness and comprehensiveness of this education needs to be further researched. This is in light of misconceptions identified in children's perception of specific hazards (e.g., when the eye of the hurricane is passing, it is safe to go outside and have a look). If we assume that the home is the main source of disaster knowledge at this age, it would therefore appear that there is significant misconception within the home environment. Therefore, the role of schools in providing appropriate and factual information for disaster loss reduction cannot be overstated. The pretest exercise suggests that caution should be exercised when interpreting the results of tools designed to evaluate levels of disaster awareness and education among children. For example, results from the administration of the pretest questionnaire survey suggest a high level of awareness among sampled children. However, results of the Disaster Awareness Game score sheet, a practical demonstration of levels of awareness, were contrary to those of the questionnaire survey. This discrepancy is likely the result of children's tendency to perceive positive responses to questions as desirable. Because the Disaster Awareness Game does not allow nonfactual responses, the children's true level of awareness is better evaluated.

Children's exposure to and experiences with hazards are major influences on their disaster education and level of awareness. Given the age range of sampled children, recent exposure to hazards in their environment affects their knowledge. For example, Grade 5 students in the TCI had a higher (72.7%) level of exposure to hurricanes than those in SVG (57.6%). This level of exposure is a reflection of recent hurricane events, such as Hurricanes Ivan, Francis, and Charlie in 2004, that have impacted the TCI. Children in the SVG sample had minimal exposure to hurricanes in their region of the country (the Marriaqua Valley). With regard to flooding, exposure of children in SVG was greater (57.6%) than those in the TCI (39.4%), which reflected the greater frequency of flooding in SVG as well as the location of the Ona Glinton Primary School from which the sample was drawn. It can be speculated that, had the sample been drawn from a primary school on the island of Providenciales in the TCI, the level of exposure would likely be higher owing to recent flood events on that island. This suggests that a more comprehensive comparison of levels of awareness, education, and exposure among Caribbean countries would require more comprehensive and representative samplings.

Children's Risk Perceptions

People's risk perceptions are a major factor in determining vulnerability because this perception informs the decisions that will either mitigate or aggravate vulnerability. This is especially relevant for school children who might be away from adult supervision when disasters occur. The risk perception of sampled children was assessed in terms of perceived vulnerability of their country, community, and homes. It is recognized that this assessment of risk perception is somewhat simplistic but was

Table 3.4 Pretest Risk Perception among Sampled Children

	Flood			*Hurricane*		
Sample Source	*Perceived Vulnera-bility of Country*	*Perceived Vulnera-bility of Commu-nity*	*Perceived Vulnera-bility of Home*	*Perceived Vulnera-bility of Country*	*Perceived Vulnera-bility of Commu-nity*	*Perceived Vulnera-bility of Home*
Turks and Caicos Islands (%)	75.8	56.3	43.8	87.9	81.8	63.6
St. Vincent and the Grena-dines (%)	90.9	81.3	68.8	87.9	93.9	78.6

deemed inappropriate given the age and level of education of the sample a more complex approach to the assessment would have been inappropriate. The assessment is confined to hazards that are common to both the TCI and SVG to allow comparison. Table 3.4 summarizes the risk perceptions of sampled children.

When compared to the TCI, the higher risk perception of children in SVG is consistent with higher exposure and real vulnerability. Not only is SVG vulnerable to a wider range of hazards, but frequencies of occurrence are also higher than in the TCI. The lowest perceived vulnerability among children from the TCI is the effect of flooding on their homes. This perception is consistent with reality because the sandy soils and gently undulating topography of much of the TCI is not conducive to flooding except in areas adjacent to salt ponds as well as low-lying areas from which drainage is impeded.

Students identified hurricanes as greater threats than flooding, even though small-scale flooding occurs annually throughout the TCI and a category 3 or higher hurricane has not directly impacted the TCI in the past 45 years. The higher threat level for hurricanes reflects experiences with the outer bands of hurricanes that pass along the northern and southern regions of the TCI. Although these bands are far removed from the hurricane's eye, they have caused wind damage and localized flooding. A major natural hazard has not affected the TCI for more than half a century and although the TCI periodically experiences flooding due to heavy rainfall or high tide, this flooding has not severely affected social or economic life and is merely seen as an inconvenience. Caution must once again be exercised in the interpretation of questionnaire results for reasons indicated earlier, and more extensive reliability check questions can eliminate such shortcomings in the data.

Children's Level of Preparedness

The assessment of children's level of disaster preparedness is essentially an evaluation of the existing levels of preparedness in their household. Therefore, the pretest assessment of children's preparedness focuses on their knowledge of the measures required to ensure their safety and minimize impact during an event. Table 3.5 summarizes household preparedness among sampled children.

The pretest results indicate relatively high levels of preparedness among children in both the TCI and SVG. If these results were reliable, it would be extremely difficult to explain the continued high levels of impact from hurricanes and floods. In that regard, the assessment tools should have been more exploratory in identifying the specifics of their preparedness. For instance, when children responded that they are aware of the measures required to reduce the impacts of floods and hurricanes on their household, the specifics of this knowledge require investigation. The challenge is in devising means of garnering such information without frustrating the responses of these young children. It is noteworthy that although most children in the study stated that they know how to prepare for floods and hurricanes, fewer than 75% have ever participated in emergency drills at home, at school, or in their community. Although most of the sampled children were aware of what the contents of an emergency bag should be, none of their households had such a bag. When probed, the vast majority were unaware of simple preparation measures such as cutting down overhanging limbs from trees, removing fruits that could become missiles during storms (e.g., coconuts), or elevating furniture and unplugging electrical items before a flood. This indicates a need for comprehensive assessment of preparedness measures that children are exposed to that can be used to inform interventions for the promotion of disaster education in schools. The formal

Table 3.5 Pretest Summary of Preparedness among Sampled Children

	Flood			*Hurricane*		
Sample Source	*Understand What To Do to Prepare*	*Understand What To Do to Evacuate Safely*	*Understand What To Do to Recover From the Damage Caused by Floods*	*Understand What To Do to Prepare*	*Understand What To Do to Evacuate Safely*	*Understand What To Do to Recover from the Damage Caused by Hurricane*
TCI (%)	87.9	69.7	66.7	81.8	87.9	66.7
SVG (%)	93.9	75.8	69.7	93.9	75.8	69.7

introduction of disaster education into school curricula would increase students' disaster knowledge and would have a trickle-down effect on the wider society.

Posttest Analysis

The posttest assessment of the Disaster Awareness Game is intended to measure the effectiveness of this technique in enhancing the disaster awareness knowledge base of targeted students and to determine whether there are significant differences in the knowledge acquisition of students in a multicultural school when compared with those in a more homogeneous setting. In that regard, the pretest process was followed by a posttest exercise in which students completed an identical questionnaire survey, played the board game, and completed an evaluation score sheet. Changes in awareness levels were assessed using the same criteria as the pretest and are discussed in the following sections.

Awareness of Local Hazards

Analyses of data generated in the Disaster Awareness Game exercise suggest that levels of awareness about local hazards increased after exposure to the Disaster Awareness Game. Although the results of the pregame survey suggested high levels of hazard awareness, especially in regard to children from SVG, the reliability check provided by the initial game exposure contradicted this. The conclusion is that children either did not comprehend the questions on which the analysis is based or deliberately inflated their knowledge of hazards and disasters in the pregame survey. The correctness of the responses on the game score sheet was used as a comparative measure of the extent to which exposure to the game enhances disaster awareness (Tables 3.6 and 3.7). In the case of children from the TCI, the level of awareness in relation to hurricanes and floods increased by an average of 20% following exposure to the Disaster Awareness Game. For SVG, awareness increased by more than 24% in relation to the five hazards to which that country is vulnerable. Although it can be concluded that the Disaster Awareness Game process significantly contributed to the enhancement of disaster awareness among the sampled children, caution must be exercised in relation to the reliability of these

Table 3.6 Turks and Caicos Islands: Comparative Levels of Disaster Awareness before and after Exposure to the Disaster Awareness Game

Hazard	*Before Workshop (%)*	*After Workshop (%)*
Flood	64.8	86.8
Hurricane	65.4	83.4

Table 3.7 St. Vincent and the Grenadines: Comparative Levels of Disaster Awareness before and after Exposure to the Disaster Awareness Game

Hazard	*Before Workshop (%)*	*After Workshop (%)*
Flood	69.2	74.6
Hurricane	75.0	83.0
Volcano	54.0	80.2
Landslide	51.0	92.0
Mudflow	43.0	83.0

results because there is a tendency of attrition of knowledge as time from the event increases. Therefore, the reliability of these results can only be confirmed through continuous evaluation of awareness levels over an extended period of time. These results are to be interpreted as preliminary and represent opportunities for further research.

Children's Risk Perception

Evaluation after the Disaster Awareness Game indicated a notable increase in risk perception for both TCI and SVG children in relation to floods (Table 3.8) and hurricanes (Table 3.9). In the case of flooding, the risk awareness of the TCI sample

Table 3.8 TCI and St. Vincent and the Grenadines Flood Hazard Risk Perception

	Turks and Caicos Islands			*St. Vincent and the Grenadines*		
Variables	*Pretest*	*Posttest*	*Increase (%)*	*Pretest*	*Posttest*	*Increase (%)*
Perceived vulnerability of country	75.8	84.8	9.0	90.9	100	9.1
Perceived vulnerability of community	56.3	90.9	34.6	81.3	90.8	9.5
Perceived vulnerability of home	43.8	60.6	16.8	68.8	80.5	11.7

Table 3.9 TCI and SVG Hurricane Hazard Risk Perception

	TCI			*SVG*		
Variables	*Pretest*	*Posttest*	*Increase %*	*Pretest*	*Posttest*	*Increase %*
Perceived vulnerability of country	87.9	93.9	6.0	87.9	95.9	8.0
Perceived vulnerability of community	81.8	90.9	9.1	93.9	98.5	4.6
Perceived vulnerability of home	63.9	72.7	8.8	78.6	90.9	12.3

increased by approximately 20%, while risk awareness of the SVG sample increased by approximately 10%. The smaller increase in risk awareness for the SVG sample is a reflection of higher levels of existing flood awareness in the pretest evaluation and is due to greater exposure to flood events when compared with the TCI sample. The analysis of TCI posttest data indicated that the most significant increase in flood risk awareness related to the vulnerability of communities that reflected lower levels of exposure to and experience with flooding in the communities from which the students originated. In the case of SVG, the most significant increase was related to the perceived vulnerability of homes because, although students have a high level of exposure to flooding, however sometimes given the specific location of their homes they are not directly affected by flooding unless it is of high magnitude.

In relation to hurricanes, the increase in risk perceptions following exposure to the Disaster Awareness Game was generally lower than flooding; hurricane risk perception increased by approximately 7.8% in the TCI and 7.6% in SVG. The main explanation for the lower increase is that the existing levels of hurricane awareness before exposure to the Disaster Awareness Game were significantly high for both samples. The similarity of the increase for TCI and SVG reflects the universal dissemination process for hurricane information throughout the Caribbean, especially within the context of the Caribbean Disaster Emergency Response Agency* participating states, to which TCI and SVG belong. Information about hurricanes is the most accessible for hazards that affect the Caribbean region; therefore, students from the TCI and SVG would be exposed to similar hurricane-related information.

* CDERA is the umbrella disaster management organization for the Caribbean region.

Children's Knowledge of Hazard Preparedness and Mitigation

The role of preparedness and mitigation in disaster risk reduction has emerged as a dominant paradigm in disaster management, particularly because of their relationship to the sustainability of development in both developed and developing countries. Increasingly, preparedness and mitigation are being promoted at the household and community levels and children's knowledge of preparedness and mitigation issues has become critical in this prevailing paradigm. There was a significant increase in children's knowledge of risk awareness and preparedness and, by extension, mitigation measures related to floods (Table 3.10) and hurricanes (Table 3.11) for both samples. In the case of flooding, knowledge of preparedness measures among students increased by an average of approximately 22% in the TCI and approximately 17% in SVG. In relation to hurricanes, preparedness knowledge increased by 16% in the TCI and 15% in SVG.

Explanations for the differences in increase between the two locations are similar to those for risk awareness. The increase in preparedness knowledge was manifested in the children's ability to list items that should be included in an emergency evacuation kit and how these can be used during an emergency. In addition, children demonstrated a better understanding of hazards and their impacts and a more comprehensive knowledge of steps that can mitigate the effects of hazards after exposure to the Disaster Awareness Game. Exposure to the Disaster Awareness Game also assisted students in identifying ways they can assist their parents with disaster preparedness activities. Most children were able to compile checklists to remind their parents of preparations required to mitigate the impact of specific

Table 3.10 TCI and SVG Flood Hazard Preparedness

	TCI			*SVG*		
Variables	*Pretest*	*Posttest*	*Increase %*	*Pretest*	*Posttest*	*Increase %*
Understand what to do to prepare	87.9	100	12.1	81.8	95.2	13.7
Understand what to do to evacuate safely	69.7	97.0	27.3	87.9	95.5	7.6
Understand what to do to recover from the damage	66.7	93.9	27.2	66.7	95.2	28.5

Table 3.11 Turks and Caicos Islands and St. Vincent and the Grenadines Hurricane Hazard Preparedness

	Turks and Caicos Islands			*St. Vincent and the Grenadines*		
Variables	*Pretest*	*Posttest*	*Increase (%)*	*Pretest*	*Posttest*	*Increase (%)*
Understand how to prepare	93.9	100	6.1	93.9	97.6	3.7
Understand how to evacuate safely	75.8	96.9	21.1	75.8	93.9	18.1
Understand how to recover from the damage	69.7	90.9	21.2	69.7	92.9	23.2

emergencies. Evacuation knowledge also increased with exposure to the Disaster Awareness Game, although this increase was more significant for children from the TCI than children from SVG. This result is primarily due to a higher incidence of pretest evacuation knowledge among children from SVG.

Conclusion

The importance of hazard awareness promotion through the provision of and access to disaster information and knowledge is increasingly being recognized as a critical strategy for the mitigation of the social, economic, and environmental impacts of disasters and, by extension, the enhancement of the process of sustainable development. Interventions for promoting disaster awareness must take the complexities in the disaster profile of the environment in which the intervention is intended into consideration. As this relates to the Caribbean region, one aspect of that complexity relates to ensuring equity of access to disaster information in the increasingly multicultural landscape, a phenomenon precipitated by the solidifying of the regional integration movement, the Caribbean Community and Common Market (CARICOM). Comparison of pretest and posttest results of the Disaster Awareness Game indicates that this game technique is effective in overcoming the barriers and challenges normally associated with communicating disaster risk in a multicultural setting. Multicultural children who participated in the Disaster Awareness Game in the TCI showed improvements in their awareness at a level comparable to those in the more homogeneous cultural environment of SVG. Given that the capacity of children to comprehend, assimilate, and apply information is different from that of adults, special design considerations must be taken into account in the development of tools that target disaster awareness promotion among children. The Disaster Awareness Game was designed with consideration of the specific needs of children in terms of the provision of disaster information and knowledge. The results from preliminary testing of the Disaster Awareness Game indicate that these design considerations are effective in promoting awareness among children. Gaps such as those related to the reliability of questionnaires in the design considerations have been addressed and the Disaster Awareness Game will undergo a second round of testing. The game can also be used to identify and prioritize interventions for promoting disaster awareness. This is especially relevant in the Caribbean region, where scarcity of economic resources dictates the need for prioritization. In that regard, the Disaster Awareness Game has the capacity to establish ranking among countries in terms of the level of awareness among children.

References

Anderson, A. W. 2000. Women and children facing disasters. Retrieved from http://www.crid.or.cr/cd/CD_Inversion/pdf/eng/doc13121/doc13121-contenido.pdf. Accessed February 3, 2006.

Article 12, UN Convention on the Rights of the Child. Retrieved from: www.2ohchr.org/english/law/crc/htm#art12 (accessed October 15, 2008).

Banks, S. P. 2000. *Multicultural public relations: A social-interpretive approach*, 2nd ed. Ames, Iowa: State University Press.

Bennett, P. 1997. Communicating about risks to public health: Pointers to good practice. Retrieved from http://www.dh.gov.uk/assetRoot/04/03/96/70/04039670.pdf (accessed February 3, 2006).

Bowling, P. A., and Shah, E. 2001. Measuring patients' preferences for treatment and perceptions of risk. *Quality in Health Care* 10(Suppl I):i2–i8.

Burrel, G., and Morgan, G. 1979. *Sociological paradigms and organisational analysis: Elements of the sociology of corporate life*. London: Heinemann.

Carney, B. 1993. Communicating risk. *IABC Communication World* 4:35–38.

Coichetto, S., and Haley, B. 1995. *How do we measure awareness: Experiences with lead-based paint survey*. Washington, DC: Bureau of Census and U.S. Department of Housing.

Edwards, A., Elwyn, G., and Mulley, A. 2002. Explaining risks: Turning numerical data into meaningful pictures. *BMJ* 324:827–830.

Enns, J. T. 1990. Attention in children. In *The development of attention research and theory*. Advances in Psychology #6. Amsterdam: Noth-Holland, Elsevier Science Publishers.

Fischhoff, B., Bostrom, A., and Quadrel, M. J. 1993. Risk perception and communication. *Annual Review of Public Health* 14:183–203.

Fothergill, A., Maestas, E. G., and Darlington, J. D. 1999. Race, ethnicity, and disasters in the United States: A review of the literature. *Disasters* 23:156–173.

Grimes, D. A., and Snively, G. R. 1999. Patient's understanding of medical risks: Implications for genetic counselling. *Obstetrics and Gynecology* 93:910–914.

Gutteling, J. M., and Weigman, O. 1996. *Exploring risk communication.* London, UK: Kluwer Academic Publishers.

ISDR. 2002. Gender mainstreaming in disaster reduction. Retrieved from http://www.unisdr.org/eng/risk-reduction/gender/rd-gender-eng.htm (accessed October 3, 2008).

ISDR. 2006. Africa regional strategy for disaster risk reduction. Retrieved from http://www.unisdr.org/africa/af-hfa/docs/africa-regional-strategy.pdf (accessed October 3, 2008).

ISDR. 2007. *Towards a culture of prevention: Disaster risk reduction begins at school: Good practices and lessons learned.* Retrieved from http://www.unisdr.org/eng/about_isdr/isdr-publications/11-education-good-practices/education-good-practices.pdf (accessed October 3, 2008).

Kobe Report. 2005. Knowledge, innovation and education: Building a culture of safety and resilience. Paper presented at the World Conference on Disaster Reduction, Kobe, Hyogo, Japan, January 18–22, 2005.

Lindell, M. K., and Perry, R. W. 2004. *Communicating environmental risk in multiethnic communities.* Newbury Park, CA: Sage.

National Research Council (NRC). 1989. *Improving risk communication.* Washington, DC: National Academic Press.

Nielsen, S., and Lidstone, J. 1998. Public education and disaster management: Is there any guiding theory? *The Australian Journal of Emergency Management* 13(3):14–19.

Press, F. 1989. Implementing the international decade for natural disaster reduction. Paper prepared for the Secretary General of the United Nations Organization by the Ad Hoc Group of Experts, April 11, 1989. New York: United Nations.

Press, F. 1996. Mitigating natural hazards. Retrieved from http://www.nwpublichealth.org/docs/wph/hazard.html (accessed October 3, 2008).

Ropeik, D., and Slovic, P. 2003. Risk communication: A neglected tool in protecting public health. *Risk in Perspective* 11(2). Retrieved from http://www.hcra.harvard.edu/rip/risk_in_persp_June2003.pdf.

Suzuki, K. 2006. Abstract — "Promoting disaster education for our children." Retrieved from www.adrc.or.jp/events/octoberforum/pdf/Abstract_Mr.Suzuki.pdf (accessed October 3, 2008).

Tierney, K. J., Lindell, M. K., and Perry, R. W. 2001. *Facing the unexpected: Disaster preparedness and response in the United States.* Washington, DC: Joseph Henry Press.

Tomlinson, J. 1999. *Globalization and culture.* Chicago, IL: The University of Chicago Press.

UNDP. 2002. *Living with risk: A global review of disaster reduction initiatives.* Retrieved from http://www.adrc.or.jp/publications/LWR/LWR_pdf/index.pdf (accessed September 22, 2008).

UNESCO. 2004. APELL for schools and educational buildings: A community-based approach for school safety and education for disaster reduction. Retrieved from http://www.preventionweb.net/files/5473_apellschools.pdf (accessed October 3, 2008).

Van Vliet, H. A. A. M., Grimes, D. A., Popkin, B., and Smith, U. 2001. Lay person's understanding of the risk of Down syndrome in genetic counselling. *British Journal of Obstetric Gynaecology* 108:649–650.

Wisner, B. 2006. *Let our children teach us: A review of the role of education and knowledge in disaster risk reduction.* Retrieved from http://www.unisdr.org/eng/task%20force/working%20groups/knowledge-education/docs/Let-our-Children-Teach-Us.pdf (accessed October 3, 2008).

Chapter 4

Rebuilding a Regional Tourist Image in the Aftermath of Disaster: New Orleans and Hurricane Katrina

Mark Hutter and DeMond S. Miller

Contents

Introduction

Tourism, with its associated industries, is one of the world's largest income generators. For example, in 2006, the World Travel and Tourism Council projected $6,477.2 billion in total demand; in the United States, travel and tourism were expected to generate $1,652.6 billion of economic activity, contributing an estimated 15,604,000 jobs or 1 in every 9.2 jobs for 2006 [World Travel and Tourism Council 2006]. A global high in tourism was reached in 2001 when 898 million people set out as travelers to tourist destinations, with the fastest arrival growth rates in the Middle East (13%) and Asia Pacific (10%) regions [Shurland and de Jong 2008]. In many developing nations with an abundance of natural and cultural resources, tourism has become an economic development strategy of choice that provides economic expansion opportunities for the small business sector in tertiary services, crafts, foods, fishing, and other ancillary services [Shurland and de Jong 2008].

These global trends occurred similarly in pre-Katrina New Orleans. Cultural structures particular to a destination, both concrete and symbolic, are influenced by underlying political, economic, sociocultural and environmental factors. These factors are commodified for public spectacle and sold. In the aftermath of Hurricane Katrina, tourist and convention planners, city leaders, cultural institutions, and private businesses such as tour guides, vied for opportunities to coordinate, control, and produce an image of the city. The resultant image ultimately emphasized the cultural revitalization of New Orleans and served to develop rules and mechanisms to assure that cultural revitalization [Beritelli 2007]. Policies governing the use of culture as the cornerstone of both the tourism industry and economic development of the New Orleans region were developed.

The commodification and selling of a place's or a region's symbolic economy is a complex process that includes a variety of stakeholders composing a network of institutional players that shape the image of a destination. Key stakeholders in the tourism industry ultimately govern how places are branded, rebranded, and how a destination's symbolic economy becomes the primary driver for rebuilding after a disaster. For the state of Louisiana and the Greater New Orleans Region, the commodification and selling of the region's symbolic economy translated into, "61% of the tourism-related expenditures for the state of Louisiana [being] spent in New Orleans and [the local] the tourism industry provided 35% of the operating budget of the city while generating $233 million in tax revenue from approximately $4.9 billion in visitor spending" [Williams 2006; Pearlman and Melink 2008, p. 58]. Pearlman and Melink go on to describe the losses associated with Hurricanes Katrina and Rita and the impact on potential tourists' perceptions of the region's image by asserting the following:

> A report by Louisiana Recovery Authority [2006] noted that roughly 81,000 businesses in 13 Louisiana parishes suffered business interruptions or direct damages as a result of Hurricanes Katrina and Rita. More

> specifically, 1,409 tourism and hospitality businesses shut down after the storm affecting 33,000 hospitality-based employees. The report also noted that the negative image of affected areas portrayed by the media has caused business and leisure travelers to choose other destinations resulting in the New Orleans region losing an average $15.2 million per day [Pearlman and Melink 2008, p. 59].*

Ultimately, this chapter contributes to the study of regional disaster recovery and tourism in the aftermath of disasters by analyzing how destination information or the imagery of a city in the media can influence destination selection and travel decisions and how the image-forming process influences the destination and regional promotion within the tourism industry. The conclusions and the main implications derived in this chapter for the tourism industry are considered.

Regional Tourism, Disaster, and the Image of a City

When Hurricane Katrina made landfall on August 29, 2005, its storm surge overwhelmed coastal communities in Louisiana and Mississippi. Of equal importance is that the hurricane's path of death and destruction was of a magnitude that was unimaginable, making it the most costly natural disaster to befall the United States in recent memory. Katrina—The Event— was widely covered by television and cable news networks as flood waters inundated 80% of the city of New Orleans [Chacko and Marcell 2007, p. 223]. America and the world watched as media images of a cityscape that was 80% underwater descended into an abyss of chaos. Vivid visuals showed cars, boats, houses, and pieces of houses floating down streets, stranded citizens on rooftops, weary soldiers, and chaos reigning in various parts of the city. All of these images created an image of a city not conducive to attracting visitors to the region. Shortly after the storm, the lower Ninth Ward section of the city, bearing the full brunt of the ruptured Industrial Canal, became the media focal point of the city's state of ruin and weakened infrastructure. Conventional wisdom suggests that visitors would want to stay away from the area. Even with evidence that tourism developments are likely to be exposed to the sudden onset of natural disasters,

* From Pearlman and Melnik [2008, p. 58]: The importance of a tourist destination's image is universally acknowledged since it affects the individual's subjective perception and consequent behavior and ultimately destination choice [Chon 1990, 1991; Echtner and Ritchie 1991]. There are a number of studies focused on perception of selected destinations. According to a previous study, "owing to differences in experiences, beliefs, and expectations, travelers tend to differ in terms of the overall perceptions they hold" regarding a certain destination [Javalgi et al. 1992, p. 46]. Illiewich [1998] stated that within tourism research, image has been recognized on a microlevel in terms of the decision-making process and on the macrolevel referring to tourism development.

in particular beach and coastal areas [Obasi and Frangialli 1998], it seems unusual to think about tourism and disaster simultaneously [Santana 2003]; this is because the thought of tourism naturally evokes feelings of enjoyment, pleasure, relaxation, and safety while conversely, disasters arouse distress, fear, anxiety, trauma, and panic in individuals [Santana 2003; Lehto et al. 2008].

As residents were rescued from their once familiar terrain by boat or plucked off their roofs by helicopters, the city once branded by the slogan *Laissez les bons temps rouler** was depicted as a place of sorrow and despair. It was a place where so much had been lost that the memories attached to the New Orleans landscape only echoed their plight to survive amid a surreal landscape of floating corpses. Media images of floating bodies amid the city's architecture became commonplace as the perception of a general state of lawlessness and violence branded the city as a wasteland. After Katrina, images of desperation, lawlessness, and looting in one of America's favorite destinations brought a flourishing tourism industry to its knees [Chacko and Marcell 2007, p. 224]. Huang and Min [2002, p. 145] note that "[t]he accuracy of media coverage is essential for shaping potential visitors' attitude toward the destination during the aftermath of a disaster; however, accurate coverage is often the exception rather than the rule during those times." Instead, the media may exaggerate the extent of the disaster and complicate the understanding of improvements in tourist locations, health facilities, security, and other attractions during the recovery, and lessen the willingness of tourists, or potential tourists, to visit [Milo and Yoder 1991; Huang and Min 2002].

With a large portion of the city's basic infrastructure—not to mention its place as a regional tourist destination—in disarray, many saw it as an opportunity for structural redevelopment with a chance to reshape the image of the city while maintaining its status as a tourist destination. However, with regard to rebuilding the tourist infrastructure, Chacko and Marcell [2007, p. 224] ask a critical question: How do you promote and attract tourists to a destination that has a relatively intact infrastructure but a completely devastated destination image? In 2004, tourism was one of the city's main economic engines and, prior to Hurricane Katrina, hospitality and leisure employment in New Orleans accounted for 80,827 jobs, generating $30 million in state income taxes [UNO Hospitality Research Center 2005b]. The hospitality industry also produced $158.6 million in state tax revenue and channeled $10 million a year to the New Orleans school system [New Orleans Metropolitan Convention and Visitor Bureau 2005]. Visitation to New Orleans peaked in 2004, when 10.1 million visitors came to the city, spending $4.9 billion [UNO Hospitality Research Center 2005b; Chacko and Marcell 2007, p. 224]. The concern of this chapter is that the constructed image of New Orleans, seen through tourist sites, has become a significant contributory factor for political and economic decisions on the city's

* A Cajun French phrase meaning "let the good times roll."

future. In contrast, and often in opposition, are the images and the symbolic meaning held by the residents.

New Orleans: A Regional Destination before Katrina

Before the storm, tourists flocked to the area of the city known as the French Quarter for food, cultural events, and entertainment. Tourism, for the arts, entertainment, and recreation accommodation and food services, is widely recognized as a staple to the New Orleans and regional economies. "Besides the attraction of its French Quarter, its internationally renowned restaurants, and its first-class accommodations, a series of celebrations, including Mardi Gras, the New Orleans Bowl, the Sugar Bowl, and the New Orleans Jazz and Heritage Festival, have drawn thousands of tourists to the city" [Dolfman et al. 2007, p. 4]. However, the area hardest hit by the storm, the lower Ninth Ward, lies beyond those sites and existed outside of the commercial tourist path. After the storm, visitors would readily go into the neighborhoods most impacted on one of the many disaster tours where people seek out destinations of calamity by their own means through a commercial tour operator [Pezzullo 2009, p. 33] or, as they are now called, a recovery tour.

The imagery of the city—a city in distress, a comeback city, or a recovery city—can be seen as a result of how people perceive the city. Urban imagery also has consequences in the shaping of city life itself. Urban images have a symbolic function that positions the city or region in the mindsets of travelers and residents alike. According to Pike and Ryan [2004, p. 334], "a major objective of any destination positioning strategy will be to reinforce positive images already held by the target audience, correct negative images, or create a new image." Positioning is a way to communicate, via symbolic language, the commodification of culture. Images help provide strong associations with a place that help facilitate interaction between people who share a common environment. Shared images of places and communities are the facilitator in the development of strong bonds between people. The urban social geographer Kevin Lynch has pointed this out:

> The landscape plays a social role as well. The named environment, familiar to all, furnishes material for common memories and symbols which bind the group together and allow them to communicate with one another. The landscape serves as a vast mnemonic system for the retention of group history and ideals [Lynch 1960, p. 126].

Social theorists have commented on how the nature of the space [or environment] informs us that specific objects, places, or attributes of the city become symbolically representative of the city or region as a whole. Thus the New York skyline, San Francisco's Golden Gate Bridge, or the French Quarter of New Orleans became identification symbols of the given city or region and serve as a source of personal identification for the inhabitants.

Reconstructing the City and Regional Images

Images of a destination are shaped in the traveler's mind through synthesis of information gathered over a period of time. Crompton [1979, p. 18] argues that a destination's image is "a sum of beliefs, ideas, and impressions that a person has of a destination." These images, both negative and positive, are crucial in determining a destination's ability to draw tourists who are likely to pay particular attention to the effect of destination image on travel intention [Leisen 2001; Lehto et al. 2008] making the image of a destination a critical factor in the tourists' destination selection process [Sirgy and Su 2000]. A positive image of destination results in more visitations [Gartner and Hunt 1987; Gartner and Shen 1992; Dimanche and Lepetic 1999], but changing a negative one requires long and costly marketing efforts [Huang and Min 2002, p. 145]. "While positioning strategies may be carefully planned and orchestrated, the major problem during a catastrophic event is that the mass media coverage of the disaster has an overwhelming impact on a destination's market position" regionally, nationally, and internationally [Chacko and Marcell 2007, p. 224]. Since the tourists' perceptions of safety concerning a destination will influence their intentions to visit [Dimanche and Lepetic 1999; Sonmez and Graefe 1998], it is important to reintroduce a destination by providing up-to-date information [Durocher 1994]. Furthermore, it is also desirable to establish a good relationship with the media [Milo and Yoder 1991; Drabek 1995; Huang and Min 2002, p. 145].

The Symbolic Economy and Rebuilding the Image

New Orleans is dependent upon both domestic and international tourism. It is a city much dependent upon what Sharon Zukin [1995] has called the symbolic economy—where culture has become an important urban economic commodification. This new symbolic economy is anchored in the economic importance of tourism, media, and entertainment for cities. She believes that with the disappearance of local industrialization and manufacturing, culture is increasingly the business of cities. The cultural consumption of art, gourmet food, fashion, music, and tourism, which often packages them together, are the foundation on which the city's symbolic economy rests. Zukin describes the symbolic economy as an economy that is created by cultural strategies of urban redevelopment and the privatization of public space. Zukin persuasively argues that the recent growth of the symbolic economy has tangible spatial repercussions: it "reshapes geography and ecology" [Zukin 1995, p. 8]. It creates new types of workplaces, commercial areas, and residences, and throws into turmoil conventional meanings of public versus private, local versus transnational, and commodity versus culture.

The symbolic economy for Zukin is not a singular thing, rather it involves a number of different ways in which the cultural symbols of a place are combined with capitalist activity: the "intertwining of cultural symbols and entrepreneurial capital" [Zukin 1995, p. 3]. She suggests that historically this economy emerges and influences how a place is seen and used as well as the subsequent kinds of investment and economic development that can be used to regenerate a city. The symbolic economy becomes dependent upon the creation of attractions, whether they be museums, entertainment zones (such as the new family-friendly Times Square), sports stadia, parks, and festival marketplaces for the more affluent groups of the city. For example, in New York City's Bryant Park, the establishment of an expensive sidewalk cafe was one means to establish the presence and control of these people over the park: "A sidewalk cafe takes back the street from casual workers and homeless people" [Zukin 1995, p. 9].

Zukin's dominating concerns are not only issues of who occupies real space, but also the utilization of symbolic spaces as well. In an allied work, Zukin states: "To ask 'Whose City?' suggests more than a politics of occupation; it also asks who has a right to inhabit the dominant image of the city" [Zukin 1996, p. 43]. However, with so many prevailing negative perceptions of New Orleans, the city's cultural value, via its symbolic economic value or brand, began to lose value and tourism marketers knew they had to develop a repositioning strategy. As stated by Ries and Trout [1993, p. 5], "The basic approach to positioning is not to create something new and different, but to manipulate what's already there in the mind, to retie the connections that already exist."

The physical space of New Orleans is redefined in terms of place by those in positions of power and economic control. In the immediate aftermath of Hurricane Katrina, the tourism industry saw an opportunity to benefit from the devastation by incorporating it into disaster tours. "As of March 2007, practically all of the tour operators in New Orleans explicitly had begun advertising a Katrina Tour, including Tours by Isabelle (which was the first), Gray Line Tours, Van from New Orleans, Dixie Tours, Airboat Adventures, Celebration Tours LCC, Big Easy Tours, Viator, Cajun Encounters, Sightseeing World, City Tours, and, on screen, the Entergy IMAX Theatre" [Pezzullo 2009, p. 33]. The lower Ninth Ward and adjacent areas have become the sites where tour buses have become a ubiquitous presence in the devastated landscape. These tours function as a "means to experience the shock and awe in a voyeuristic reality-style manner through a packed tour bus—complete with cameras flashing and souvenirs for those seeking to get as close to the real experience as they can ..." [Miller 2008, p. 118]. While it has been more than five years since the hurricane, these buses continue to run even while the area has been largely cleared and new buildings are now being constructed. In the long run, most conceivably, tourism will once again become focused on the traditional sites of the French Quarter, the Garden District, Audubon Park, etc. However, the continued presence of disaster tourism is in itself symbolic evidence of the importance of the symbolic economy in the economic future of New Orleans.

Many cities have used tourism as a strategy for urban redevelopment and revitalization, forcing local governments and the tourist industry to forge close institutional and fiscal alliances to sell the city. Such alliances by government and business interests to market imagery and themes usually positively impact the economy of cities and regions by capitalizing on the cultural aspects of the area. Such a proactive stance is necessary to challenge negative media and to reflect the reality of the undamaged areas of the city. However, residents contend that this is an inaccurate picture of the area as they struggle to regain a slice of normalcy. Nonetheless, "various tourism industry stakeholders, both private and public, pooled their resources to open The New Orleans Media Center to put out messages that reinforced the fact that tourism infrastructure was still viable" [Chacko and Marcell 2007, p. 231]. Using glossy advertisements and armed with a new message, the public relations strategy included the production and distribution of a free DVD titled "Make Way for the Rebirth" providing visual evidence that the tourist areas, such as the French Quarter, were intact and that many businesses were open [Chacko and Marcell 2007, p. 231].

Former Mayor Nagin's plan for rebuilding the city focuses on a cultural redevelopment component seeks to reestablish the city's musical, visual, culinary, architectural, literary, and graphic arts traditions. This plan invests approximately $648 million over a three-year period to jumpstart the redevelopment of the city's culture over the next few decades. Facets of the plan include rebuilding artistic talent pools, repairing and rebuilding cultural venues, teaching the city's cultural traditions to young people, attracting national and international investment, requiring shared public and private investment in the city's infrastructure, and "marketing New Orleans as a world-class cultural capital" [City of New Orleans 2006, p. 14]. By adhering to the plan proposed by the mayor's office, many leaders believe that:

> The proposed public and private investments will revive the city's cultural base, benefit businesses and residents of every neighborhood, endure the return of displaced artists and cultural workers, restore leading cultural facilities and create new cultural venues that celebrate the city's unique musical history and the cultural traditions of its diverse neighborhoods, revitalize street life and performances, increase tourism, and lever other investments many times over [City of New Orleans 2006, p. 14].*

Although turnaround and recovery from a natural disaster is a complex issue, the role of marketing communication in regaining tourist confidence is undeniable. Subsequent to a disaster, destinations are faced with not only the daunting tasks

* For more information, see City of New Orleans [2006].

of rebuilding infrastructure, facilities, and communities, but also image recovery [Lehto et al. 2008, p. 30].

Since the storm, the loss of New Orleans's distinctive culture is a primary concern of the city's residents. Culture is significant to the survivors because it has shaped a sense of place and place attachment. However, with an emphasis on tourism and the outside workers that are bound to follow, how significant will the renewing of New Orleans culture be to them? There is no doubt that those material components and festive aspects of New Orleans' culture will remain significant to the workers who will flock to the city for employment opportunities in the tourist industry. Yet, the city's economic revival rests in the commodification and selling of the traditional New Orleans experience—complete with jazz, food, and fun—as opposed to its cultural value just for its traditional practitioners. Gotham explained that the "concept of commodification refers to the dominance of commodity exchange-value over use-value and implies the development of a consumer society where market relations subsume and dominate social life" [Gotham 2002, p. 1737]. Gotham's work [2006, 2007a–c] emphasizes the role of commercial tourist industry post-Katrina in New Orleans, theorizing how the rebranding of the city is transforming the city.

Therefore, in the context of selling the city's culture and sense of place, local customs, rituals, festivals, and ethnic arts will become tourist attractions performed on cue when tourists want to be entertained [see Gotham 2006, 2007a, 2007b], thus devaluing the traditional experiences in the eyes of traditional practitioners. Devaluation of traditional culture by residents will occur as traditional events become trivial and less significant in relation to annual events with cultural meaning; traditional events will give way to more financially motivated activities that can be packaged and sold as culture on demand. Local concerns about the establishment of a dominant tourist sector are exacerbated by a more commodified tourist industry that will be centered on less culturally significant activities rooted in and connected to the landscape. Gotham [2007b] also noted the commodification of Hurricane Katrina including the production of souvenirs and tours of flooded neighborhoods.

In the pursuit of economic success, the city and private businesses will attempt to exploit the city's image; however, this may have a detrimental effect on the city's culture in the long-term. Because culture will be an economic stimulus, and possibly the most relied on sector in the region, business and city planners will have to avoid the occurrence of Disneyfication* and the manufacture of a nonauthentic urban experience and culture. With the occurrence of Disneyfication, local traditions, famous buildings and landmarks, and other heritage sights and events become

* Gotham [2002] explained "the transformation of public spaces into privatized 'consumption' spaces and the latest attempts by tourism entrepreneurs and other economic elites to provide a package of shopping, dining, and entertainment within a themed and controlled environment—a development that scholars have called the 'Disneyfication' of urban space" [p. 1738]. For more information, see Eeckhout [2001], Reichl [1999], and Sorkin [1992].

hyper-real, causing people to lose the ability to distinguish between the real and the illusion or between original culture and marketing-manufactured culture.

Additionally, Harvey [1992] and Holcomb [1993] explain that with the expansion of the tourist industry into a dominating sector, place promotion no longer concerns itself with informing or promoting in the ordinary sense (i.e., selling the original cultural characteristics of New Orleans) but rather is geared toward manipulating desires and tastes through imagery that may or may not have anything to do with the product being sold—New Orleans' culture [Harvey 1992; Gotham 2007b, 2007d]. In essence, the tourist industry ceases to sell the authentic or true culture of New Orleans; instead, city leaders are manufacturing a new New Orleans by adapting, reshaping, and manipulating images to remake a sense of place for visitors from afar.

Functioning within this economic and power environment, New Orleans' culture and demographics may change because New Orleans' culture will develop directly in conjunction with tourism market forces. Moreover, Gotham [2007b] has noted that shifts toward tourist-dominated economics in an urban setting have historically coincided with population decline, white flight to the suburbs, racial segregation, poverty, and a host of other social problems, including crime, fiscal austerity, poor schools, and decaying infrastructure, all of which were observable in New Orleans prior to Katrina. Therefore, in looking toward to the future, one can foresee that an expansion of the service sector may aid the economic redevelopment of New Orleans in the immediate short-term, but in the long-term, social conditions may revert back to what they were prior to Katrina in addition to losing the cultural significance, place attachment, and sense of place that many of the residents held for New Orleans and the region prior to the storm. The future of New Orleans and the region is now in the hands of the government and increasingly in the hands of the growth machine—land speculators and investors, real estate developers, commercial banks, corporate property owners, and politicians who also employ the image-makers. Decisions will be made on how to rebuild the city both structurally and through imagery and rebranding. Shurland and de Jong [2008, p. 10] argue that "[a] successful tourism and destination experience is therefore predicated on sound government policy and strategy and the efficacy of the tourism supply chain."

Conclusion

Some tourism regions have a relatively high probability of particular crisis events, and while "earthquakes, hurricanes, or avalanches cannot be accurately predicted; it is certainly possible to learn from past crisis, to pre-plan emergency procedures, and to mitigate the verity of such events by adopting appropriate building codes, providing alternative escape routes, and develop crisis management strategies" [Laws and Prideaux 2005, p. 5] to address these disasters. Power and economic factors combined with concerns for the symbolic meaning of space interact and have consequences for how people interpret their built environment. The point that we must emphasize in our understanding of urban imagery and the future of New Orleans is that the labeling or seeing process is influenced by people in positions of power who can influence and control the symbolic meaning given to a community. Real estate developers, leaders in the tourist industry and other power brokers including bankers and big businesses, through the consent and support of government agencies, will largely shape the region's redevelopment. These agencies have the power to transform the urban space through the manipulation of images, symbols, and rhetoric. For those most affected by Hurricane Katrina, the people who lived in that region, what role they will play and how effective they will be in shaping the urban imagery and the building environment of the future New Orleans and the region remains to be seen.

The way regional disasters and recovery are portrayed in the media has fundamental implications for strategic destination image management, which can assist in designing strategies to enhance a region's image and to correct negative images. But just as the media's presentation of the disaster can provide negative images, the media can also provide an opportunity to show the redevelopment in a positive light, highlighting the region's rebuilt infrastructure. Laws and Prideaux [2005] echo the sentiment that mass media can play a large role in presenting positive images. These images blend to make a coherent, yet contradictory, whole New Orleans that continues to be reshaped, redefined, and rebuilt.

An individual's behavioral intention as it relates to travel is influenced by a destination's image. Weakened or negative images of a destination can be a result of the degree of increased risk associated with the destination. It can significantly alter the perceived benefits to be derived from an intended travel experience. As a result, image restoration is a critical part of destination recovery. A practical consequence of this research draws attention to the marketing implications of the study findings for the destination management organizations and governing bodies in charge of national tourism policy formulation. It is important to recognize that the impact of such a natural disaster affects not only the immediate communities, livelihoods of local people, and the natural environment, which are all important stakeholders at any given destination, but also the minds of the potential travelers [Lehto et al. 2008, pp. 40–41]. The control of images is important for the region; destination image and how a region is positioned and promoted impacts the demand and decision-making process for travelers in the market for a vacation to the Gulf Coast region. For industry leaders and tour promoters, Molina et al. [2010, p. 727] argue "that the use of information sources [i.e., brochures and guides] as promotion tools has a strong influence on the formation of tourist destination images ... [and] that the image generated by tourist brochures is firmly related to actual destination image." These businesses and municipal authorities promote the public good and indirectly reflect and promote economic prosperity via tourism. The question remains whose image will dominate in the rebuilding and rebranding of New

Orleans and the region—images of the media or those working to put forth a positive regional image? Durocher [1994 as cited in Huang and Min 2002: 152] maintains that "it may take several years to rebuild the industry to pre-disaster levels for any tourism destination after a natural disaster." The government's policies and attitudes play a vital role in the recovery process. The prompt and effective efforts of New Orleans city government and other regional stakeholders contributed greatly to the recovery that has taken place and enhanced awareness of the region around the globe.

References

Campbell, D. 2005. From the ruins, an opportunity for rebirth: Planners face huge challenge to rebuilding city: But it could take years. *The Guardian*, Friday September 2, 2005.

Chacko, H., and Hawkins Marcell, M. 2007. Repositioning a tourism destination: The case of New Orleans after Hurricane Katrina. *Journal of Travel and Tourism Marketing* 23(2/3/4):223–235.

Chon, K. 1990. The role of destination image in tourism: A review and discussion. *Revue due Tourisme* 2:2–9.

Chon, K. 1991. Tourism destination image modification process. *Tourism Management* 12(1):68–72.

City of New Orleans. 2006. Rebuilding New Orleans. Retrieved from http://www.bringneworleansback.org/Portals/BringNewOrleansBack/Resource (accessed January 21, 2007).

Cooper, M. 2005. Japanese tourism and the SARS epidemic of 2003. *Journal of Travel and Tourism Marketing* 19(2/3):117–131.

Crompton, J. L. 1979. An assessment of the image of Mexico as a vacation destination and the influence of geographical location upon that image. *Journal of Travel Research* 17(4):18–23.

Dimanche, F., and Lepetic, A. 1999. New Orleans tourism and crime: A case study. *Journal of Travel Research* 38(1):19–23.

Dolfman, M. L., Wasser, S. F., and Bergman, B. 2007. The effects of Hurricane Katrina on the New Orleans economy. *Monthly Labor Review,* June 2007, http://www.bls.gov/opub/mlr/2007/06/art1full.pdf (accessed March 19, 2010).

Durocher, J. 1994. Recovery marketing: What to do after a natural disaster. *Cornell Hotel and Restaurant Administration Quarterly* 35(2):66–71.

Drabek, T. E. 1995. Disaster planning and response by tourist business executives. *Cornell Hotel and Restaurant Administration Quarterly* 36(3):86–96.

Dyson, Michael E. 2006. *Come hell or high water: Hurricane Katrina and the color of disaster.* New York: Basic Civitas Books.

Echtner, C. M., and Ritchie, J. R. B. 1991. The meaning and measurement of destination image. *Tourism Studies* 2(2):2–12.

Eeckhout, B. 2001. The 'Disneyfication' of Times Square: Back to the future? In *Critical perspectives on urban redevelopment,* ed. K. F. Gotham, 379–428. New York: Elsevier Press.

Fall, L. T., and Massey, J. E. 2005. The significance of crisis communication in the aftermath of 9/11: A national investigation of how tourism managers have re-tooled their promotional campaigns. *Journal of Travel and Tourism Marketing* 19(2/3):77–90.

Fischer, W., and Sard, B. 2006. Bringing Katrina's poorest victims home. *Center for Budget and Policies Priorities*. Retrieved from http://www.cbpp.org (accessed: March 19, 2010).

Gartner, W. C., and Hunt, J. D. 1987. An analysis of state image change over a twelve year period. *Journal of Travel Research* 26(2):15–19.

Gartner, W. C., and Shen, J. 1992. The impact of Tiananmen Square on China's tourism image. *Journal of Travel Research* 30(4):47–52.

Gotham, K. F. 2002. Marketing Mardi Gras: Commodification, spectacle and the political economy of tourism in New Orleans. *Urban Studies* 39:1735–1756.

Gotham, K. F. 2006. *Authentic New Orleans: Tourism, culture, and race in the Big Easy*. New York: New York University Press.

Gotham, K. F. 2007a. Fast spectacle: Reflections on Hurricane Katrina and the contradictions of spectacle. *Fast Capitalism* 2(2). Retrieved from http://www.uta.edu/huma/agger/fastcapitalism/2_2/gotham.html (accessed August 22, 2008).

Gotham, K. F. 2007b. (Re)Branding the Big Easy: Tourism rebuilding in post-Katrina New Orleans. *Urban Affairs Review* 42(6):823–850.

Gotham, K. F. 2007c. Critical theory and Katrina: Disaster, spectacle and immanent critique. *City* 11(1):81–99.

Gotham, K. F. 2007d. Selling New Orleans to New Orleans: Tourism authenticity and the construction of community identity. *Tourist Studies* 7(3):317–339.

Gottdiener, M. 1986. Culture, ideology, and the sign of the city. In *Capitalist development and crisis theory: Accumulation, regulation, and spatial restructuring*, eds. M. Gottdiener and N. Komninos. London and New York: Macmillan.

Gottdiener, M. 1997. *Theming of America: Dreams, visions, and commercial spaces*. Boulder, CO: Westview Press.

Gray, C. 2005. Saving cultural treasures: Restaurants, musicians struggling after the storm. *The Philadelphia Inquirer* 12 September 2005: A1, A7.

Harvey, D. 1992. *The condition of postmodernity: An inquiry into the origins of cultural change*. New York: Blackwell.

Holcomb, B. 1993. Revisioning place: De- and re-constructing the image of the industrial city. In *Selling places: The city as cultural capital, past and present*, eds. G. Kearns and C. Philo. Oxford, England: Pergamon Press.

Huang, J.-H., and Min, J. C. H. 2002. Earthquake devastation and recovery in tourism: The Taiwan case. *Tourism Management* 23:145–154.

Illiewich, S. 1998. Deconstructing destination-image: A conceptual model of image formation, empirically tested in the context of European city tourism. In *29th TTRA Annual Conference Proceedings* (pp. 287–294). Ft. Worth, TX: TTRA.

Javalgi, R. J., Thomas, E. G., and Rao, S. R. 1992. US pleasure travelers' perceptions of selected European destinations. *European Journal of Marketing* 26(7):45–64.

Khudhairy, D. A., Annunziato, A., Best, C., Bettio, M., Caravaggi, I., Confalonieri, R., Christou, M., Ehrlich, D., Ketselidis, M., Louvrier, C., et al. 2005. Katrina event information report. Seville, Spain: European Commission Joint Research Centre.

Knabb, R. D., Rhome, J. R., and Brown, D. P. 2005. Tropical Cyclone Report Hurricane Katrina 23–30 August, 2005. National Hurricane Center. Retrieved from http://www.nhc.noaa.gov/pdf/TCR-AL122005_Katrina.pdf (accessed September 20, 2010).

Lehto, X., Douglas, A. C., Park, J. 2008. Mediating the effects of natural disasters on travel intention. *Journal of Travel and Tourism Marketing* 1540–7306, 23(2):29–43.

Laws, E., and Prideaux, B. 2005. Crisis management: A suggested typology. *Journal of Travel and Tourism Marketing* 19(2/3):1–8.

Leisen, B. 2001. Image segmentation: The case of a tourism destination. *Journal of Services Marketing* 15(1):49–66.

Lofland, L. H. 2003. Community and urban life. In *Handbook of symbolic interaction*, eds. L. T. Reynolds and N. Herman-Kinney. Lanham, MD: Alta Mira Press.

Louisiana Recovery Authority. 2006. *Proposed action plan amendment number 2 for disaster recovery funds disaster recovery initiative.* Baton Rouge: Office of Community Development.

Lynch, K. 1960. *The image of the city*. Cambridge, MA: MIT Press.

Mele, C. 2000. *Selling the Lower East Side: Culture, real estate, and resistance in New York 1880–2000*. Minneapolis, MN: University of Minnesota Press.

Miller, D. S. 2008. Disaster tourism and disaster landscape attractions after Hurricane Katrina: An auto-ethnographic journey. *International Journal of Culture, Tourism and Hospitality Research* 2:115–131.

Miller, D. S., and Rivera, J. D. 2008. *Hurricane Katrina and the redefinition of landscape.* Lanham, Maryland: Lexington Books.

Milo, K. J., and Yoder, S. L. 1991. Recovery from natural disaster: Travel writers and tourist destinations. *Journal of Travel Research* 30(1):36–39.

Molina, A., Gómez, M., and Martín-Consuegra, D. 2010. Tourism marketing information and destination image management. *African Journal of Business Management* 4(5):722–728.

Molotch, H. 1969. The city as a growth machine. *American Journal of Sociology* 82:309–330.

Mooney, C. 2005. Thinking big about hurricanes: It's time to get serious about saving New Orleans. *The American Prospect.* Retrieved from http://www.prospect.org/web/page.ww?section=root&name=ViewWeb&articleId=9754 (accessed March 19, 2010).

National Oceanic and Atmospheric Administration and National Climatic Data Center. 2005. *Climate of 2005 summary of Hurricane Katrina*. NOAA/NCDC. Retrieved from http://lwf.ncdc.noaa.gov/oa/climate/research/2005/katrina.html (accessed December 12, 2005).

National Oceanic and Atmospheric Administration and National Weather Service. 2005. *Monthly tropical weather summary*. NOAA/National Weather Service. Retrieved from http://www.nhc.noaa.gov/archive/2005/tws/MIATWSAT_aug.shtml (accessed December 12, 2005).

Obasi, B., and Frangialli, F. 1980. Preface. In *World Tourism Organization and World Meteorological Organization handbook on natural disaster reduction in tourist areas.* Madrid: WTO.

Park, R. E. 1925/1967. The city: Suggestions for the investigation of human behavior in the city environment. In *The city*, eds. R. E. Park et al., 1–46. Chicago: University of Chicago Press.

Park, R. E., and Burgess, E. 1970/1921. *Introduction to the science of sociology*. Chicago: University of Chicago.

Pearlman, D., and Melnik, O. 2008. Hurricane Katrina's effect on the perception of New Orleans leisure tourists. *Journal of Travel and Tourism Marketing* 25(1):58–67.

Pezzullo, P. C. 2009. Tourists and/as disasters: Rebuilding, remembering, and responsibility in New Orleans. *Tourist Studies* 9:23–41.

Pike, S., and Ryan, C. 2004. Destination positioning analysis through a comparison of cognitive, affective, and cognitive perceptions. *Journal of Travel Research* 42:333–342.

Reagan, M., Web, G., Harkleroad, C., Brown, C., Reagan, L., and Murphy, D. 2005. *Katrina: State of emergency*. Kansas City, MO: Lionheart Books.

Reichl, A. 1999. *Reconstructing Times Square: Politics and culture in urban development*. Lawrence, KS: University Press of Kansas.

Ries, A., and Trout, J. 1993. *The 22 immutable laws of marketing*. New York: HarperCollins Publishers.

Santana, G. 2003. Crisis management and tourism: Beyond the rhetoric. *Journal of Travel and Tourism Marketing* 15(4):299–321.

Short, J. R. 1999. Urban imagineers: Boosterism and the representations of cities. In *The urban growth machine: Critical perspectives two decades later*, eds. A. E. G. Jonas and D. Wilson, 37–54. Albany, NY: State University of New York Press.

Shurland D., and de Jong, P. (lead authors). 2008. *Disaster risk management for coastal tourism destinations responding to climate change: A practical guide for decision makers*. Paris, France: United Nations Environmental Program ISDR/CAST.

Sirgy, J. M., and Su, C. 2000. Destination image, self-congruity and travel behavior: Toward an integrative model. *Journal of Travel Research* 38(4):340–352.

Sonmez, S. F., and Graefe, A. R. 1998. Determining future travel behavior from past travel experience and perceptions of risk and safety. *Journal of Travel Research* 37(2):171–177.

Sorkin, M. 1992. *Variations on a theme park: The new American city and the end of public space*. New York: Hill and Wang.

Strauss, A. 1961. *Images of the American city*. New York: The Free Press.

Strom, E. 1999. Let's put on a show! Performing acts and urban revitalization in Newark, New Jersey. *Journal of Urban Affairs* 21:423–435.

Thevenot, B., and Russell, G. 2005. Rape. Murder. Gunfights. *The Times Picayune* [New Orleans]:A1, A4–A5.

University of New Orleans Hospitality Research Center. 2005. *2004 New Orleans area visitor profile: Annual report*. New Orleans, LA: Author.

Vale, L., and Campenalla, T. 2005. *The resilient city: How modern cities recover from disaster*. New York: Oxford University Press.

Williams, J. 2006. The state of the hospitality/visitor industry one year after the hurricanes of 2005. Paper presented at the meeting of the New Orleans Metro Convention and Visitors Bureau, New Orleans, LA.

Wohl, R. R., and Strauss, A. L. 1958. Symbolic representation and the urban milieu. *American Journal of Sociology* LXIII:523–532.

World Travel and Tourism Council. 2006. *World travel & tourism climbing to new heights: The 2006 travel & tourism economic research*. Retrieved from http://www.wttc.org/ (accessed November 22, 2006).

Zukin, S. 1995. *The culture of cities*. Cambridge, MA: Blackwell.

Zukin, S. 1996. Space and symbols in an age of decline. In *Re-presenting the city*, ed. A. King. London: Macmillan.

Chapter 5

Addressing the Future Urban Water Crisis in Southern Ontario: An Ethnographic Look at the Promise of Expert–Lay Collaboration in Water Management Research

Benjamin W. Kelly

Contents

This chapter explores the recent activity of a group of academic environmental engineers and scientists in southern Ontario, Canada, who have been organizing themselves around an emerging democratic model that advocates for social and environmental change within the water and sanitation sector since 2006. As part of the larger worldwide environmental movement, these scientists have committed themselves to incorporating local end-users and community stakeholders in the development and maintenance of ecologically sustainable technologies and strategies. By recognizing the value of diverse lay knowledge systems, these scientists claim they can more adequately address the latent risks associated with urban water management than more conventional models. Although there are numerous multinational initiatives designed to tackle the problems associated with water conservation, the engineers and scientists I have been following expand and modify several environmental narratives as a means to bring together various individual and institutional stakeholders such as scientists, engineers, local businesses, and municipal officials into a process aimed at identifying and implementing solutions to the problems of urban water conservation within their own local districts. Persuaded by what they see as the failures inherent in conventional top-down models of knowledge translation, they are creating what they call the Learning Alliance and seeking an integrative, multistakeholder, transparent, and open-ended research approach to water resource innovation [Moriarty et al. 2005]. My interest is in tracking these engineers and scientists, the Learning Alliance they have created, and the overall social world in which their work is embedded.* The Learning Alliance seeks to transcend the interests of the researcher and implement new effective knowledge translation strategies to build greater end-user capacity and overall uptake of innovation within localities that are susceptible to potential poor water quality, scarcity, infrastructure degradation, and technological failure. Using interviews and participatory observations, I looked at the rhetorical strategies of these environmental engineers as they struggle to challenge conventional research practices, which they view as neglectful of community members' needs and dismissive of their lay knowledge. Exploring the claims-making activity of these environmental scientists and engineers as they advocate on behalf of end-users in the hopes of averting future environmental crises sheds light on the challenges and promises associated with expert initiated efforts in democratizing environmental science.†

* Although this chapter does not include the voices of those who are most vulnerable to disaster and risk, the group of environmental scientists and engineers I follow are struggling to redefine conventional pollution prevention research and give these communities a voice in the design and implementation of technoscience as a model of best practices in sustainability and effective knowledge translation within diverse water management arenas.

† Rather than look specifically at one example in which the Learning Alliance achieved successful sustainability with lay involvement, I take a more generic approach to the social definitional work invoked by science activists [see Prus 1996]. I ask, what is involved when experts speak back to science? How do science activists go about problematizing environmental, social, and scientific discourses to further legitimize their commitment and alternative vision in claiming that they in fact can advert or solve environmental crises with the assistance of lay expertise?

It has been argued that science may be unable to resolve a host of environmental problems because its design and application are replete with social and political consequences that it alone cannot accommodate [Beck 1999]. For this reason, members in the public sphere, cognizant of the uncertainties and risks inherent in technoscience and environmental problems, are advocating for a more open and democratic approach to research and design [Funtowicz and Irwin 1995; Kleinman 2000; Ravetz 1993; Sclove 1995; Vanderburg 2000]. Because of the complex, indeterminate, and transdisciplinary nature of environmental factors, many argue that the black box of technoscience and expertise must be opened [Wynne 1992]. As an institution that plays an active role in constituting material and social worlds, technoscience can ill afford to deny the public inclusive debate regarding its development and overall impact on society and the environment.*

Over the past two decades, we have witnessed the rise of a few social movements that are critical of science governance and its consequences [see Frickel and Moore 2006]. On some occasions, experts are implicitly criticized for questioning the knowledge and involvement of end-users in technical matters, as a result further limiting them from active involvement in the governance of problems that most affect their lives [Wynne 1996]. This, according to the critics of science, perpetuates an ever-increasing risk society and deeply ingrained ambivalence in which technoscience not only benefits society, but also whose failings and uncertainty have drastic and far-reaching consequences [Beck 1992; Giddens 1990]. Despite the impact science has on the well-being of average citizens, its governance is placed instead in the hands of technical experts and their bureaucracies [Habermas 1971]. Many feel the doors of the scientific enterprise are closed to public participation and thus susceptible to corporate and bureaucratic influences and constraints that negatively impact the well-being of all citizens. A majority of democratizing movements are lead by nonexperts who demand that the legitimacy of science be reconstructed within the transparent public sphere rather than mystified and detached from lay involvement [McCormick 2007]. In this case, the Learning Alliance, as a democratizing environmental movement, is being constructed not by end-users but rather by experts who are frustrated with conventional water management research and its blatant neglect of lay knowledge contribution.

Drawing on the interactionist approach to group life as accomplishing shared meaning [Blumer 1969; Prus 1996; Strauss 1993] and the constructionist perspective on environmental problems as claims-making activity [Hannigan 2006; Spector and Kituse 1977], this chapter looks at the Learning Alliance as a group

* Nowhere is the need to consult those citizens who are most impacted by these issues more apparent than in the case of water quality within First Nation communities. Continuing to ignore the water challenges Canada's indigenous peoples face is nothing short of environmental injustice [Mascarenhas 2007].

of academic environmental engineers and scientists who are struggling to further democratize science as a means to solve problems in water conservation and sustainability and to bring about positive social change. Redefining conventional research to empower end-users is not easy and the group is in the midst of legitimizing and clarifying their role as both activist and expert. Through members' experienced tensions and frustrations with conventional methods of science and engineering practices, the Learning Alliance is reconfiguring their own position, working out their professional space, and slowly beginning to articulate their alternative strategies regarding urban water conservation. The discourse of my participants implicates research practices as flawed, corrupt, and in need of dramatic overhaul because these conventional practices are ill-equipped to respond to or avert an environmental crisis. The ethnographic data gleaned in this chapter demonstrates that a group of science activists' efforts to legitimize their alternative vision of environmental science and empower lay citizens is an important factor in the overall social processes involved in creating more sustainable and resilient communities that would otherwise be more vulnerable to environmental risk.*

Facing the Challenges of Urban Water Management

With the onset of climate change and intense urbanization, cities worldwide face several water management problems. In 2008, the world's urban population (3.3 billion) rose above the rural population (3.2 billion). This trend is projected to continue into 2030, with 81% of the world population living in urban regions [Schlein et al. 2008]. Increased demands for quality freshwater place a great deal of stress on antiquated infrastructure and conventional water management approaches that are unable to handle and satisfy the future water needs of many major urban centers. Al Radif [1999, p. 146] defined water stresses as "unprecedented demands for agriculture, the provisions of domestic water supply and sanitation, industry, energy production, environmental amenity, changes in the patterns of consumption as a result of industrialization, rural/urban shifts, migration, and unaccounted for water." These factors all contribute to several potential environmental, political, social, and economical crises that conventional water management science is unprepared to handle.

Urban water management is not restricted to the confines of the city. A holistic approach requires that we acknowledge the link between urban population growth

* Data was collected by way of a qualitative multimethod approach. I attended meetings, workshops, conferences, carpooling to industrial locations, and other gatherings in which these science-activists organized their social world over a period of 2 years. My primary sources of data are the field notes I generated via participant observation. These notes are supplemented by 16 open-ended informal interviews.

and agriculture. With an increase in population comes the need for more water for overall food production. Water used for agricultural purposes plays a major role in environmental damage and water depletion. With these contributing factors in mind, scientists, governments at al levels, and water management personnel must give this potential crisis top priority on their policy and research agenda [Molden et al. 2007, p. 39]. Protecting freshwater as a limited resource, understanding the risks involved in ignoring nature's limits, and securing sustainable ecosystems that can handle increased stresses from population and agriculture expansion demand a new mindset in water management [Postel 2003].

Canada is one of six countries (including Brazil, China, Columbia, Indonesia, and Russia) that hold more than half of the world's freshwater [Sterling and Camhi 1997]. Even with its abundant freshwater reserves, Canada is not immune to the various economic and environmental problems associated with water safety and sustainability. In 2000, Walkerton, Ontario, was the site of a devastating freshwater contamination. The water foremen, who were not formally trained, did not properly maintain the cities chlorinator. Intense rains allowed manure to seep into the town's wells causing, along with neglected bacteria testing, a massive *E. coli* O157:H7 contamination. In a town of 5,000 people, 2,300 people were gravely sick and 7 people died [Morris 2007]. Due to poor federal legislation and water quality regulation, water treatment failure also occurs in several First Nations communities across Canada. As of 2008, there were as many as 93 First Nations communities living under boil-water advisories or do-not-consume orders [Eggerston 2008]. A recent report published by the Federation of Canadian Municipalities states that the national infrastructure deficit is approximately $123 billion and there is sure to be radical water system failures in the near future if Canada continues to ignore the impending crisis [Mirza 2007]. Although many water planning and management projects in Canada appear to resemble an orientation to sustainability, the conventional water management and sanitation ethos is entrenched in practices that remain outdated and lack approaches that recognize future problems. The number one roadblock appears to be an overburdened bureaucratic system and complacent research culture that is neglectful of efficient knowledge translation, a process that belittles end-users and stunts authentic long-term innovation [Kelly and Farahbakhish 2008].

The Learning Alliance believes that they have a vision and an approach that can counterbalance risk and ineffective knowledge translation and help water management research break free from the barriers that deny end-user participation. They argue that a Learning Alliance approach can implement widespread replication and sustainability as a means to prepare for the economic and environmental stresses that will most impact freshwater in the near future. Alliance members strive for, in their terms, a paradigm shift in urban water management science. Their ongoing efforts represent a transition from conventional water management science to what Gibbons et al. [1994] call Mode 2 knowledge production.

Mode 2 Knowledge Production

Recent dramatic transformations in scientific research practices have challenged the way scientific knowledge is produced [Nowotny et al. 2001, 2003]. Nowotny et al. [2003] summarize these trends by way of a comparison between what has come to be known as Mode 1 and Mode 2 science. Mode 1 science is characterized as the old paradigm of scientific discovery. Knowledge production under Mode 1 is associated with abstract and experimental science. It is discipline-specific and isolationist at its core. Mode 1 science has recently been challenged by a new paradigm, which Nowotny et al. [2003] refer to as Mode 2 knowledge production. This new paradigm is "socially distributed, application-oriented, transdisciplinary, and subject to multiple accountabilities" [Nowotny et al. 2003, p. 179]. The ever-changing landscape of scientific research and the emergence of Mode 2 science are significant for several sociological reasons. Many sociologists of science and technology are struggling to understand how best to study the dizzying pace of technoscience and its relation to industry, government, academia, and the public. Moreover, many who are interested in the collaborative interaction between different groups of scientists are curious as to how they form and are sustained given their diverse and often clashing interests. Nowotny et al. [2001, p. 144] use the term transaction spaces to describe the social interaction that occurs when one scientific world intersects with another or with nonscientific actors in the hopes of achieving some type of mutual understanding or to participate in a collaborative venture.

Much of the research about Mode 2 science is abstract and lacks in-depth ethnographic analysis. The few empirically based studies that do exist have attempted to either refute or confirm the reality of Mode 2 trends [Gulbrandsen and Langfeldt 2004; van de Kerkhof and Leroy 2000]. Those who write about Mode 2 science are concerned with questions such as the following: Is science changing? Is it becoming more sensitive to the nonscientific other? If not, should it? What are the best practices for science to incorporate if it wants to be relevant in the twenty-first century? How can science be made more transparent and open to lay perspectives?

As an ethnographer, my concern is not so much with questions such as these as it is with those who ask them and with efforts within the scientific community to bring Mode 2 science about within urban water management. I am not interested in challenging or defending either Mode 1 or Mode 2 science. Using a concept drawn from Goffman's [1959] dramaturgical approach, I want simply to understand the back region or backstage social processes underlying the efforts of a new breed of scientists to transform how water research is done.* To this end, this chapter examines how Learning Alliance members manage to construct an alternative scientific vision in contrast to more conventional environmental research practices.

* In this study, the back region refers to generic places where Learning Alliance members can let their guard down, compare notes with one another, anticipate other's issues, and adapt to potential or eminent barriers to achieve their goals.

In addition to being scientists and engineers, Learning Alliance members are also committed advocates for social–environmental change. This makes them ethnographic fodder as I investigate how they perceive, interact, and collaborate with other actors in the arena of urban water management. The goal of this chapter is to better understand how members of the Learning Alliance create distance between themselves and traditional water research, ultimately generating a new type of scientist that can acknowledge uncertainty and complexity and adapt to alternative perspectives as a means of adapting to local contingencies. Through their demarcation rhetoric, Learning Alliance members manage to construct a common purpose around the discourse of both advocacy and scientific expertise. The chapter aims to examine how Learning Alliance members understand their roles and responsibilities as both scientists and advocates for social–environmental change. They express their vision in the terms of a paradigm shift in the way environmental science and engineering research should be done. A paradigm shift can be considered an umbrella term used to orient members toward innovative ways of organizing collaboration among experts and end-users. From the perspective of the Learning Alliance, a paradigm shift is not a matter of revolutionizing science [Kuhn 1970]. Rather, Learning Alliance members would like to see a shift in the way scientific research practices contextualize knowledge, elevate lay perspectives, and attend to process within urban water management. When asked what a paradigm shift means to them, the following was a response that all members agreed on during a Learning Alliance meeting:

> We are tearing down an old system and replacing it with another that is bottom-up, not top-down. End-users are not being talked at but listened to. We are building a new system, one that empowers the people and actually assists experts who have good intentions and want to make a difference but keep banging their head against the wall as nothing they try seems to work. They mean well but have no avenue to express change; they are stuck in the old ways of doing things. The learning alliance will provide them with new strategies. . . [and] a great deal gets done when you bring in experts who are willing to move beyond their comfort zone. That is what we are doing, forcing each other to grow. Like the rest of society, we are stuck in our own inertia, but we have a responsibility to lead the way out. . . (interview)

Alliance members feel society is on the precipice of environmental and economic collapse. Involving stakeholders and refashioning research to accommodate their involvement is a matter of life or death. At times, Learning Alliance members use dramatic, almost apocalyptic, language to get their point across regarding the existential importance of their efforts. One engineer felt that humanity is in a state of denial. When I asked him if things are really that bad, he replied, "Yes, indeed, we are that giant ocean liner, the Titanic. . . we think we are indestructible. We party late into the night, with no care in the world, unaware that in a few short hours we have a date with an iceberg. . . It's the final hour, are we going to change course?" Another member expresses the same metaphor for urgency:

> Humanity is delusional in that it thinks it is flying high. "Look, I'm flying" it says, when in reality we are accelerating to the ground in a free fall. Humanity is like a guy who jumped out of a 30-story building and at the fifteenth floor says "So far, so good" . . . we've got kids, damn it, we have to do something, we have to save this place. (interview)

Learning Alliance members contrast their view of science and their goals as scientists with the more prototypical practices and characteristics of academic scientists and engineers. At the heart of the Learning Alliance's meaning-making activities are their understanding of Mode 2 change within an environmentally sensitive context—a context that is making a social space for nonexperts.

Extended Peer Communities

One way in which I conceptualize how such expert-activists are refashioning Mode 1 science and empowering end-users within urban water management is through their promotion of extended peer communities [Funtowicz and Ravetz 1993; Ravetz 1999]. Extended peer communities bring both expert and lay members to the table to consult on several pertinent issues and thus "positively enrich the processes of scientific investigation" [Funtowicz and Ravetz 1993, p. 753]. A consultative model that tends to recognize indigenous knowledge may provide several safeguards that protect end-users from the expert's cold distant gaze and the downplaying of environmental risks and hazards. The emergence of an extended peer community that will address water issues in Southern Ontario is being built up by several scientific, political, and social narratives. The Learning Alliance orients its members and their selected audiences toward extended peer communities that subscribe to Mode 2 definitions and practices of research and knowledge production as a means of addressing future problems in urban water quality and scarcity. Learning Alliance members feel strongly that the fate of the world depends on learning how to think differently about environmental science and how to practice it. Analysis of the data indicates that Learning Alliance members use three strategies to establish extended peer communities and achieve a level of Mode 2 knowledge production that manages to separate what they do from the Mode 1 practices of traditional water management. These Mode 2 strategies are (1) contextualizing knowledge, (2) elevating lay expertise, and (3) constructing an incremental approach. I organize each of the three Mode 2 strategies by first looking at how members come to understand and express the issues they have with how things are currently done in urban water management and then follow this with a discussion of how they imagine things could be done. I conceive conceptually of these Mode 2 strategies on the part of Learning Alliance members in terms of the efforts at legitimating their newly forming scientific social world.

Contextualizing Knowledge

Learning Alliance members hope to build extended peer communities by separating themselves from conventional colleagues with respect to how they think about knowledge. The Learning Alliance believes that scientific knowledge cannot be divorced from how it is used and what consequences it brings about. Contextualizing knowledge is about uncovering mode 1 science as being oblivious to its context; it is divorced from the real world and its problems. In a highly technical and uncertain world, authentic expert knowledge can no longer be associated with isolated disciplines and their prestigious researchers. Rather, the production of knowledge is situated along many knowledge dimensions [Gibbons 1999, p. 15]. These dimensions are both scientific and moral. As far as Learning Alliance members are concerned, facts are not developed in a vacuum because they are often associated with values [Shalin 1986]. They believe that the researcher has responsibilities because all research has consequences. In this case, the issues are environmental. Alliance members make it clear that they do not want to divorce knowledge production from the problems for which science should be developed. One Learning Alliance member recalls a series of situations in which he argued with colleagues over how to spend money more practically:

> I know professionals who are now more administrative than scientists. They want to keep putting money into infrastructure. Enough with the upgrade of buildings. This is how they want us to write major proposals. . . I always ask them why? They say we need money for a new science building on campus; this process is a juggernaut that out of control [sic]. I often bring this up at our department meetings. My colleagues just nod their head. . . they pretend to agree with me but they are still in this old mindset and just dismiss me. . . they may suggest some kind of mobile lab to take to the field but this isn't going to solve real life problems. . . its just the same old, same old. I tell them it's not about taking a lab into the field because ultimately it's still a lab. With the amount of funding they get to create buildings, you could identify, say four problems, give them each five million dollars and go solve the problem. Can you imagine the amount of change for the better? (interview)

During a meeting with end-users, two Alliance members demonstrated a performance of commitment to practical problems by distancing themselves from the iron cage of bureaucracy. Fully aware that end-users are not appreciative of manipulative researchers and their concern for publication, Learning Alliance members construct the image of humble expertise. The following excerpts represent a conversation two environmental engineers had during a Learning Alliance meeting.

> If you look at academia as a place to solve problems. . . some researchers find user-led research threatening. I ask my colleagues why does the university continue to build infrastructure, millions and millions spent on labs. The university is writing up a 30 million dollar proposal, I told them this is ridiculous, takes a few years to get the funding then a few more to build the lab, once the infrastructure is ready 5 years and millions of dollars later, then what? 30 million dollars can solve many problems in society right now but we are expecting the problems to knock on the university door for the experts to solve rather then spend the money for us to go out there and meet the people and the issues they face. . . (field notes)
>
> I couldn't agree more. . . they say we need more labs and more space to solve problems, when we do this we are turning the university into a complex that can no longer manage itself. Researchers are becoming too engaged in nonscience, they are busy solving bureaucratic problems rather than humanities problems. What if we took the university to the problems? In the process of attempting to solve the problem we are gonna engage them (end-users) in learning. Solve a problem and monitor how solutions work in the real world. That can be a lab we can use for years to come without adding additional bureaucratic layers to the university structure. We often complain, oh, those ideas work in the lab but they don't apply in the real world, so there is something fundamentally wrong with that model, its not user friendly, it's not really meant to solve problems. It's now just about getting the university more money to pay for their adult play pens (labs). (field notes)

Learning Alliance members do not feel that researchers and the university are serving the public good. They have intimate understanding of how research at universities is not conducive to real world solutions. A Learning Alliance member makes a distinction between those who talk and those that do:

> We want to take all the necessary measures to make sure the Learning Alliance does not fall into this same old trap. . . we need to DO, and not TALK. It's all about action. We must empower those who are DOING, not just sitting around thinking. At this stage in humanity's development, we need doers, we have enough thinkers. (field notes)

Another Learning Alliance member speculates about the potential of a well-governed practical Mode 2 science enterprise:

> I realize you need to fund fundamental research but can you imagine if funding agencies and tenure committees said, most of the funding and acceptable publications are now going to go toward research that is dedicated to solving real world problems. Not just some solution devised with no practical application. At the end a problem must be solved or else. That's the difference! (field notes)

One Learning Alliance member suggests that many of her fellow researchers are not interested in the alliance approach. "I see plenty of research that is extremely abstract and theoretical . . . promising great things, but nothing practically useful in which meaningful work can come from . . . They do not want anything to do with us." (interview). She goes on to talk about the stagnant nature of research on display at many academic conferences:

> Everyone (researchers) at these conferences and workshops talk only about successes but no one talks about obstacles and failures, so no real learning takes place. . . The whole science dialogue is about avoiding failure and inconvenience. I don't go to conferences anymore because no one is really challenging people to improve. (interview)

The Learning Alliance commitment to contextualized knowledge production is something they claim is worthy of their efforts. This belief separates them from the Mode 1 science they feel is unable to provide environmental solutions. The Learning Alliance belief in elevating the contextual and pragmatic aspects of research results in an unexpected form of legitimation: the downplaying of technology.

It seems counterintuitive for a group of engineers and scientists in water research to downplay the importance of technology, but placing technology secondary to the ultimate good of science (in this case the social–environmental issues they are trying to address) plays a prominent role in keeping research practical and realistic for Learning Alliance members. A closer look at how Learning Alliance members downplay the role of technology in collaborative projects with lay end-users paints a more pragmatic picture than one would expect. During my participant observation of the Learning Alliance, I was struck by how they contextualized knowledge by limiting their discussion of technological issues. I was initially reluctant to study environmental engineers because I feared that much of the social world would be sociologically irrelevant and limited with group discussion revolving around technical matters that I was not well versed in. It turns out that this apprehension became an ethnographic strength. It allowed me to recognize that technology did not play the central role I assumed it did, and with this interesting turn I was able to question my participants further as to why they were not consulting on technical matters more. It was soon apparent that social, political, and economic behavior was just as relevant to these science-activists because these contextualized narratives were an interpretive resource to redefine the priorities of science.

For Learning Alliance members, redefining priorities takes on a rationalization that society must recognize the earth as a closed system with limited resources that demand to be managed more responsibly. The quote below was indicative of members attempting to justify constraint, moderation, and redefining manipulation, perhaps implicitly legitimating their social world ideology.

> How do we as researchers convince people to live with constraint and moderation, so we can get through the hard times ahead. . . It's time to recognize we live in a closed system and require constraints so we can start to curb our outrageous wasteful consumption. Standard practices bias environmental interests. They delay infrastructure repair, avoid costs, and keep putting things off. Sure, we may try to limit growth, but even if we just maintain the same level of production we will fall short. . . we need to realize we will run out of resources. As we grow in population and wealth our consumption increases. . . we must change the criteria and introduce constraint so we can conserve. We must force conservation or we won't have anything left. . . We have manipulated the environment to allow maximum use, rather than design products to fit the environment. (interview)

Frustration with the status quo is a focal point for Learning Alliance members. It not only plays a prominent role in member's attempts to redefine priorities, but it also emerges in the narrative work that constitutes a justification for designating conventional collaborative efforts in environmental science as fundamentally defective in contrast to their egalitarian efforts.

According to Learning Alliance members, scientists should be more reflective about how their knowledge is used and the consequences it has for others. In terms of environmental science more specifically, this translates into the argument that Learning Alliance members tend to view ever-advancing technology as merely a band-aid solution to urban water management and environmental concerns because it provides society further distraction from deeper societal problems. The Learning Alliance stresses a redefinition of priorities to change the status quo in water research so as to reflect the urgency to alter environmentally adverse behavior. The following excerpt demonstrates how frustrated members are with the public's neglect of underlying issues and their mistaken assumption that technology is the holy grail of Western problems.

> Our opportunity to act is diminishing each day. We don't want to do anything to change the status quo. So we are willing to fund research that looks on the surface as different, but it's not. We are not thinking

> about changing the way we behave or view the world and our place in it. . . but we are obsessed with changing technology so we can stay the same. . . like nonfossil fuels that still demand just as much energy procurement. . . One scientist I know is working day and night on technology that changes CO2 into diesel. But in order to do this you must raise the temp 1500°C. Of course this is not efficient. I asked her: "Why not just use all that energy for electricity?" Her answer was, "Because people want to drive gasoline cars." This is poor science because it only perpetuates the problem. (interview)

At the heart of the Learning Alliance's prioritization of lifestyle over technology is the belief that innovation must take into account sustainability and the fact that we live in a world of limited resources. The following quote demonstrates that it is not just science that needs to change, from the perspective of Learning Alliance scientists, but society as well. Members blame the scientific profession for not involving itself more in the issues of climate change on behalf of the public good. This provides justification for their involvement so as to stop the continual decline of western society and its obsession with technology. One member suggests that it is not an overhaul of technology that is required, but rather a fundamental change in lifestyle.

> The biggest failure of scientists in the twentieth century is not connecting the science of global warming with changes in behavior, no one says, "Drive less". . . but rather we are saying, "Why don't you just maintain your levels of consumption and lifestyle and rely on technology?" After 9/11, there was no pause to check our behavior, it was more like, things are ok, support the economy and the American way of life, go shopping. Stay a consumer. . . we will make sure your behavior doesn't impact our way of life. Our habits are so bad that the North American life expectancy and standard of living will drop in the twenty-first century for the first time. . . as we begin our continual decline we will turn to technology rather than the underlying problems. (field notes)

Another member talked about people being a more important resource than technology as a means of achieving greater sustainability:

> You can spend all the capital you want, but if people are not well educated on water conservation and sustainability. . . if you don't get them on board it will never work. . . technology will fail, but if people are committed they eventually always find a way to fix it. (interview)

Technological development and design that is neglectful of people and limited resources has Learning Alliance members deeply troubled, and so they often downplay the importance of technology when consulting with each other. When they do discuss the importance of technology, it is always placed within a local context.

Downplaying the importance of technology not only provides contrast for members to focus their attention on societal reluctance to buck the status quo and the lack of innovation in conventional research, but it also serves the Learning Alliance members' mandate to elevate local knowledge systems and emphasize practical solutions within their proper context. The following passage demonstrates how vital context is to the learning alliance:

> Usually end-users laugh at engineers and are not asked for input... "Ok, we will build it if that is the way you want it" and they end up having to go back and modify it. The key is not to force technology on the problem... but look at the problem first and foremost. Get the total context. Don't focus on a new technology, we already have what we need, but come up with integrated solutions to a problem... keep it to a local site. Usually it isn't one technology per se, but a battery of many sources to solve the local problem... (interview)

Again, we are seeing the limited role of technological design and the elevation of what is more sustainable for the needs of end-users. "In the labs people are designing new technology, but there is already lots of good technology sitting on the shelf gathering dust no one is using... why not use what we already have and apply it, integrate it, and modify it?" (interview). Members found that one of their major stepping stones was the recognition that technology is not the only barrier to collaborative success. One researcher, discussing the roadblocks that emerged with rainwater barrels, makes this clear:

> Realizing that adequate technology already exists, the project team has begun to acknowledge that other barriers, social and political in nature, stand in the way of the projects successful realization... the team is realizing that a lack of capacity looms as the main barrier and, hence, developing capacity has become a key objective of this project. (internal memo)

Another Learning Alliance member makes this point by contrasting the developed world with the underdeveloped world:

> Five billion people just need simple technology to live... basic water treatment, food. The other billion have high technological needs and concerns. We need to recognize that what works well over there can work here. Why do we have to put so much money and expertise into our research... it is overly complicated and wasteful. (interview)

With this form of contextualized knowledge, technology is not an end but a means of serving local problems. This redefinition of technology redefines the whole research process as a pragmatic innovation. A process that is reflexive and because it differs from traditional research goals in its democratic outlook, demands high levels of participation on behalf of all stakeholders.

Knowledge for Learning Alliance members is contextualized by speaking to the urgent environmental problems that face humanity. Issues can be practically solved by building up capacity through the local adaptation of current urban water technology. Learning Alliance members build their social world around the belief that society does not need to keep developing more complex technology but can modify already existing technology to address the social, economic, and political barriers that past water research tends to neglect. Learning Alliance members contextualize knowledge because they believe that science and society can no longer afford to keep producing knowledge that is abstract and not generated for specific purposes. The problems we face, according to the Learning Alliance, call out for applied knowledge and specific solutions produced with end-users in mind.

Elevating Lay Expertise

In addition to contextualizing knowledge, Learning Alliance members further advocate for change in urban water research by participating in discourse that downplays their own expertise and elevates the tacit knowledge and practical skill set of their nonscientific collaborators. The Learning Alliance vision of how science should be done reconceptualizes not only the relationship among and between scientists, but the relationship with end-users as well. They reject the conventional view that end-users have no place in knowledge production. Because many of the environmental problems they are dealing with are complex and require cooperation with a wide diversity of stakeholders, the Learning Alliance is building a social world in which they see their own knowledge as limited. Gibbons [1999, p. 15] remarked:

> Experts must respond to issues and questions that are never merely scientific and technical, and must address audiences that never consist only of other experts. The limits of competence of the individual expert call for the involvement of a wide base of expertise that has to be carefully orchestrated if it is to speak in unison.

Members believe that urban water problems are too complicated for just one group of stakeholders to solve; everyone has to work together in the spirit of true interdisciplinary action. Learning Alliance members wish for their research to be practical. They see knowledge not as pure abstraction for ivory tower consumption, but always with an end goal in mind of being of service to humanity and providing end-users with sufficient capacity-building so that innovation is sustainable. Learning Alliance members recognize that stakeholders do not want to be overly dependent on experts. To be independent, Learning Alliance members must bring in and manage stakeholders in ways that allow them to feel as equals in the development and implementation of water conservation technology and strategies. The Learning Alliance attempts to impress on end-users that lay perspectives and involvement are vital for the overall success of water conservation projects. They believe knowledge and its application should be practical, simple, local, and diverse. To do otherwise is to make stakeholders overly dependent on experts while making experts insular and ignorant of potentially valuable lay insights. They recognize that without total commitment from stakeholders, their vision of a Mode 2 water management science is nonviable.

Mode 2 knowledge is what Nowotny et al. [2001] call "socially robust." This type of knowledge is different than Mode 1 knowledge in three ways. First, the validity of this knowledge is not dependent on the laboratory. Although it has legitimacy in the laboratory context, it also has value outside the laboratory. Second, the legitimacy of Mode 2 knowledge is achieved by an extended group of experts. This includes lay expertise as well. Third, due to its acknowledged origins in society, such knowledge is more likely to weather the storm of intense critique [Gibbons 1999]. Mode 2 science is made more robust due to its heterogeneous organization.

> Since expertise now has to bring together knowledge that is itself distributed, contextualized and heterogeneous, it cannot arise at one specific site, or out of the views of one scientific discipline or group of highly respected researchers. Rather it must emerge from bringing together the many different knowledge dimensions involved. Its authority depends on the way in which such a collective group is linked, often in a self-organized way [Gibbons 1999, p. 15].

For the Learning Alliance, stakeholders are the most important factor in carrying forward their paradigm shift. The following excerpt demonstrates that socially robust knowledge is dependent on end-user participation and perspectives. One engineer describes the importance of end-users in participatory terms that allow for real needs.

> User-lead design processes help us recognize the importance of inclusion and the potential contribution of lay expertise. This of course implies that stakeholder involvement is not solely based on technical or scientific knowledge. No doubt, technical and scientific knowledge is necessary during the latter stages of the design process. . . inclusion of lay expertise ensure that real needs are identified and simpler solutions are given priority. (interview)

Another Learning Alliance member spoke about the progress the group achieves when they allow for a nonthreatening environment for the expression of lay knowledge.

> Once experts recognize that their attitudes toward lay are wrong and that they are not intellectually superior but actual facilitators in educating them. . . that they just take their own understanding of things for granted and that end-users don't have this knowledge, once they take the time to get to know their end-users, they begin to build trust and the lay no longer feel defensive or looked down at, they open up and then, only then can experts recognize the roadblocks to communication. There is a moment when the light goes on for everyone and all benefit. (interview)

These inclusive strategies move Learning Alliance members beyond monocultures. The lay standpoint is considered valuable because they are more familiar with the everyday problems on the ground. Opening channels of communication between lay and experts makes research practical.

> We must allow end-users a voice in the whole research process. . . provide them with the opportunities to say what information they need and what they have to offer practical research. Researchers can then use this information to get more money to promote innovation. . . so it's a reciprocal relationship, information needs to flow freely between those on the ground. . . yeah, the front line workers and those on top. (field notes)

Learning Alliance members claim that if urban water management science wants to create more sustainable environments, truly generate collaboration, handle risk, and build capacity, it must make research more practical and relevant to all parties involved. To do this, they further suggest that water research must become more diverse in its execution, more democratic in its process, and more sensitive to the needs and expertise of stakeholders. The following quote was recorded during a Learning Alliance meeting. Here, the engineer was explaining the importance of practical solutions and the relevance of stakeholders as doers being essential to the overall success of research.

> The future of research is doing! We can't wait around for groundbreaking research. . . that is passive. We can't afford it financially or environmentally. We need to fix things here and now while we are studying them with the people most involved and most impacted. . . we learn through doing. When we funnel money toward practical solutions. . . this is not a waste. These projects would keep up on monitoring because all relevant stakeholders are on board. They help in the maintenance and are therefore key to success. (field notes)

Of course, to make research practical and diverse, that means bringing several different stakeholders into all aspects of design. Learning Alliance members feel strongly about challenging the status quo to make the whole research enterprise more inviting for stakeholders. The local knowledge and practical application of end-users are paramount to successfully producing knowledge collaboratively.

> When I am collaborating with them [end-users] I'm not barking orders. I'm listening and observing. . . their local knowledge is vital. We can't just barge into their environment and tell them what to do. . . that's not collaboration, that's a dictatorship. I remember one project where we noticed mangos dropping from trees were wasted because there were not enough people in the community to eat them before they spoiled. We found a few locals who dried them. We then worked together to build a mango drying plant to expand the efforts they already started. We are merely helping them do what they are doing already. (interview)

This egalitarian format makes a socially robust space for the proper development of Mode 2 knowledge production.

Empowering end-users by elevating their knowledge and humbling scientific expertise is not just a moral imperative for Learning Alliance members; they also want to establish open participation as a means to further advance effective research and overall urban water project development. A Learning Alliance member makes this connection clear:

> Why not experiment with a number of tools that can be used by everyone, not just experts; people can take these tools and ideas, try them out in their communities, then come back and reflect on them with the experts, share the challenges and successes and go back into the field for another round. . . we don't want to make end-users dependent on experts, we want to empower them, allow them to start improving and measuring their own progress. . . being forced to commit to top-down regulations can be disheartening. (interview)

We see in the above excerpt that, according to Learning Alliance members, measured outcomes do not need to be divorced from authentic collaboration and can serve the Learning Alliance goals of empowering stakeholders.

Constructing an Incremental Approach

Learning Alliance members contrast their activities with traditional research practices that overemphasize ultimate solutions to environmental problems. Rather than occupy themselves with an end product that putatively solves everything, members make a point of viewing water research as incremental steps that are more successful in the long run. Members know that outcomes are still important but process is the avenue to liberate traditional research and society from its solutions that are not always applicable to local contexts. The narrative involved in constructing an incremental approach to Mode 2 urban water collaboration is composed of taking small, less ambitious steps and diversity of solutions, paying attention to process, making sure there is forward movements, trusting that problems will be solved, and not investing in the dramatic big discovery of quick fixes. The following excerpt links the acknowledgment of process with innovation as a form of socially robust knowledge production.

> If the process is right the outcome will be right . . . with every project I am involved in, I try to take the extra time and energy needed to understand and build a process, in doing so the outcome is often a far better quality. It's very difficult to turn a goal-oriented culture to a process culture. (interview)

Another Learning Alliance member refers to the importance of steps and its relation to empowering people.

> Complex solutions disempower us all. The Learning Alliance allows clarity, not only to see the solution, the product at the end, but more importantly to see clearly the steps, the process needed to do it. That is allowing the end-user and public witness the black box of science. That is progress . . . (field notes)

Knowing that their concern with process differentiates them from traditional pollution prevention protocols, Learning Alliance members rationalize their unique activities as an end in themselves, rather than striving toward external recognition. One scientist states, "The NOBLE prize for us is in the Learning Alliance process" (internal memo). In the following excerpt, a Learning Alliance member describes their vision of a paradigm shift as incremental rather than what engineers and scientists are socialized into producing-grand flashy solutions.

> Research is required to be sexy or it's not noteworthy, it's dismissed . . . everyone is looking for that 100% silver bullet solution, the mega treatment, but what about 10% solutions building on each other over time; this would be more successful in the long run, this would be a massive paradigm shift for science and engineering. (interview)

A processual Mode 2 research paradigm includes a need for Learning Alliance members to make an effort to accommodate the end-users that assist in adapting technologies and pollution prevention strategies to fit incremental solutions to local problems.

The Learning Alliance believes that many of the problems with Mode 2 knowledge production are due to participants being overly focused on outcomes at the expense of developing a sufficient grounding in cooperative processes and knowledge translation. The process for science-activists is defined as building stakeholder capacity through efficient knowledge transfer among all parties involved in the project. The following excerpts explain why this is a vital aspect of the learning alliance: "The goal is to improve the quality of life or address a real need or problem rather than new product development. . . this is impossible without user-lead processes" (interview). To achieve these goals, members envision this collaborative process through a series of steps. The members' concerted efforts to establish knowledge production in a cyclic fashion rather than through the traditional linear format are an important aspect of the Learning Alliance dedication to the model (see Figure 5.1). Attempts to produce knowledge collaboratively involve stakeholders and science-activists in what they describe as a cyclic approach that is characterized by four stages: (1) defining the problem; (2) identifying the solution; (3) implementation; and (4) reflection [Kelly and Farahbakhish 2008]. Figure 5.1 represents the template Learning Alliance members utilize to help guide them and their end-users in the multistakeholder collaboration process [Farahbakhsh and Kelly 2008].

Members speak of process as discovering diverse solutions within collaborative circles. Diversity is recognized as a defining feature of producing knowledge collaboratively, further legitimizing the Learning Alliance's claim that local knowledge

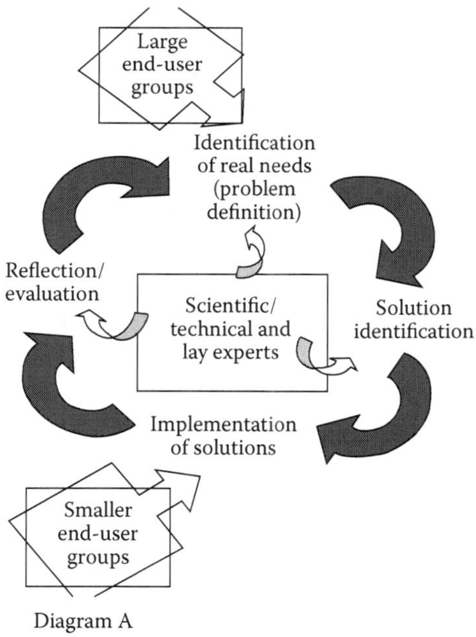

Figure 5.1 The Learning Alliance user-led design process.

and its practical application constitutes a more innovative process than traditional Mode 1 research formats. During a meeting, one member tells the group of a project in Holland that was successful due to its commitment to local diversity:

> There is a massive architectural reconstruction project in Holland. All disciplines are working together, not building empires. No one over there is giving all the money to just one firm . . . they are ensuring diversity. You don't have all these researchers and companies competing for funding and contracts, no one is wasting public money. It is a massive project facilitated with participants at the ground level and forcing people of all walks of life to recognize their weaknesses and integrate ideas and strengths together. (field notes)

This member's statement demonstrates a strong narrative of diversity. Diversity, according to science-activists, appears to be a mechanism by which they attempt to avoid the pitfalls and traps inherent in more traditional forms of urban water research and yet still manage to produce knowledge collaboratively.

> We should always be turning to a diversity of solutions and not putting all our eggs in one basket. The key is integrated teams working toward a diversity of solutions with a common vision . . . not attempting to shove some single solution into some foreign context in which it has no place being . . . we have all seen that. (field notes)

Members often spoke of the possibility of alienating nonexperts during collaboration. Diversity of solutions was a method in which to generate a form of knowledge transfer that was not alienating because it differed from conventional research. The following excerpt was recorded during a heated discussion among Learning Alliance members and their struggle to include lay perspectives.

> We must change current approaches so they promote diversity of solutions. Rather than solutions being typically predetermined by a group of experts, and neglecting knowledge transfer, we commit ourselves to collaboration with stakeholders during all phases of research . . . not alienating them. This will allow us to avoid the monoculture of science and engineering. (field notes)

Another engineer concerned with alienating others in the project, expressed his fear of naming what they do overly esoterically.

> Sure, we found a funding agency that wants to give funding to us as long as we call what we do green chemistry, but if we use that term we will get twenty experts knocking down the door, if all of this is plausible,

> lets just call it pollution prevention, because renewable resources is more than green chemistry, both green and chemistry are alienating others, it is like saying, "We are green and you are not" . . . this will just make it more difficult to bring others to the table and play the game . . . (field notes)

With this acknowledgment of process comes further legitimatization that the Learning Alliance perspective on collaboration is a more sustainable enterprise in the long run compared to traditional methods.

Concern for process and an overall incremental approach to sustainability constitutes members' orientation toward user groups' expertise, solution identification, and reflection. At the heart of these processes are elements of diversity, democracy, practicality, and local application of design. It is people-centered and user-lead perspectives that drive the group's efforts rather than research-centered and expert-lead practices that are so characteristic of Mode 1 science.

Conclusion: The Promise of Mode 2 Innovation

Rather than focus on lay-expertise and end-user's struggles with experts in their attempts to alter technoscience within urban water management, this chapter explored what Moore [2006] terms a "professionally initiated participatory science." By analyzing the democratic claims of the Learning Alliance, I demonstrated how scientific moral entrepreneurs [Becker 1967] attempt to bring together and manage expert–lay collaboration around issues of urban water conservation as a means to prevent future environmental crises. The science-activists, who are my key informants, seek to empower lay members of the public by recruiting and encouraging their involvement in knowledge production and overall scientific processes as a means to achieve greater success in ecological footprint reduction within several Canadian urban centers.

Efforts such as the Learning Alliance are possible because their meaning-making activities draw on multiple normative narratives that address eco-friendly agendas and more scientifically guided institutional realities. The Learning Alliance must generate science that is both effective and morally sound from an eco-democratic perspective if it is to carve out a viable solution to urban water quality and scarcity. The three interrelated Mode 2 strategies of contextualizing knowledge, elevating lay expertise, and constructing an incremental approach constitute the intersubjective reality needed for the Learning Alliance members to distance themselves from conventional research and ultimately supply audiences and themselves with the interpretive resources necessary to justify their desire to construct a more democratic environmental science without completely alienating themselves from the public or science in general. Boundaries that are built between competing universes of discourse [Mead 1934] take on rhetorical forms that justify why members participate in nonconventional urban water research and why they believe their way of doing things is better than the status quo. Learning Alliance members attempt to legitimate their research ideology in a more favorable light than the status quo embodied in their colleague's style of research. Rather than further legitimating their work through scientific ideological resources, Learning Alliance members are struggling to establish their vision of innovative water research by breaking down the scientific barriers that separate or close off professional expertise from end-users and lay perspectives. Accomplishing Mode 2 strategies on behalf of the Learning Alliance acts as a demarcation device that provides members with a narrative to legitimate involvement in and construction of a new scientific subworld committed to democratic ideals. Moreover, this narrative is built on an interpretive foundation of critiquing other research activities and ideologies that are distinct and in conflict with their own. As scientists, members are still forced to construct rhetoric that is within the confines of instrumentality. In addition to such rationality, I also found that solutions to an urban water crisis are established through a moral context and are advocates for social and environmental change.* I found that the Learning Alliance's egalitarian practices and representative ideology not only served the public good but also provided end-users with practical innovative designs and strategies within pollution prevention research. As the Learning Alliance struggles to organize and distance itself from Mode 1 science, we see that they draw on both instrumental and moral rhetoric to justify what I term Mode 2 innovation. As a newly forming research group in urban water management, Learning Alliance members developed a shared understanding of what they are doing in relation to conventional science as both morally and instrumentally superior. Implicit in the demarcation and distancing that occurs between Mode 1 and Mode 2 urban water management are value systems that are both value-rational and instrumental. More specifically, it is these normative value systems that members draw on to legitimize how they envision a diversity of solutions to handle future freshwater stress.

The Mode 2 strategies invoked by Learning Alliance members constitute a delicate balance that allows them to straddle two different research ethos in urban water management as science-activists. The concept of Mode 2 innovation helps describe how the Learning Alliance manages to keep one foot in the world of advocacy for social–environmental change while the other remains firmly entrenched in the traditional world of scientific water research. Mode 2 strategies are not only more equitable for incorporating other knowledge systems (rational–value) but also make for better science (instrumental) according to Learning Alliance members.

* This dichotomy is inspired by Max Weber's [1947, p. 185] formal and substantive rationality. The former is measured quantitative calculation while the latter is concerned with social action that is associated with "a given set of ultimate values . . . or to the content of the particular given ends." In this case, formal activity represents the Mode 1 scientific logic of instrumentality while substantive action orients members toward Mode 2 knowledge production associated with moral-environmental advocacy.

The manner in which the Learning Alliance drew on several collective normative resources to find solutions that are both instrumental and moral demonstrates the complexity involved when a scientific social world intersects with several other worlds that house competing logics of interest. The following Learning Alliance member's explanation mimics what I have seen from many members in this study:

> . . . We can show that these alliances in the long run are much more cost-effective and that is what people want to see, these quantifiable outcomes, yet we are doing so much more than this . . . We are building both community and capacity. It's about time we recognize the two are interrelated. (interview)

Members of the Learning Alliance feel that their brand of scientific practice is innovative because it strives to attain a more sincere form of collaboration between experts and end-users within water conservation projects. Although they have yet to apply this model in a crisis scenario, they continue to modify it as they work on less vital water conservation projects, seeing these endeavors as building networks and lessons for future crisis management they believe is just around the corner.

References

Al Radif, A. 1999. Integrated water resources management (IWRM): An approach to face the challenges of the next century and to avert future crisis. *Desalination* 124:145–153.

Beck, U. 1992. *Risk society: Toward a new modernity*. London: Sage Publications.

Beck, U. 1999. *World risk society*. Oxford: Polity Press.

Becker, H. 1967. Whose side are we on? *Social Problems* 14:239–247.

Blumer, H. 1969. *Symbolic interactionism: Perspective and method*. Englewood Cliffs, NJ: Prentice Hall.

Clarke, A., and S. Star. 2003. Science, technology, and medicine studies. In *Handbook of symbolic interactionism*, eds. L. Reynolds and N. Herman-Kinney. New York: AltaMira Press.

Eggertson, L. 2008. Despite federal promises, First Nations' water problems persist. *Canadian Medical Association Journal* 178(8):985.

Farahbakhsh, K., and B. Kelly. 2008. Managing complexity and promoting usability: A user-led design approach. Paper presented at the annual International Design Engineering Conference, July 27–29, 2008, Halifax, Canada.

Frickel, S., and K. Moore. 2006. *The new political sociology of science: Institutions, networks, and power*. Madison: The University of Wisconsin Press.

Funtowicz, S., and R. Ravetz. 1993. Science for the postnormal age. *Futures* 25:735–755.

Gibbons, M. 1999. Science's new social contract. *Nature* 402:81–84.

Gibbons, M., C. Limoges, H. Nowotny, S. Schwartzman, P. Scott, and M. Trow. 1994. *The new production of knowledge: The dynamics of science and research in contemporary societies*. London: Sage Publications.

Giddens, A. 1990. *The consequences of modernity*. Cambridge: Polity Press.

Goffman, E. 1959. *The presentation of self in everyday life*. New York: Anchor.

Gulbrandsen, M., and L. Langfeldt. 2004. In search of "Mode 2": The nature of knowledge production in Norway. *Minerva* 42:237–250.

Habermas, J. 1971. *Toward a rational society*. London: Heinemann.

Hannigan, J. 2006. *Environmental sociology*, 2nd ed. New York: Routledge.

Irwin, A. 1995. *Citizen science: A study of people, expertise and sustainable development*. London: Routledge.

Kelly, B., and K. Farahbakhish. 2008. Innovative knowledge transfer in urban water management: An attempt at democratizing science. *International Journal of Technology, Knowledge and Society* 4(3):73–83.

Kleinman, D. 2000. *Science, technology and democracy*. New York: State University of New York Press.

Kuhn, T. 1970. *The structure of scientific revolutions*, 2nd ed. Chicago: University of Chicago Press.

Mascarenhas, M. 2007. Where the waters divide: First Nations, tainted water and environmental justice in Canada. *Local Environment* 12(6):565–577.

McCormick, S. 2007. Democratizing science movements: A new framework for mobilization and contestation. *Social Studies of Science* 37:609–623.

Mead, G. 1934. *Mind, self and society*. Chicago: University of Chicago Press.

Mirza, S. 2007. *Danger ahead: The coming collapse of Canada's municipal infrastructure: A report prepared for the Federation of Canadian Municipalities*. Ottawa, Ontario.

Molden, D., C. De Fraiture, and F. Rijsberman. 2007. Water scarcity: The food factor. *Issues in Science and Technology* 23:39–48.

Moore, K. 2006. Powered by the people: Scientific authority in participatory science in *The new political sociology of science: Institutions, networks, and power*, eds. S. Frickel and K. Moore. Madison: The University of Wisconsin Press.

Moriarty, P., C. Fonseca, S. Smits, and T. Schouten. 2005. Background paper for the symposium: Learning alliances for scaling up innovative approaches in the water and sanitation sector. IRC International Water and Sanitation Center. Retrieved from http://www.irc.nl/content/download/16138/208040/file/Background_paper_symposium_.pdf.

Morris, R. 2007. *The blue death: The intriguing past and present danger of the water you drink*. New York: Harper.

Nowotny, N., P. Scott, and M. Gibbons. 2001. *Re-thinking science: Knowledge and the public in an age of uncertainty*. Cambridge: Polity Press.

Nowotny, N., P. Scott, and M. Gibbons. 2003. "Mode 2" revisited: The new production of knowledge. *Minerva* 41:179–194.

Postel, S. 2003. Securing water for people, crops, and ecosystems: New mindset and new priorities. *Natural Resources Forum* 27:89–98.

Prus, R. 1996. *Symbolic interaction and ethnographic research*. New York: State University New York Press.

Ravetz, J. 1999. What is post-normal science. *Futures* 31:647–653.

Schlein, L., J. De Capua, and S. Krüger. 2008. For humanity's sake, developing world must prepare for soaring urbanisation. City Mayors Society. Retrieved from http://www.citymayors.com/society/urban-population.html.

Sclove, R. 1995. *Democracy and technology*. New York: The Guilford Press.

Shalin, D. 1986. Pragmatism and social interactionism. *American Sociological Review* 51:9–29.

Spector, M., and J. Kitsuse. 1977. *Construing social problems.* Menlo Park, CA: Cummings.
Sterling, E., and M. Camhi. 1997. Sold down the river. *Natural History Magazine*, 119.
Strauss, A. 1993. *Continual permutations of action.* New York: Aldine De Gruyter.
Van de Kerkhof, M., and P. Leroy. 2000. Recent environmental research in the Netherlands: Towards post-normal science? *Futures* 32:899–911.
Vanderburg, W. 2000. *The labyrinth of technology.* Toronto: University of Toronto Press.
Weber, M. 1947. *The theory of social and economic organizations*, eds. A. Henderson and T. Parsons. New York: Free Press.
Wynne, B. 1992. Misunderstood misunderstanding: Social identities and public uptake of science. *Public Understanding of Science* 1(3):281–304.
Wynne, B. 1996. May the sheep safely graze? A reflexive view of the expert-lay knowledge divide. In *Risk, environment and modernity: Towards a new ecology*, eds. S. Lash, B. Szerzynski, and B. Wynne. London: Sage Publications.

AFRICA AND THE MIDDLE EAST

II

Chapter 6

Critique of Stakeholder Participation and Decision-Making Processes Affecting the Design and Implementation of Transboundary Water Governance Projects Bordering Vulnerable/High Risk States: Lake Chad and Basin, Danube and Tisza River Basin

Filip Aggestam and Stephanie Hodge

Contents

Introduction

Over the past century, the demand and variety of ways by which humans have modified the biological, physical, and chemical characteristics of ecosystems has increased dramatically. With regards to water, technological development has enabled societies to improve and provide services such as sanitation, food production, fisheries, power generation, drinking water, and recreation—services that we today consider an essential dimension of human welfare and development. This ever-increasing pressure on the utilization of an often shared and finite resource also exemplifies the pressing need to improve ecosystem management services, collaborative planning, and cooperation between different institutions, states, and individuals. On the whole, the growing stress on our natural environment, coupled with the uncertain impact of climate change, has in recent years resulted in an increase in efforts to develop and apply alternative management strategies for the use of natural resources [Azqueta and Delacamara 2006; Xevi and Khan 2005].

But despite great leaps forward in our understanding of how to manage limited resources, present day water governance constitutes an increasingly complex web that integrates technical and scientific knowledge, legal requirements, socioeconomic aspects and multistakeholder participation [Baggett et al. 2006]. One dimension of this development has, for example, been the growing international acceptance that environmental projects need to be inclusive and that relevant stakeholders are incorporated into the decision-making process [HarmoniCOP 2005; Hare and Pahl-Wostl 2002; Brugha and Varvasovszky 2000]. In addition, in our efforts to comprehend the effects of climate change and the ever increasing pressure on finite land resources—frequently causing overexploitation and land degradation—we have also seen an intensification in research on societal resilience to natural and manmade disasters. This particularly in the aftermath of disasters such as the Indian Ocean Tsunami [Clark 2007] and Hurricane Katrina [Baker and Refsgaard 2007]—catastrophes that have drawn attention to many of our societal vulnerabilities. These events have once again illustrated how low- and middle-income countries and marginalized sectors of society are particularly at risk.

With this background in mind, resilience is seen as a desirable property of natural and human systems. A development that has encouraged several fields of research to refine and further develop strategies relevant for community disaster recovery and resilience. Science rooted in studies of resilience is consequently increasingly derived from multidisciplinary research from a broad range of theories and methodologies [for more details, see Baker and Refsgaard 2007; Klein et al. 2003]. For that reason, resilience and related concepts, such as vulnerability and adaptation, are receiving renewed consideration in connection to measures to counter future climate change [Klein et al. 2003]. Perhaps more importantly for this chapter, efforts have also been made to better understand underlying social patterns that define resilience and human coping strategies when faced with a disaster.

As disasters often carry a high cost in terms of human lives (on a global, regional, or local scale) and the potential damage to ecological and socioeconomic systems is great, catastrophic events provide an opportunity for better understanding development efforts. This is because such events represent a situation in which institutional systems have been unable to prevent damage. A condition from which we can draw valuable lessons for the future management of our natural environment.

Community Disaster Recovery and Resilience

While it is important to have a common understanding of the concepts involved in this chapter, we will not provide an overview of the conceptual development of resilience. In this case, the meaning of resilience will be generic. For example, as defined by the Oxford English Dictionary, a resilient individual is "able to withstand or recover quickly from difficult situations." In a broader sense, resilience will refer to a socioecological system's ability to tolerate disturbance and recover to its earlier state. In theory, resilient communities should be able to withstand extreme events and recover rapidly from disasters whenever they occur. Resilience is as such dependant on careful planning and organization of society, both to limit the impact of a disaster and to facilitate the recovery process. But such comprehensive planning would require an all-embracing strategy to reduce risk and exposure. A strategy which would have to take into consideration factors such as age and family structures, spatial patterns, socioeconomic factors, gender issues, education, cognitive factors, social and ethnic networks, etc. [Linnerooth-Bayer and Mechler 2007].

Generally speaking, these elements are all interrelated and will have a significant influence on any community's capacity to recover from a disaster. A truly comprehensive strategy would therefore have to include all the elements from our natural environment, as well as human-use systems. This is complicated further by how systems constantly change over time, representing a network of interactions that is significantly dependent on the context. This dynamic aspect of communities requires the near impossible, namely, a complete and continuous understanding of the interactions taking place in a given system [Tobin 1999]. But regardless of the difficulties associated with managing and planning for disasters in real-life, an important development in recent years has been the increasing recognition in policy that ecological and human systems are interlinked. It is becoming clear that resilience relates to the function and interaction of the system rather than to the stability of the components, or the ability to maintain or return to a state of equilibrium [UN/ISDR 2004].

Thus, reducing vulnerability to disasters represents an ongoing and iterative process that incorporates planning, design, implementation, and monitoring, as well as awareness-raising and information collection. At the community level, this means, among other things, being able to improve current living standards in the face of changes in the environment that may have an impact on peoples' livelihoods. This is particularly important in relation to climate change in high-risk states, where reducing vulnerability requires having the correct mechanisms in place, access to technology, resources, and expertise to cope with the process of implementation. In relation to the management of vulnerable ecosystems, part of the solution to this problem is offered by active stakeholder participation as a measure for reducing risk and strengthening recovery mechanisms.

The question now arises as to how participatory management relates to community disaster recovery and resilience when developing strategies to reduce vulnerabilities to natural hazards. The next segment will be dedicated to discussing the value of stakeholder participation as a means to improving community-based disaster risk recovery, management, and planning.

Public and Stakeholder Participation

> "[a]ll too often still silos of knowledge get produced that fail to help make systems and communities more robust to extremes and to change" [Vogel et al. 2007, p. 352].

To successfully manage and plan for disaster recovery is not only dependent on costs, available technologies, and relevant expertise, but it is also dependent on understanding and identifying the social context in which a given institution will operate. Unless the public and all relevant stakeholders are engaged in the decision-making process, project managers might find it difficult to be the facilitator of knowledge and information. This would have a direct impact not only on efforts to manage natural resources, but also on a community's capacity to recover from a disaster [Bagett et al. 2006]. Hence, the implementation of a project dealing with disaster recovery at the community level is dependent on the overall acceptance of its strategy by all affected parties, including the public.

Participation may therefore provide decision-makers with important knowledge if there is a structured and transparent mechanism that supports open dialogue with stakeholders [Johnston and Soulsby 2006; Wattage and Mardle 2005]. For example, engaging stakeholders in a two-way communication process could (at any stage) have a positive influence on planning, implementation, and management. The application of a stakeholder analysis and participation strategy could help to ensure that the right people are brought into a project at the right time to allow for discussion, learning, and agreement in connection to the use and management of a limited and/or shared resource.

In the cases that will be presented in this chapter, the emphasis is on stakeholder participation within the water sector. This choice has predominantly been made as this field is often characterized by unpredictable events, such as flooding or severe droughts. It is consequently suitable for studying the processes underlying community-based disaster recovery and resilience. However, to begin, it may be appropriate to briefly consider a key area in this field, namely, risk management. Imagine, for example, a water expert who might characterize flooding risk in terms of public health. In a situation such as this, you would most often find that the public's definition of risk would represent a much more complex construct. For instance, as mentioned above, the facilitation of information with regards to risks associated with flooding is dependent on how believable the public considers the information to be. As a result, the institutions credibility is significantly linked with the level of trust exhibited by stakeholders and the public. With this in mind, engaging in superficial and inconsistent participatory management may in fact have long-term counterproductive effects, as well as damaging socioeconomic consequences [Sadoff and Grey 2002]. Consequently, social trust constitutes one significant argument in favor of participatory management at the community level. Not only is it relevant for the long-term management of limited natural resources, but also the erosion of social trust can seriously limit an organization's ability to reduce natural hazards.

The purpose of this discussion is simply to emphasize that evaluating and engaging elements of the public and stakeholders may help to assess and guide knowledge requirements and risks associated with disasters. If the process of managing natural hazards is perceived as being under joint control, it seems more acceptable than if managed by others. This is particularly important within water resource management, as it is typically a technology- and expert-driven field. Besides the unpredictability of future events and the quality of information, the recognition and analysis of these social limitations increases the usefulness of stakeholder participation, as well as the potential it possesses to effect a community's readiness for a disaster.

Using stakeholder participation and transboundary cooperation as a medium for analysis, this chapter aims to explore their relevance to the concept of community-based disaster recovery and resilience. A brief overview of the relevant concepts employed throughout this study can be found in the appendices. The following section will present two cases focusing on linking various aspects of participatory management with resilience and disaster recovery. The third section will review the results presented in the case studies and draw conclusions in the context of international development interventions and the management of shared natural resources in risky ecosystems. Finally, the fourth section will present a set of policy recommendations related to participatory management and transboundary cooperation.

Case Studies

A case study approach is employed to compare stakeholder participation and transboundary cooperation efforts across different countries. The approach is based on the following:

- Review and comparative analysis of publicly available project documents
- Semistructured interviews and contributions from an online forum conference.

The combination of interviews and document analysis bolsters validation by allowing triangulation of data sources. The chapter relies considerably on the inside views of program officers and practitioners—individuals who are in a good position to describe their views and understanding of the participatory process, incorporation of stakeholder input, and policy decisions.

Case 1. Engaging Stakeholders in Practice: Managing the Danube and Tisza River Basin

One dimension of the changing water sector in Europe has been the democratization of the decision-making process and policy-making at national, regional, and international levels. The European Union has, for example, integrated this consideration into its Sixth Environmental Action Programme.* As a consequence, European Union member states are now required to continuously introduce scientific input into the environmental policy process [European Commission 2001].

With this in mind, we have seen an increased emphasis on the incorporation of stakeholder interests and the promotion of public participation when planning and implementing projects within the water sector. It is becoming regular practice to characterize and categorize levels of interest and power relations among stakeholders and the public. Yet in many cases we still find that project managers and decision-makers apply a soft (or deliberative) decision-making approach in which a key individual is trying to perceive the whole picture, observing it from various angles, and finding the best alternative through expert intuition [Wierzbicki et al. 2000].

Research on decision-making in the face of difficult choices suggests that project managers are still likely to use a range of simple strategies to address a complex decision problem/process, such as simple heuristics [Gigerenzer and Tood 1999] or cognitive rules of thumb [McDaniels et al. 2003]. Despite some beneficial aspects of simplifying the decision-making process, this is not applicable to all environments, particularly regarding individual decisions for the following reasons: (1) decisions are seldom optimized given situations of uncertainty; (2) access to information is limited; and (3) individual optimization could lead to conflicts in situations with a limited number of stakeholders (as is often the case in water governance) [Wierzbicki et al. 2000]. As expressed by Castelletti and Soncini-Sess [2007, p. 18], "Ignorance is being unaware that our awareness is imperfect. An imperfect awareness implies uncertainty and uncertainty generates apprehension." Furthermore, it is argued that decision-makers often have a tendency to remove the problem of uncertainty by promoting deterministic scenarios and models to provide exact estimates. In essence, hiding uncertainty by trying to make their evaluations perfect.

Besides these issues, project performance is not only influenced by individual decision-making strategies, but also by behavioral biases (or contextual preferences).

* See http://ec.europa.eu/environment/newprg/index.htm.

A dimension that may have a significant impact on the decision-making process [Chae et al. 2005]. For instance, Wierzbicki et al. [2000] has suggested that decision-makers form their own aspirations adaptively through a learning process that is satisfied if the aspiration levels are attained for the decision. Instead of optimizing, this subjective aspect of decision-making may make it difficult for the decision-maker to distinguish between facts and value judgments—an issue that could greatly influence the perception of a problem and the use of input from stakeholders and communities [Castelletti and Soncini-Sess 2007].

There is, consequently, a need for exploration of the implications of these biases in real-life decision-making and their effects on stakeholder participation. An improved understanding of the behavioral and cognitive processes at work will provide a better position from which to truly judge how stakeholder participation is working at project level.

Aim and Scope

The aim of this case study will be to discuss what constitutes a stakeholder and participation according to project managers and practitioners. An emphasis is placed on projects run through the United Nations Development Programme (UNDP), the International Commission for the Protection of the Danube River (ICPDR),* and related organizations. Within this framework, the purpose will be to analyze variations in perspectives, definitions, and approaches toward stakeholders and participation among executive decision-makers in water-based environmental projects along the Danube and Tisza River Basin, as well as the Caspian Sea. Three main objectives have been identified:

- Define how project managers and decision-makers perceive stakeholder participation and how they fit into the context of project management;
- Determine whether there are any prevalent behavioral biases among project managers that may influence (1) how participation is carried out and (2) how stakeholder input is interpreted;
- Evaluate whether any applied stakeholder participation strategy has had an influence on decision-making and project implementation.

The independence of the selected organizations and projects make a strong argument for comparing the common elements that appear in them, particularly as both UNDP and ICPDR are involved in projects concerning the Danube region and the Caspian Environment Programme (CEP).† As such, a comparison between

* ICPDR is based on the Danube River Protection Convention. ICPDR operates to ensure the sustainable and equitable use of waters and freshwater resources in the Danube River Basin region. For more information see http://www.icpdr.org.

† See http://www.caspianenvironment.org.

perspectives and approaches toward stakeholder participation and subsequent use of input during decision-making can help to clarify whether current participation strategies are effective. The purpose will not be to compare different stakeholder participation methodologies or to determine their effectiveness. The emphasis will rather be on a different dimension of participation, namely, how project managers employ input from stakeholders in the decision-making process. Therefore, descriptions of the different perspectives are used to learn about participation in practice, and to extract some of the common features among those involved in stakeholder participation. As such, it will serve a purpose by highlighting problems and benefits with integrating stakeholders in decision-making.

Method

To enable a comparison between organizations, projects, and people, this study combined a review of project documents and semistructured in-depth interviews. The first phase, a review and comparative analysis of publicly available project documents, was conducted to gain a basic understanding of the applied participation strategies, results, and to evaluate how stakeholder input was used during decision-making. The interviews were aimed at exploring the level and perceived success of stakeholder participation. An approach that enables a comparison between involved decision-makers and helps to clarify their experiences, arrangements, and thoughts regarding stakeholders [Dearnley 2005; Smith et al. 2001]. The selected participants worked at various levels in the United Nations Development Programme (UNDP), International Commission for the Protection of the Danube River (ICPDR), Regional Environmental Centre for Central and Eastern Europe (REC),* International Institute for Applied Systems Analysis (IIASA), Nimfea Environment and Nature Conservation Association,† Caspian Environment Programme (CEP), and the Tisza Biodiversity Program.‡ In addition to the interviews a forum on the topic of new approaches to stakeholder management was held over a full day during an online UNDP conference (more information available on http://www.waterfair.org). A range of water experts, primarily from UNDP, the academic community, and nongovernmental organizations (NGOs) contributed to the forum. Some of the information provided has been extracted to support the discussion on stakeholder participation.

Methods and Data Analyses

The choice to use semistructured interviews is based on the ability to elicit relational data across participants and to provide the interviewer with the flexibility

* See http://www.rec.org.
† See http://www.nimfea.hu.
‡ See http://www.elotisza.hu (*only in Hungarian*).

to probe and openly explore the participant's thoughts and experiences [Dearnley 2005]. To this end, the interviewer is able to seize upon opportunities to elaborate on values expressed by the interviewee without being restricted to the rigidity of a structured interview [Smith et al. 2001].

There was also a strong element of comparison within the research design. For instance, the interviewer often contrasted stakeholder interests with water resource management issues. This was done to add the depth of comparative research [Triandis 1976; Rose 1991; Lidstrom 1999], but it also allows the participants to obtain a broader view of the difficulties surrounding genuine participation.

The applied methodology is predominantly qualitative but incorporates some elements of quantitative research characteristics. As such, although great care was taken to apply a systematic process, it is reasonable to assume that a certain degree of subjectivity is unavoidable in a study of this nature [more information available in Aggestam 2007].

Engaging Stakeholders and Transboundary Cooperation

Terms such as stakeholders and participation are rather vague when considering the many purposes for which they are being employed. Even more, they are concepts that are becoming increasingly diluted and applied within a wide range of different professional fields [Harmoni 2005; Herman 2005; Hare and Pahl-Wostl 2002]. This becomes particularly apparent when discussing the rationale of engaging stakeholders with project managers and practitioners. The defined approaches (e.g., purpose, scope, indicators, level, and definition of participation) and the context in which stakeholder participation is carried out vary significantly. Within the context of water resources management, this diversity in definitions and participation approaches constitutes a noteworthy obstacle for comparison of project performance and results. But more important, these loose and varying definitions have implications for projects attempting to achieve participatory management. Consider, for example, the impact different interpretations of participation may have on community-based projects striving to reduce risk toward natural hazards.

Against this background, it is apparent that the inclusion of stakeholders, transparency, and community participation is open for interpretation, despite their being essential and required components for environmental projects operating inside as well as outside the borders of the EU. As stated by one of the participants, "The law on participation can only be guided by what project managers intend stakeholder involvement or engagement to achieve."

In relation to the Danube region, there is a history of 15–20 years of interventions. This context represents a clear benefit with regards to the establishment of relationships and building trust within communities [UNDP/GEF 2003, 2005]. As a result, on the part of the UNDP and ICPDR, there is a strong emphasis on stakeholder participation for the Danube River Basin. But these organisations also rely on the assumption of having a sufficient understanding of relevant stakeholders in the region (an issue this study is unable to confirm).

Regarding public participation strategies for ICPDR, a general roadmap provides details on what needs to be done in connection with their river basin management plans. Part of this process includes milestones for the integration of the communities and stakeholders. One aspect of this integration was, for example, workshops with participants from each Danube country, flood defense, navigation, hydropower, NGOs, and the scientific community [ICPDR 2003, 2006]. But despite this emphasis on participation, the level of stakeholder participation is to some extent still determined by the projects management unit. As noted by one participant, "Stakeholders have an influence on the direction of the projects, if you seriously involve them," illustrating that genuine stakeholder participation depends on the individual project manager's beliefs. As such, within the context of the reviewed projects, most participants seem to agree that engagement is vital. Especially if the stakeholders do not agree with stated project objectives. On the other hand, given the organization's size, what is defined as participation by ICPDR does not necessarily equate to actual engagement on the ground. Even so, by pushing for participation at a national or regional level, it is possible that ICPDR can positively influence community-based participation strategies in regions where this is needed.

Another dimension of the complexities surrounding participation can be demonstrated by a World Wildlife Foundation (WWF)* project that was conducted along the Tisza River Basin in Hungary [WWF 2004]. This particular project investigated gravel abstraction in the Upper Tisza River and its contribution to flooding. As this is not a problem that is immediately apparent for many regional communities, the question is how you can successfully convey the problem to the general public? Moreover, how can you influence the situation so that the public actually becomes concerned about these activities? As this project demonstrated, managing to influence or engage stakeholders became a serious obstacle. As expressed during one interview, "You are lucky if any stakeholder turns up to your workshop." A common statement that illustrates that the successful engagement of communities is not only subject to the relevance allocated by project managers, but also the communities themselves.

In line with this argument, and in connection with participation through ICPDR, one participant expressed that the agricultural sector was considered to be the least successful sector in terms of providing input, while other industries only had limited input. The relevance of this issue concerns the apparent increased interest from sectors such as hydropower and navigation. Stakeholders have become more engaged as a result of the perceived threat from the Water Framework Directive (WFD) [UNDP 2003]. This demonstrates an additional problematic aspect of actually realizing the engagement of vital stakeholders. Only when a project or, in

---

* See http://www.panda.org/about_wwf/where_we_work/europe/what_we_do/danube_carpathian/blue_river_green_mtn/danube_river_basin/middle_danube/tisza_river/index.cfm.

this case, institution has a certain level of perceived importance would stakeholders make the effort to engage or commit themselves. This issue of perceived relevance is of equal importance within the context of community-based participation.

One explanation, as elaborated by a participant external to ICPDR, is that many NGOs and organizations do not appreciate working with the organization. It should be noted that the reasons for this were not stated explicitly and, furthermore, this view might be a result of individually differing opinions. It does, nevertheless, emphasize the real-life problem created by the perceived distance in-between project managers and practitioners at one end and stakeholders and communities at the other.

In support of this, it became quite clear while speaking to project managers and practitioners that most projects do not actively involve communities in decision-making. This despite the fact that a decision-making body in the Danube region has an obligation to provide adequate information as well as consult with stakeholders. In this case, a literal interpretation of the WFD, as it requires that stakeholders should be heard and considered, while actively involving them as part of the decision-making process is only encouraged—an opinion that seems to be reflected among project managers themselves. As stated during one interview, "At the onset you don't get all the stakeholders, you do the stakeholder analysis, but you don't get all stakeholders on board besides maybe one or two NGOs." But, on the other hand, managers also express the wish to instill a sense of ownership and accountability among stakeholders. The question is whether this can be accomplished without engaging stakeholders in the decision-making body.

As suggested by several participants, it further appears as if project managers often do not even know who the relevant stakeholders are. As a consequence, relevant stakeholders have nothing to do with project implementation up until the point when the course cannot be altered. Therefore, a fundamental problem appears to be that stakeholders are not getting actively involved early enough. Additionally, the process of engaging stakeholders (within the context of UNDP) can be an "often not very satisfactory way of doing it" (interview excerpt). The reasons offered for this were (1) difficulties in persuading governments that it is worthwhile; (2) difficulty finding a good and representative list of stakeholders; (3) problems with engaging stakeholders once they have been identified; and (4) stakeholders having a very narrow and local view of problems.

In contrast with the previous examples, a successful illustration of participation concerns the application of a stakeholder analysis conducted through the Caspian Environment Programme (CEP). In this case, the stakeholder analysis moved from being an investigative tool to a participation strategy. Within the context of the initial analysis for the Caspian Sea, "The notion of stakeholders was expanded from the standard focus, given the level of government permeation into the social structures of the region" [Matthews 2004, p. 10]. The analysis focused on exploring whether there were any "conflicts among stakeholders that may constrain effective interventions" [CEP 2002, p. 3] and the manner in which the stakeholders and local communities prioritized eight major perceived problems identified by experts [UNDP 2005; CEP 2002, 2005, 2006]. As such, the process (including elements of consultation) managed to generate a reprioritization of project objectives. "Stakeholder groups rated some concerns much higher than experts, whereas others that the experts believed to be the most prominent were ranked far lower than expected" [Matthews 2004, p. 10]. As a consequence, subsequent to the stakeholder analysis, the project changed direction as the original objective turned out to be of low importance to the public. This supports incorporating stakeholder input much earlier in the planning process, as relying explicitly on expert-driven input is clearly not sufficient to adequately formulate the problem environment and structure.

Related to the governance of shared water, it is of interest to note some difficulties in association to the transboundary management of river basins in the Danube region. It has been noted that attendance across countries differs significantly. A problem that primarily appears to be due to different value structures, institutions, and perspectives on the democratic process in the various regions. Since stakeholder participation is a vague term, it could be asked whether participation should be more concretely and strictly defined, so as to make its implementation more homogeneous. Additionally, given these regional and national differences, it raises the question of how you should analyze, engage, and compare communities across national and regional borders. This is especially relevant in connection to the transboundary management of water bodies within the European Union.

Why Engage Stakeholders and Communities?

Questions of why we should pursue community-based participation follow naturally in light of all the obstacles highlighted above. After all, a project manager might say he or she will listen to stakeholders, they may say they will reply to written comments, they may even say they will circulate the decision for comment, but there are many ways in which to illustrate where project managers' (and governments' and institutions') willingness to engage stakeholders and incorporate input ends. "The outcome of many examples of participation is a tragedy because it is a conflict and not cooperation" (interview excerpt). When projects, for example, do not define participation, stakeholders often begin to suspect each other's motives and the process becomes one of conflict rather than cooperation. Moreover, successful stakeholder participation (of any kind) requires trust. As a result, many projects often rely on the specific abilities of a talented facilitator to make things happen. But if the intended engagement is not a sustained effort, there can (in the long term) be a deterioration of trust, causing communities to become uncooperative.

On the whole, there are a multitude of context-specific barriers that might arise during the development of a community-based participation strategy. Nevertheless, during the course of this study, there were a number of issues that were highlighted as being particularly important to take into consideration. These issues relate to the following:

1. Consultative versus active participation. This signifies that projects most often only engage in one-way channels of communication. Stakeholders are, as a result, rarely participating actively in decision-making.
2. Invited versus voluntary participation. During project formulation/planning, it is vital to consider what stakeholder groups are represented, particularly as many project managers often do not make an effort to extract input from the least interested (but often equally important) parties.
3. Obstacles related to stakeholder participation as a protest. Several cases were mentioned in which stakeholders participated during project implementation solely to slow down or prevent the projects from progressing.
4. Alternative motivations of stakeholders. Similar to the third point in terms of what stakeholders are participating. For example, it is most often the most vocal participants, such as an NGO, that provide input, which might provide a skewed perspective of the problem environment.
5. Risks involved with only inviting/engaging already known stakeholders in the project: "My feeling is that a good deal would be won when stakeholder analysis and consultation would be done even before projects start, when there is still flexibility to fully account for their concerns. In existing projects, most budgets have already been earmarked, experts identified, and if stakeholders come up with something new that doesn't fit in directly with existing project structures, it is difficult to include those issues for many practical reasons. Also, their concerns may call for expertise that is traditionally not well represented in the water sector." (interview)

Despite engagement being potentially biased and even partial, it should not be forgotten that it can not only help communities reduce risks but also improve resilience toward natural and manmade disasters. This will require not only that a structured and transparent strategy is applied to support engagement and a continuous dialogue with stakeholders, but also that project managers and practitioners react and adjust according to received input [Bagett et al. 2006; Johnston and Soulsby 2006; Hare and Pahl Wostl 2002; Wattage and Mardle 2005]. Stakeholder participation should be a two-way process. In such cases, if employed correctly, engaging stakeholders and communities can greatly contribute to the successful sharing of limited natural resources through ensuring discussions, learning, and a certain degree of agreement in connection to its use and management [Sadoff and Grey 2002; Failing et al. 2004].

Case 2. The Lake Chad Crisis

> "It will be a puddle. You'll get crops and drinking water out of it, but you'll have no ecosystem left to speak of" [Coe and Foley 2001].

This case illuminates key areas of concern and describes the central argument—the case for enhanced community and stakeholder participation in water and basin governance in order to reduce poverty and vulnerability to disasters. Issues include the following:

1. Need for attention to formal and informal education and identifying strategies nationally to scale up the benefits of the cross sector benefits of such strategies in regional poverty reduction, conservation, and development policies.
2. Need for strategies for local innovation and knowledge inputs in regional ecosystem policy development, as well as strengthening institutions to do so.
3. Need for real community and stakeholder participation and the development of platforms for dialogue, resource, and knowledge exchanges and joint strategy development in lake and basin governance.

Lake Chad is disappearing for two reasons: (1) natural (increasing drought and climatic change conditions) and (2) human factors (lack of regulation and massive bad irrigation practices). It is estimated that about one-third of the stream flow today is being diverted from the Chari River before its flow even reaches Lake Chad. Between 1983 and 1994, irrigation water use increased fourfold [Glantz 2004]. About 50% of the decrease in the lake's size since the 1960s is attributed to human water use, with the remainder attributed to shifting climate patterns [Davit 2001; Coe and Foley 2001].

In response to these changes, on May 22, 1964, the Lake Chad River Basin Commission (LCBC) was established by the Fort Lamy (N'djamena) Convention and Statutes by the governments of Cameroon, Chad, Niger, and Nigeria. In 1994, the Central African Republic was admitted as the fifth signatory. Sudan was admitted during the tenth Head of State summit held on July 28, 2000, increasing the LCBC jurisdiction from 966,955 square kilometers to 1,035,000 square kilometers [Lake Chad Basin Commission 2003]. The Fort Lamy Convention recognizes the sovereign rights of each member state over basin water resources within its own territory but forbids unilateral exploitation of lake water where such use detracts from the interests of other states. Member states are required to abstain from measures likely to alter the water budget, water quality, integrated water and resources management health, or water access by other member states. The convention recognizes the right of member states to plan projects within the conventional basin in consultation with the LCBC.

Limitations

Despite the broad mandate of LCBC, which includes transboundary water and land, economic integration, and peace and security issues, the existence of institutional legal frameworks has not translated into effective action. For example, years after the creation of the LCBC, national water policies within member states still remain deficient or nonexistent. The absence of regional and national standards or guidelines to govern the monitoring of water quality, quantity, and cost-sharing mechanisms has led to continued environmental degradation within the basin. As a result, LCBC's efforts at regional water resource management have failed. Its failure to take action to achieve active community participation and wide stakeholder consensus have not provided any results.

Significant operational issues at the LCBC have arisen from the overall fragmentation of scientific effort, responsibility, and authority, leading to fiduciary concerns, the substantive lack of accountability, and a weakening of capacities at all levels. As a consequence, the organization's ability to function on a day-to-day basis is impeded.

Timeline for Reform

It is obvious that the current system is failing: lake and basin governance is lost; management is faulty; Lake Chad is endangered; and time and opportunities are running short. The next question is of imminent importance: What is the timeline for reform? The management of Lake Chad is already having a negative impact on every life system being sustained by the lake. The estimated 22 million people who depend on the lake and its resources for their livelihoods are at serious risk. It is already possible to observe the consequence of expanding extreme poverty in a troubled region already featuring the driest regions (Eastern Chad, Northern CAR, and Western Sudan), which will intensify the ongoing conflict and increase social disruption. Estimates of a worst-case scenario in which Lake Chad dries up with no alternative livelihoods or social safety net in place are dire.

Principal Barriers to Knowledge Diffusion and Technological Innovation and the Need for Strengthening Community Involvement

The Lake Chad River Basin Commission rules and protocols were built on age-old conventions that are not necessarily relevant for the current ecosystem and the adaptive management needs of the lake and river basin. However, before initiating reforms at LCBC, governments must assess the broader economic environment including the potential for regional development, infrastructure, technology, innovation, and basic human capital stocks. Overcoming national and regional structural barriers may not be possible under the auspices of a LCBC restoration project, as macroeconomic structural economic adjustments are called for to generate a "fundamental shift in the business as usual model of governments" to achieve regional economic development [Murenzi and Hughes 2005, p. 43]. In addition, constant political instability perpetually undermines regional economic progress and stunts the development of national and local learning innovation systems. However, at the institutional level, actions can be taken in order to create a Lake

Chad Basin Commission Learning System. This would involve more investment in knowledge creation, knowledge transfer, and a governance structure suited to ecosystem management and innovation approach [Murenzi and Hughes 2005, pp. 43, 52].

The following sections also illuminate the problem and provide concrete suggestions concerning the preconditions needed for an effective learning system to evolve (regional integration, infrastructure, and nonformal and formal education).

Regional Development (Macroeconomic Policy)

Romaine Murenzi and Mike Hughes believe that the recent African effort at regional economic development reflects recognition of the importance of science and technology in development [2005, p. 51]. They suggest that new efforts to import technologies must be combined with homegrown institutional innovation in order for the development to be successful. In addition, they posit that new strategies for international cooperation in Africa consider major trends such as globalization and regional integration initiatives [Murenzi and Hughes 2005, pp. 48–59]. For example, they state, "World Markets are a source of technology and capital; it would be silly for developing countries not to exploit these opportunities. But globalization is not a short cut to development. Successful development strategies have always required a judicious blend of imported practices with domestic institutional innovation" [Murenzi and Hughes 2005, p. 51].

Infrastructure (Technological Learning)

Furthermore, Murenzi and Hughes suggest that the ability of infrastructure development to diffuse technical skills into the economy has been overlooked. In connection to this position, they argue that Rwanda's growth policies since 1997 stipulate that new regional infrastructure projects should be tailored to the transfer of skills and facilitation of the learning process in order to be effective [Murenzi and Hughes 2005, p. 62], which, according to Ridley and Lee [2005], support that "the absence of infrastructure services is a serious problem hindering efforts to develop Africa."

Juma [2005] posits that the technological learning process can provide organizations with an opportunity to acquire and diffuse new knowledge and skills. During a presentation to the leaders at the African Development Bank in April of 2006, he stated, "Development and infrastructure literature often overlooks infrastructure's dynamic nature. Every stage of an infrastructure project, from planning and design through to construction and operation, involves the application of a wide range of technologies and associated institutional and management arrangements" [Juma 2006b]. Thus, infrastructure constitutes a major part of the development of a dynamic learning system.

Education (Human Development Policy)

Oyelaran-Oyeyinka and Barclay [2003] argue that the base of human capital and the institutions inherited by African states have set the tone for the development in national systems of innovation. They consider the African development context as a "non-dynamic system of innovation" and suggest that limited human capital development explains the relatively difficult processes of implementing science and technology institutions on the continent. They suggest that in order to be effective, national and international innovation systems must include important factors such as basic education, economic growth, and industrialization.

In 1981, Easterlin expanded on this argument [Oyelaran-Oyeyinka and Barclay 2003]. He stated, "There is a direct correlation between schooling of the relevant context and countries' ability to master new technologies." Oyelaran-Oyeyinka and Barclay [2003] elaborated, stating that "the combined rates of technological and human capital (transmitted through educational attainment) are connected, and the existing supply of human capital such as a mixture of skills at the onset of the industrialization process is an important prerequisite for economic growth." When considering the elements necessary for progress through the lens of the disappearing Lake Chad situation, one question raised is whether existing national structures of innovation and human capital base are sufficient to provide capacity to support a standalone model institution aimed at providing knowledge services to achieve integrated water resource management.

Multifaceted Cooperation

Cooperation with donors also represents an issue that has presented unique challenges for organization of good regional ecosystem governance structures in the Basin area, especially as donor interest has been waning in recent years. This was not always the case. According to the UNDP report on restructuring [1988], 1964–1989 was a period characterized by donor vitality, an organizational vision, and positive cooperation. Donor support, particularly from UNDP, was significant and included the establishment of LCBC and maintenance of many of its activities. Today, with a growing international interest in climate change and international waters, the international community has started to work with organizations such as LCBC again. However, incumbency and related poor management within the UN dynamic and older governance structures. LCBC's lack of internal technical and administrative capacity, as well as staff incentives to embark on a complex regional work plan and, finally, the poor stakeholder participation strategy threaten the renewed relationships.

Within this context, the review of institutional documentation revealed that donors tended, in the past, to grant assistance in accordance with their own mandates and often included a preset criterion that has changed depending on the development context. Dealing with regional problems of the Lake Chad Basin, the reliance on donor-led planning support has ultimately failed. It will be vital to learn from these experiences. While donors have practical interests in the organization goals, they should not be relied on for follow-through after the initial input had been made. As such, regional structures such as LCBC must plan and supplement their activities with donor inputs only after considering where donor criteria fit within their own strategic planning mandate.

Building Community Resilience and Ensuring Food and Water Security for All

Within the context of participation, the LCBC is responsible for a diverse set of stakeholders. However, stakeholder participation in LCBC activities varies according to each different stakeholder, individual, or organizational mandate. Engagement usually occurs through informal arrangements and on an ad-hoc basis, depending on the nature of the project in question.

Moreover, LCBC is characterized by the paradox of being an internationally recognized organization that is virtually invisible inside its own jurisdiction. This is, in part, attributed to the absence of two critical stakeholder groups in the Basin governance process, namely, subregional planners and local end-users.

Benefits of Expanded Stakeholder Participation

The benefits of involving stakeholders especially those at the community level in the regional policy and problem-solving processes are as follows: (1) the process appears more democratic and the engagement of stakeholders may legitimize the final policy; (2) stakeholder groups (particularly industry and NGOs) may have additional data that may inform policy deliberation; (3) stakeholders bring diverse technical expertise; and (4) stakeholders bring diverse viewpoints.

The lack of intra- and international coordination between the LCBC and various tiers of government, the private sector, and civil society severely inhibits good regional planning, policy construction, and relevant local actions. The LCBC does, however, have the opportunity to build upon the output from years of informal stakeholder activity that has contributed to the identification of key Basin management issues. These informal discussions can provide the foundation for a more formal set of decentralized structures that can govern the regional planning mandate and, thereby, influence potential outcomes. The GEF Project is in the process of formalizing many of these relationships, as these new structures need to be integrated into the existing LCBC structure in order to be effective.

Barriers to effective, systematic stakeholder participation in LCBC activities include the following [LCBC 2005]:

- Inconsistent interregional coordination
- Lack of goals and benchmarks

- Lack of an implementation plan
- Lack of strategic management planning
- Poor communication networks
- Lack of detail-orientation
- Lack of reliable data related to monitoring and assessment
- Disparities in resources
- Complex and highly technical scientific data and geographic data requiring expensive proprietary software
- Lack of ease finding a common place and time for discussion and negotiation

With this in mind, LCBC management must identify the key stakeholders and the nature of their engagement in organizational activities. Participation should be planned in relation to the LCBC mandate and outcome goals, and monitored against a strategic LCBC annual work plan. LCBC must, therefore, work in conjunction with its stakeholders and clearly define the nature of these relationships (possible examples include LCBC–members, employer–employee, constituent–beneficiary, and professional–donor association, LCBC–end-user).

In relation to this, national authorities need to evolve mechanisms necessary for effective coordination and consultation with stakeholders to ensure awareness of and involvement in policy formulation, implementation, monitoring, and decision-making. Environmental education is also critical.

Stakeholders interviewed confirmed there is a lack of regional and local stakeholder participation in LCBC decision-making [Hodge 2006]. During a consultation with Lake and Basin stakeholders held in Cameroon, the group highlighted Basin management and governance concerns:

- General lack of knowledge concerning technical LCBC activities. No publications or information updates were available.
- Synergistic networks exist but are very task-related and usually unsubstantiated.
- Lack of a formal network to discuss Basin issues.
- Lack of communications and IT infrastructure that inhibit email correspondence.
- Data monitoring and assessment deficiencies.

A way to facilitate these important stakeholder linkages at the local administrative levels can be through, for instance, a memoranda of understanding (MoU). This would broaden the knowledge base of LCBC, provide better communication flow, and, ultimately, enhance cooperation. There are precedents (for example, the Mekong River Basin Commission in Thailand) that demonstrate that river basin working groups can function effectively when a bottom-up development process is incorporated.

Proposed Evolutions

Vision—Regional Integrated Water Resources Management

The current LCBC 2025 Vision that has been approved by the Member States,* is summed up as follows:

> . . . land, water and all natural resources are conserved, sustainable exploited, managed in an integrated manner and shared equitably, in order not only to eradicate poverty and improve living standard of the people in the Lake Chad Basin, but also to ensure peace, security, cooperation and sound economic developments of the region."†

This vision represents a significant departure from LCBC's original mission, which was to "develop general regulations, ensure effective application of regulations, coordinate research activities of Member States, study projects prepared by Member States, recommend plans for the implementation of surveys and other projects in the Lake Chad Basin, and, generally, maintain liaison among the Member States [Odada, Oyebande, and Oguntola 2005].

During the tenth summit meeting held in July 2000, N'djamena, the LCBC members declared Lake Chad a wetland of international importance (Chad Wet). This sparked increased awareness and a conscious revision of the importance of integrating the principles of conservation and sustainable management of the hydrographic Basin. The original mandate focuses on political and economic development and Integrated Water Resources Management (IWRM), emphasizing the regulatory function and placing less emphasis on its regional function. The current orientation, therefore, is a conscious focus on the regional basin management function.

A set of unifying policy-related outcome goals and an overall comprehensive LCBC vision statement are urgently needed. There have been several activities attesting to a new focus and vision that need consolidation with achievable goals. For example, recently the RAMSAR Secretariat and LCBC enacted a MoU‡ during

---

* Managing the Water Resources of the Lake Chad Basin, Power Point Presentation, Muhammad Sani Adamu–Executive Secretary LCBC, GTZ Workshop, LCBC N'djamena, August 31, 2005.

† Editorial of the LCBC Executive Secretary, Lake Chad Bulletin, No. 0. November 2001.

‡ A Memorandum of Cooperation in November 2002 between the Bureau for the Convention on Wetlands in accordance with the Ramsar Convention of 1971 and the LCBC spelled out the following objectives:

- Reinforce the role of wetland integrated water resources management in sustainable development
- Reinforce organizational partnerships at the Basin and national levels among all stakeholders, governmental entities, intergovernmental organizations (IGOs), nongovernmental organizations (NGOs), and other stakeholders concerned with conservation and sustainable wetland use
- Establish coherent national and regional networks of Ramsar sites at the Basin level as the basis for their sustainable management
- Foster innovative approaches to transboundary wetland management in the Lake Chad Basin by promoting partnerships between stakeholders, conventions (CBC, UNCCD, UNFCCC), etc.

the second World Water Forum in 2000 (UNEP/DEWA) that highlighted a number of important issues identified within the Lake Chad Basin and presented three key objectives:

1. Maintenance of Lake Chad and other wetlands in the region at sustainable levels to provide economic security, sustainable biodiversity, and the equitable use of aquatic resources of the Basin for the alleviation of poverty
2. Acceptance of responsibility for freshwater, IWRM, and biodiversity conservation and integrated River Basin management by regional and national authorities
3. Equitable access by member states to safe and adequate water resources in order to meet their needs and rights

Partnership Strategy

The ability to build coalitions is an essential component in order to facilitate regional IWRM. In this regard, there is a need for a comprehensive partnership strategy (including all relevant stakeholders, scientific and academic institutions, and donors). The role of informal agreements cannot be underestimated in order to define roles and responsibilities of the different stakeholders that have a vested interest in the management of the Basins resources.

Among the core values to target are the following:

- Development of collaborative opportunities
- Facilitate access to external expertise
- Share cross-organizational best practices
- Better serve clients and partners
- Increase visibility of LCBC as a best practice advisory and knowledge-based institution

Traditional assessments of partnership strategies emphasize cooperation over competition. Knowledge is sought and achieved through teaming with complementary individuals (e.g., cross-networking). Typical issues to be assessed include the following:

- As a team, how well does LCBC understand its clients?
- As a team, what value does LCBC demonstrate to the client?
- How well does LCBC use organizational knowledge to improve work efficiency?
- What contribution does the LCBC team make to the client's knowledge networks? Is knowledge management effective?
- How active is the LCBC team in the development of new services?

Other Regionally Relevant Knowledge-Based Organizations

A sustained effort should be made to establish, or reestablish, formal and informal relations with key research organizations involved with resources management in the region. LCBC should, for example, begin by conducting an institutional mapping exercise of potential partners.

For instance, two key research institutions in the Lake Chad area include the Lake Chad Research Institutes at Maiduguri (Nigeria) and N'Djamena (Chad). The Institute in Maiduguri, for example, has established 10 stations (experimental sites). These and other research institutions can be considered as potential LCBC allies or partners and properly equipped to deliver results that will lead to breakthroughs in knowledge and management practices. Another good example of a good partnership is the North East Arid Zone Development Programme (NEAZDP), which was established by the European Development Fund of the Lome Convention, to promote integrated rural development in the area north of Latitude 12°N in northeast Nigeria. It has achieved much and with its EU funding fully restored could be effectively used as a strategic partner of LCBC.

Partnership Strategy (Getting to Know the Clients)

As mentioned earlier, regional institutions such as LCBC must develop a firm partnership strategy that will delineate donor institution relationships in a manner conducive to LCBC outcome goals. These revamped relationships must be characterized by defining strategic partnership strategies. For this reason, a partnership strategy between LCBC and its primary stakeholders must be developed internally. The development of key relationships has a strategic intent and must therefore be properly resourced. Since the principles of knowledge networking, for example, can support the sharing of comparative experiences that can be used to develop an LCBC relationship agenda and related set of actions.

Such a strategy will (1) support enhanced understanding of member states internal agendas; (2) develop knowledge of client (government) profiles; (3) identify strategies designed to strengthen relationships with existing or build relationships with new partners; and (4) share comparative experiences and identify best practices for influencing partners.

Building relationships with these clients can be facilitated by a number of activities including promoting staff exchanges between LCBC and country-sector-specific offices, establishing a regional community of practices centered around problem-solving on issues of relevance to the Basin, building a yellow pages directory of LCBC staff expertise and experience, making staff available for programming- or policy-related missions to member states, and providing the target knowledge communities with funding for workshops, research, and other local activities.

Regionally Relevant Knowledge-Based Organization

For LCBC to undergo a process of renewal (2025 Vision), the objective must be to ensure that the Secretariat has at its disposal a correct mixture of scientific and technical expertise as well as partnerships and other internal capabilities for effective problem solving and implementing regional IWRM. This will require considerable internal refocusing, reengineering, and behavioral change (management style, operational and administrative structure, work processes, budgeting, etc.). Through this process, policy capacities can be strengthened, partnerships emphasized and expanded, and LCBC staff competencies upgraded. Such changes are required in order to enhance instruments for measuring performance and to provide a platform for an expanded knowledge resource base.

Consequently, there is a pressing need for LCBC to restructure in order to become a service-oriented, professional, and streamlined regional development organization that is capable of providing policy support where and when it is most needed. LCBC must reflect a flexible management structure that facilitates this objective. In addition, the organization's transformation must entail enhancement, streamlining, and rationalization of all staff roles in order to produce a decentralized, networked, hands-on and service-delivery-oriented operation. This action would enable LCBC staff to work effectively, thereby securing future success in achieving organizational goals.

With this in mind, LCBC might need to develop a number of key service lines and tailor internal capacities to do so. For example, LCBC can adopt a strategic program approach in order to manage activities relating to key conservation and development objectives. Existing staff must be provided with the tools and management guidance to be able to perform new knowledge facilitation and partnership-development-related duties. In order to accomplish this, the organization will inevitably need enhanced technical capacities (e.g., management support and technical guidance) in the short term. Nevertheless, any measures for change will ultimately prove futile without effective management, technical support, and leadership to make the transition possible.

Stakeholder Involvement, Knowledge-Partnerships, and Organizational Structure

In order to provide more opportunities for stakeholders to participate, own, and build coping mechanisms, LCBC must develop a mechanism for gathering all stakeholder input into policy and decision-making activities. By doing so, LCBC would learn of the gaps related to the lack of stakeholder participation, particularly as achieving active and consistent stakeholder engagement is critical in terms of creating better policies and decision-making structures.

Transparent and participatory problem formulation, clarification, and linkages would help to improve understanding of the characteristics of data collection and dissemination methods. It would serve to garner regional and local participation in LCBC's policies and planning. Consequently, it is essential that LCBC functions are updated in order to reflect the new functions implied by the 2025 Vision such as problem identification and solving approach, ability to solicit active stakeholder participation, more strategic communications, knowledge management, data collection, and methods of data analysis.

Therefore, for a management plan to be sustainable, LCBC must establish strategic partnerships with key academic and scientific institutions within member states. In order to deal with the complexity of IWRM, it is crucial to develop relations with credible scientific and academic institutions that have an interest in LCBCs mandate as this will support the technical, human resources requirement, and geopolitical issues. As an example, several regional academic and scientific institutions have professional interests in the LCBC mission and would be capable of providing support. Moreover, obtaining access to relevant scientific expertise will foster effective Basin management and enhance LCBCs credibility as a knowledge-based organization.

For this to be probable, LCBC must establish a lead scientist or a similar quality control mechanism. For example, it could revitalize the LCBC technical committee within its Department of Water and Environment. This mechanism would be responsible for reviewing LCBC policies and programs against the scientific integrity in relation to the Basin's overall management objectives. This would, in turn, support monitoring policies and decision processes in line with the IWRM approach. Especially as LCBC must maximize the use of science and technology as the basis for future planning and policy.

On the whole, the integration of science into the decision-making structures at LCBC must be the central point through which the organizational changes occur, an issue that cannot be underestimated. Appropriate scientific input, supportive technology, and infrastructure would facilitate better decisions concerning the Basin. As such, LCBC must address the blatant communication problems within the organization, and develop mechanisms that support knowledge-sharing and information flow throughout the region. By integrating this function within its organizational structures, LCBC could support and significantly improve decision-making concerning Lake Chad.

Conclusion

Although the literature on stakeholder- and community-based participation and development continues to bring in new ideas and methods for exploring development agendas, there is little evidence that one approach is better than the previous one [Herman 2005; Harmoni 2005]. As any concept evolves, debates over definitions, fundamental principles, and policy implications proceed. But, as this chapter has demonstrated, the discussion surrounding participation is not only academic and neither is it restricted to those scientists, professionals, and policymakers who work directly with water resources management. Together with defining how we can reduce risks concerning natural and manmade disasters, as well as improving the resilience of communities, there is a parallel process of redefining the role and responsibilities of humans in nature.

In relation to this issue, the assessment of a project (or program) requires the consideration of the complete context in which it is being implemented. It is not enough to discuss the credibility of a project, policy, or program, but it also requires a joint understanding of the established objectives and the methods by which they will be achieved. This is of particular importance from a development perspective (for example, in the case of Lake Chad and the Danube region), as many of the environmental problems associated with community disaster recovery and resilience involve broader social and economic concerns.

Due to this lack of commitment and insight, there seems to be a need to develop a strategic understanding of the mechanisms required for developing consensus, capacity, and relationships within and outside projects. It is not enough to fund a diagnostic analysis and develop a strategic action plan, since that does not build capacity and resilience (stakeholders are, in this sense, one element of capacity/resilience). A potential benefit of making sure that stakeholder input is taken seriously is that it helps to stimulate adaptability and the development of local solutions. Such an approach might generate solutions that work effectively by being inclusive and innovative. It is difficult to see how resilience and recovery capacities can be effectively improved without active participation of those most directly concerned.

As stated in the Introduction, this chapter aims at addressing some of the key challenges facing the development of participation strategies. This was with the intent of discussing its relevance in terms of community-based risk reduction. An interesting context in which to put these issues in perspective can be illustrated by risk mitigation surrounding glacial lake outburst floods (GLOF). This, particularly as one of the consequences of climate change, is the continued acceleration of glacial retreat and melting—a process that results in the generation of glacial lakes and, consequently, an increase in flash floods and general flooding events. This intensification, in terms of climate variability, is having a significant impact on the lives and livelihoods of mountainous communities and also introduces new risks for downstream areas.

As in many other fields, the traditional approach toward GLOF risk mitigation has mostly been an overwhelming focus on engineering and structural measures [Lewis et al. 2009]. Such measures can, for example, involve (1) drawing or diverting water from hazardous glacial lakes; (2) generating an inventory of glacial lakes together with monitoring the formation and expansion of glacial lakes using geographic information systems (GIS). Within this framework, the challenges to structural/engineering approaches include the inaccessibility of glacial lakes, labor intensity, accidental triggering of a GLOF during preventative measures, high costs, and limited long-term contribution to risk mitigation. It is increasingly becoming apparent that risk reduction requires a much more holistic approach—an approach that addresses all dimensions of disaster management and the engagement of all stakeholders. As such, efforts toward involving communities and local administration have not found much acceptance and are rarely practiced. Solutions have most often been limited to expert-driven input from selected technical or research institutions and organizations. As a consequence of this emphasis on engineering solutions, most past projects have engaged in very limited involvement of communities and other relevant stakeholders. As a result, insufficient consideration has been given to the following:

- Establishing an efficient, inexpensive, and straightforward community-based warning system
- Improving a community's preparedness in terms of responding to a flooding event
- Encouraging land-use planning in hazard prone areas
- Advancing and developing strategies for mitigating the impact of disasters through awareness-raising and basic risk mitigation measures

Therefore, within the context of sustainable land-use planning and IWRM, communities that are at risk could be involved through the process of conveying knowledge as to what land is available and its suitability for a given activity. This may be achieved through a process that simply asks the communities themselves where the high risk areas are and what activities need to be halted (or relocated) to reduce risks.

To present an example within the framework of community-based participation, a community's local knowledge and understanding of their environment could (in the case of GLOF) be obtained through a simple community mapping exercise. As such, the community would not only provide valuable information, but they could also help in the process of defining problems and objectives—approach that ultimately also provides a sense of ownership toward the project. The exercise should therefore consist of straightforward guidelines. For example, a sketch map prepared by the community that includes the involvement of any local institutions may be available. Their engagement should not characterize a formal planning process, but rather a simple approach. Within this context, little external support should be provided and some elements of awareness-raising and capacity-building could be incorporated into the overall process.

On the whole, to provide a sustainable solution to GLOF (or any other natural hazards), community-based mitigation and preparedness need to be part of the planning process. Local institutions, NGOs, administration, volunteers, community elders, etc. should all be involved in the process, so as to strengthen relationships and ownership of the intended measures. Moreover, the correct type of participation provides a degree of understanding (for project managers and practitioners) in terms of how stakeholders in the given community interpret local hazards. Thus, it is essential to employ a holistic and iterative approach as well as to develop and modify applied methods to take into consideration a community's resilience needs. This would require consideration of, for example, (1) feasible actions a community can take to prepare, mitigate, and respond to flooding events; (2) sustainability of these actions at the community level; (3) acceptance by local administration and communities; (4) available expertise; (5) training and capacity-building initiatives to prepare communities and local administrations; and (6) keeping risk reduction practices evident and tangible. Achieving this development of a risk reduction strategy requires active community-based participation and commitment from local administrations as well as enabling relevant national institutions and organizations to address challenges and provide genuine input. After all, the sharing of knowledge and experiences in this regard concerning different countries, regions, and institutions can promote the development and formulation of more effective methods and risk mitigation strategies.

Lessons Learned

The focus of this chapter has been on comparing various perspectives on the use of stakeholder input and community-based participation strategies. This question has been addressed by presenting the manner in which various project managers, practitioners, and institutions (such as LCBC and ICPDR) have chosen to interpret stakeholder participation, integrate input, and engage communities. Primarily, the purpose of this has been to stimulate a discussion on the significantly diverse definitions of what constitutes a stakeholder or community and the actual relevance accorded within the context of project management. This is particularly important as the above-mentioned case studies represent a context that supports policy implementation in the field of transboundary water management. As such, an improved understanding of this environment can help decision-makers to understand and decide on objectives, policy development, implementation, and participation strategies within the context of community-based improvement of resilience and risk reduction.

- It is fundamentally important to employ a regional approach to instill an updated IWRM approach, particularly as it describes the patterns of the use of natural resources as well as provides guidance in accordance with the principles of sustainable use.
- A country acting individually cannot bring about an integrated regional approach. This requires collaboration among the affected countries and cooperation has to be present at all levels of government.
- A successful strategy to increase the extent of an effective and efficient institution must make provisions to build its capacity, to maintain an effective and efficient Secretariat, to react quickly and well to crises, and to anticipate

important issues and initiate the formulation of timely advice and recommendations for action.

- In terms of LCBC, the Commission must become product-oriented. This means focusing on the delivery of high quality, scientifically sound products that will revive donor and member state confidence and that can prove it to be a relevant and indispensable institution in the region.
- Competence is demonstrated by a commitment to a process that will ensure recruitment of the best available scientific and other talent, as well as the establishment and application of clear rules and procedures for staff.
- International donors will support institutions that are competent, effective, inclusive, efficient, and flexible.
- Effectiveness is demonstrated through provision of useful services, such as the appropriate distribution of accurate and helpful reports to stakeholder/communities and production of reports that inform the global community of the importance of the intended work/project.
- An organization's efficiency can be achieved through the development of a process that ensures a systematic, continuous, and planned review of staff structures and monitors the number of staff; ensures the delivery of high quality output for the benefit of its members, and maintains the ecological integrity of the Basin.
- In relation to appearing efficient, flexibility can help to ensure that staffing patterns are (or can be) adjusted to meet new and changing priorities. This is particularly important at the professional level so as to ensure the ability to be proactive.
- Stakeholder and community-based participation (or inclusiveness) is the key to the future success of building resilience and contributing to risk reduction toward natural and manmade disasters. Any project or program must be seen to be inclusive on the part of all the affected stakeholders in the Basin. As such, these stakeholders should include groups and individuals at international, regional, national, and community-based levels.

Policy Recommendations

- Balanced use of expert- and stakeholder-based input during project design and conceptualization. Participation is significantly easier to stimulate if the stakeholders' real-life connections to stated project objectives are concrete and tangible.
- Highlight the importance of stakeholders and community participation (in particular) in decision-making processes for disaster risk reduction.
- Promote transboundary cooperation in the management and protection of key areas such as cross-border exchanges between local organizations.

- Move problem formulation from strictly technical concerns into the area of social orientation. This should not mean that the advantages associated with technical aspects of project development are lost.
- Develop clear guidelines to appraise the moral implications of project objectives. To address these issues, project managers should examine the moral dimensions of problem formulation, systematically represent stakeholders, engage in critical self-reflection to avoid cognitive biases, and challenge current problem assumptions.

References

Aggestam, F. 2008. Stakeholder analysis and engagement: Reviewing projects of the United Nations Development Programme (UNDP) and International Institute for Applied Systems Analysis (IIASA). SERI background paper. Vienna: SERI.

Azqueta, D., and Delacamara, G. 2006. Ethics, economics and environmental management. *Ecological Economics* 56:524–533.

Baggett, S., Jeffrey, P., and Jefferson, B. 2006. Risk perception in participatory planning for water use. *Desalination* 187:149–158.

Baker, D., and Refsgaard, K. 2007. Institutional development and scale matching in disaster response management. *Ecological Economics* 63:331–343.

Barry, J., and Proops, J. 1999. Seeking sustainability discourses with Q methodology. *Ecological Economics* 28:337–345.

Brugha, R., and Varvasovszky, Z. 2000. Stakeholder analysis: A review. *Health Policy and Planning* 15:239–246.

Catelletti, A., and Soncini-Sess, R. 2007. *Topics on system analysis and integrated water resource management.* Amsterdam: Elsevier.

Caspian Environment Programme (CEP). 2002. Transboundary diagnostic analysis for the Caspian Sea: Executive summary and environmental quality objectives. Retrieved from http://www.caspianenvironment.org (accessed February 9, 2011).

CEP. 2005. Public participation strategy for the Caspian Sea: Final draft for comments. Retrieved from http://www.caspianenvironment.org (accessed February 9, 2011).

CEP. 2006. Communication and public participation case study for the Caspian Sea. Retrieved from http://www.caspianenvironment.org (accessed February 9, 2011).

Chae, B., Paradice, D., Courtney, J. F., and Cagle, J. C. 2005. Incorporating an ethical perspective into problem formulation: Implications for decision support systems design. *Decision Support Systems* 40:197–212.

Clark, N. 2007. Living through the tsunami: Vulnerability and generosity on a volatile earth. *Geoforum* 38:1127–1139.

Coe, M. T., and Foley, J. A. 2001. Human and natural impacts on the water resources of the Lake Chad basin. *Journal of Geophysical Research* 106:3349–3356.

Daniel Baker, D., and Refsgaard, K. 2007. Institutional development and scale matching in disaster response management. *Ecological Economics* 63:331–342.

Dearnley, C. 2005. A reflection on the use of semi-structured interviews. *Nurse Research* 13:19–28.

Devitt, T. 2001. *Under human pressure, Africa's Lake Chad disappearing.* Retrieved from www.news.wisc.edu/5846.html (accessed February 9, 2011).

Failing, L., Horn, G., and Higgins, P. 2004. Using expert judgment and stakeholder values to evaluate adaptive management options. *Ecology and Society* 9:13–30.

Gigerenzer, G., and Tood, P. M. 1999. *Simple heuristics that make us smart.* New York: Oxford University Press.

Glantz, M. H. 2004. Lake Chad and the Aral Sea: A sad tale of two lakes. Retrieved from www.fragilecologies.com/sep09_04.html (accessed February 9, 2011).

Hare, M., and Pahl-Wostl, C. 2002. Stakeholder categorisation in participatory integrated assessment processes. *Integrated Assessment* 3:50–62.

Harmoni, C. O. P. 2005. *Learning together to manage together: Improving participation in water management.* Germany: University of Osnabruck.

Herman, L. M. 2005. Actor analysis for water resources management putting the promise into practice. PhD Thesis. Technical University of Delft.

Hodge, S. 2006. Knowledge innovation systems and technology diffusion strategies for ecosystems management in Africa. *African Policy Journal,* 2, 1–26.

International Commission for the Protection of the Danube River (ICPDR). 2003. UNDP/GEF Danube Regional Project. Retrieved from http://europeandcis.undp.org/WaterWiki (accessed February 9, 2011).

ICPDR. 2005. Danube River Basin stakeholder conference—Final conference report. Budapest, June 28–29, 2005. Retrieved from http://europeandcis.undp.org/WaterWiki (accessed February 9, 2011).

ICPDR. 2006. ICPDR response to the stakeholder consultations process 2005. Retrieved from http://europeandcis.undp.org/WaterWiki (accessed February 9, 2011).

Johnston, E., and Soulsby, C. 2006. The role of science in environmental policy: An examination of the local context. *Land Use Policy* 23:161–169.

Juma, C. 2005. Biotechnology in a globalizing world: The co-evolution of technology and social institutions. *Bioscience* 55:265–272.

Juma, C. 2006. *Entrepreneurship and development: Opportunities for private sector participation.* Submission to the International Development Select Committee. London: United Kingdom Parliament.

Juma, C., et al. 2005. Forging new technological alliances: The role of south-south cooperation. *The Cooperation South Journal* 59–71.

Klein, R. J. T., Nicholls, R. J., and Thomalla, F. 2003. Resilience to natural hazards: How useful is this concept? *Environmental Hazards* 5:35–45.

Lake Chad Basin Commission. 2003. *Compilation of decisions from various summits of the heads of states of the Lake Chad Basin Commission.* N'djamena: LCBC headquarters.

Lewis A. O., Thackray, G., Anderson, R. S., Briner, J., Kaufman, D., Roe, G., et al. 2009. Integrated research on mountain glaciers: Current status, priorities and future prospects. *Geomorphology* 103:158–171.

Lidstrom, A. 1999. The comparative study of local government systems: A research agenda. *Journal of Comparative Policy Analysis: Research and Practice* 1:97–115.

Linnerooth-Bayer, J., and Mechler, R. 2007. Disaster safety nets for developing countries: Extending public-private partnerships. *Environmental Hazards* 7:54–61.

Lynch, H. J., Hodge, S., Albert, C., and Dunham, M. 2008. The Greater Yellowstone ecosystem: Challenges for regional ecosystem management. *Journal of Environmental Management* 41:820–833.

Matthews, M. M. 2004. Stakeholder inclusion in Caspian Basin natural resource management. Woodrow Wilson International Center for Scholar: Occasional Paper #288.

McDaniels, T. L., Gregory, R., Arvai, J., and Chuenpagdee, R. 2003. Decision structuring to alleviate embedding in environmental valuation. *Ecological Economics* 46:33–46.

Murenzi, R., and Hughes, M. 2005. Africa in the global knowledge economy. In *Going for growth: Science, technology and innovation in Africa,* ed. C. Juma. London: The Smith Institute.

Odada, E. O., Oyebande, L., and Oguntola, J. A. 2005. *Lake Chad: Experience and lessons learned brief.* Kusatsu, Japan: International Lake Environment Committee Foundation.

Oyelaran-Oyeyinka, B., and Barclay, L. A. 2003. *Systems of innovation and human capital development in African development.* New York: United Nations University, Institute for New Technologies.

Plapp, T. 2001. *Perception and evaluation of natural risks: Interim report on the first results of a survey in six districts in Germany.* University of Karlsruhe.

Ridley, T., and Lee, Y.-C. 2005. Infrastructure, innovation, and development. In *Going for growth,* ed. C. Juma, 62. London: The Smith Institute.

Rose, R. 1991. Comparing forms of comparative analysis. *Political Studies* 39:446–462.

Sadoff, C. W., and Grey, D. 2002. Beyond the river: The benefits of cooperation on international rivers. *Water Policy* 4:389–403.

Smith, J. A., Harre, R. and Van Langenhove, L. 2001. *Rethinking methods in psychology.* London: SAGE.

Tobin, A. G. 1999. Sustainability and community resilience: The holy grail of hazards planning? *Environmental Hazards* 1:13–25.

Triandis, C. H. 1976. Methodological problems of comparative research. *International Journal of Psychology* 11:155–159.

UN/ISDR. 2004. Living with risk: A global review of disaster reduction initiatives. Retrieved from http://www.unisdr.org (accessed February 9, 2011).

UNDP. 2005. Public participation strategy for the Caspian Sea. Retrieved from http://europeandcis.undp.org/WaterWiki (accessed February 9, 2011).

UNDP/GEF Danube Regional Project. 2003. Stakeholder analysis workshop. Retrieved from http://europeandcis.undp.org/WaterWiki (accessed February 9, 2011).

UNDP/GEP. 2005. *UNDP/GEP Environmental Governance Project. Reducing Trans-boundary Degradation of the Kura-Aras River Basin through Public Involvement and Stakeholder Inclusion in Governance.* Kura-Aras Qualitative Stakeholder Analysis. Draft version.

Van Aalst, M. K., Cannon, T., and Burton, I. 2008. Community level adaptation to climate change: The potential role of participatory community risk assessment. *Global Environmental Change* 18:165–179.

Vogel, C., Moser, S. C., Kasperson, R. E., and Dabelko, G. D. 2007. Linking vulnerability, adaptation, and resilience science to practice: Pathways, players and partnerships. *Global Environmental Change* 17:349–364.

Wattage, P., and Mardle, S. 2005. Stakeholder preferences towards conservation versus development for wetland in Sri Lanka. *Journal of Environmental Management* 77:122–132.

Wierzbicki, A. P., Makowski, M., and Wessels, J. 2000. *Model-based decision support methodology with environmental applications.* The Netherlands: Dordrecht. Kluwer Academic Publishers.

Wittmer, H., Rauschmayer, F., and Klauer, B. 2006. How to select instruments for the resolution of environmental conflicts? *Land Use Policy* 23:1–9.

WWF. 2004. Living with floods: Achieving ecologically sustainable flood management in Europe. Retrieved from http://assets.panda.org/downloads/livingwithfloodswwfpolicybriefingfinal.pdf (accessed February 9, 2011).

Xevi, E., and Khan, S. 2005. A multi-objective optimisation approach to water management. *Journal of Environmental Management* 77:269–277.

Acronyms

CEP Caspian Environment Programme
EU European Union
GEF Global Environment Facility
GIS Geographic Information Systems
GLOF Glacial Lake Outburst Flood
ICPDR International Commission on the Protection of the Danube River
IIASA International Institute for Applied Systems Analysis
IWRM Integrated Water Resources Management
LCBC Lake Chad River Basin Commission
MoU Memoranda of Understanding
NEAZDP North East Arid Zone Development Programme
NGO Nongovernmental Organization
REC Regional Environmental Centre for Central and Eastern Europe
SAP Strategic Action Plan
TDA Transboundary Diagnostic Assessment
UNDP United Nations Development Programme
WFD EU Water Framework Directive
WWF World Wildlife Foundation

Appendix 6.1 Defining Terminology and Concepts

Many researchers still attempt to extrapolate individual elements from context-specific projects to a much larger stage. But as environmental initiatives are increasingly interdisciplinary, ecological and social systems are becoming more and more inseparable. As a result, even with clear project objectives, a difference in understanding relevant concepts may generate disagreement not only during implementation but also influence the evaluation process. An issue that emphasizes not only the impact the evaluator's personal and professional background may have but also the impact from other spheres of interest.

This appendix will aim to provide a generic definition of some key concepts that have been employed throughout this chapter.

Ecosystem Management

The development of ecosystem management is a response to the deepening global biodiversity crisis. As such, it is an approach to natural resource management, which aims to sustain ecosystems so as to meet both ecological and human needs. In essence, it aspires to maintain and/or restore the function, composition, structure, and services of natural as well as modified ecosystems under the basis of sustainable development. It should represent an adaptive, iterative, and participatory vision of a desired future state that integrates ecological, socioeconomic, and institutional perspectives, applied within a geographic framework and defined primarily by natural ecological boundaries.

Integrated Water Resource Management

Integrated Water Resource Management (IWRM) is an approach to resource management and decision-making that considers multiple perspectives and viewpoints on how water should be managed. In essence, "IWRM is defined as a process that promotes the coordinated development and management of water, land and related resources, in order to maximize the resultant economic and social welfare in an equitable manner without compromising the sustainability of vital ecosystems" (taken from Global Water Partnership). Hence, IWRM uses both structural and nonstructural measures to control natural and human-made water resources systems. In contrast to more traditional and sectoral approaches, IWRM provides a more holistic and interdisciplinary approach to managing all factors that affect water resources.

Stakeholder Participation and Adaptation

A general definition for public and stakeholder participation can be defined as a project element that (1) analyzes public and stakeholder interests; (2) helps to plan and provide input during complex decision-making situations; and (3) as an integrated part of project and conflict prevention/management. Participation should, if genuinely employed, represent a set of tools that is able to define dynamic relationships between stakeholders and the public. This, particularly in connection to project problem definitions, objectives, and boundaries, determines an organization's potential to influence stakeholders; define stakeholder attributes, social and cultural context; actively seek out those who will normally not be able to become involved in a stakeholder process.

Transboundary Cooperation and Sharing Water

Achieving a balance between the conservation and development of natural resources represents one of the key challenges facing societies today. This problem becomes more complex as resources (in particular water) are most often shared by more than one community or even countries. Cooperation across boundaries and borders is of key importance, a context within which shared water resources require a mechanism to become more efficient and productive. Thus, transboundary water cooperation should operate within a wide range of fields, such as crisis and conflict prevention/management, poverty reduction, and resource protection. With water resource protection in mind, transboundary cooperation should be concerned with the sustainable protection of water resources as well as the protection and viability of biodiversity, ecological integrity, and surrounding ecosystems.

Community-Based Disaster Risk Reduction

The community is central to supporting government planning and projects focusing on benefiting urban and rural communities living with hardship. As such, the aim of community-based disaster risk reduction is to empower resource-strapped communities to cope and prepare for increasing disaster events in a sustainable manner. By, for example, developing local risk identification and planning methodology that links to planning functions, a community-based development project can (1) support any national development strategy; and (2) reduce risk for vulnerable groups. In this context, community-based disaster risk reduction should serve as the main tool for the strategy and policy formulations in relation to disaster risk reduction, resilience, and recovery.

Relevance of Defining General Terminology and Concepts

Wise decision-making concerning the management of water resources requires knowledge and experience from different disciplines to identify alternatives for action as well as to assess their effects. For example, engineering knowledge might focus on infrastructure development, while a social scientist might focus on human development issues.

As the water sector is becoming increasingly interdisciplinary, project managers need to be able to comprehend and utilize many disciplines. Moreover, there needs to be a joint understanding concerning the available methodological approaches, terminology, and concepts. Particularly, as complex problems within water resources management requires the identification of alternatives as well as the assessment of the various impacts both within the sphere of human and environmental interests.

Chapter 7

Global Food Crisis and African Response: Lessons for Emergency Response Planning*

Sheryl L. Hendriks and Scott Drimie

Contents

* This paper draws extensively on the African Union Commission/New Partnership for Africa's Development's Comprehensive Africa Agricultural Development Plan's Framework for African Food Security.

Africa has experienced an average of 20 food emergencies per year since 1998 [Economic and Social Council 2005]. Most of the world's hungry live in Africa, yet this continent has made the least progress in meeting the United Nation's Millennium Development Goal of halving hunger by 2015 [Von Grebmer et al. 2008]. Global food aid has increased over the past decade, and Africa remains the largest recipient of international food aid. In 2007, sub-Saharan Africa received 67% of global emergency food aid deliveries (the next largest region was Asia at 22%), amounting to 2,444, 960 tons of food [World Food Program 2009].

Food aid in Africa has been predictable, pervasive, and persistent—an important element of national food security in many countries for more than 30 years. However, the 2008 global food crisis put African countries in an even more precarious position as staple crop commodity prices climbed rapidly, reaching the highest levels since the 1970s [Food and Agriculture Organisation 2008a]. The World Bank's [2009] provisional estimates show that, in 2007, 100 million more people were added to the total number of undernourished relative to 2003–2005. Although several factors are responsible, high food prices are driving millions of people into food insecurity, worsening conditions for many who were already food-insecure and threatening long-term global food security. The crisis was, in essence, manmade rather than a typical covariate shock or conflict typical of food emergencies over the past three decades. No major international production crisis existed, but global food stocks fell to the lowest level in 25 years, exposing the fragile nature of the global food economy.

Therefore, the 2008 food crisis came as a surprise to a world (over)confident in its ability to produce, distribute (over considerable distances across the globe); import commodities (even staples) including those demanded by changing consumer preference for manufactured goods and disregard for local seasonality; export strategic commodities and surpluses and call on international food aid when necessary.

Not all food emergencies are caused by short-term shocks. Food crises occur when deeply rooted vulnerability is confronted with various hazards and disruptions. The countries affected the worst by the food price crisis were those with deep-rooted vulnerability, which exposed their populations to the sudden and determined increases. This reiterates the broader definition of food security as the ability to acquire (by production or purchases) enough food for an active, healthy life—both now and in the future. Food security definitions usually include elements of access to the foods households prefer and the right to procure such food in a dignified manner. The right to food is a basic human right, demanding that governments protect and provide access to individuals and households to ensure food security. More recently, food security literature has shifted attention to the vulnerability to food insecurity. This focuses on the element of the definition of food security that refers to security of access to food in the future (or lack of worry as some authors indicate) as a fundamental requirement for food security. Although vulnerability is a bundle of interrelated vulnerabilities (e.g., poverty, resource deprivation, health, income, and climate), the concept provides a useful consideration for emergency response planning, as indicators of vulnerability are essentially early warning indicators of impending issues related to the food economy.

Food crises can be prevented or their effects significantly muted if underlying vulnerabilities are confronted and addressed. However, few African countries possess the required physical, human, institutional, and financial capacities to do so. The capacity gaps that constrain adequate crisis prevention and preparedness also limit the depth and breadth of effective action during crises. The principle challenges during food emergencies center on coordination, logistics, and information management. National governments are not always able to take the lead in setting the broad framework for emergency response or in prioritizing intervention modalities and locations. This is particularly true in the current crisis, where constant review of policy choices is required to respond to rapid and multiple changes in the global and local economies [African Union Commission (AUC) and New Partnership for Africa's Development (NEPAD) 2009].

The 2008 crisis challenged many traditional approaches to emergency and contingency plans and strategies. Effective emergency food responses ensure that the needs of the most vulnerable are met. Emergency prediction, preparedness, and response management are crucial for mobilizing resources and capacity to meet immediate and dire needs. However, response programs rely on the availability of food to purchase, import (if necessary), and distribute. This chapter reflects on the responses of African governments to the 2008 food crisis and suggests ways of improving emergency planning to mitigate the impact of atypical shocks in a multifaceted global crisis.

Africa's Food Emergency Legacy

The global food crisis highlighted and deepened perpetual food security problems in Africa. Food insecurity is a systemic problem in that a single shock, such as a flood or drought, creates a food security crisis and predictable year-on-year food assistance is required to fill the consumption gap of many populations. Food supply is inadequate and erratic.

African population growth rates are the highest in the world, which suggests a massive food supply challenge. Over the past 40 years, the African population has grown at an average of 2.7% per year, compared to 2% in developing Asia and 2.2% in Latin America [Wik et al. 2008]. Therefore, African economies must grow faster than the rest of the world just to keep up with the continent's rapidly growing population. However, low agricultural productivity and rapid population growth have left Africa as the only continent where per capita food production has fallen over the past 45 years. Cereal yields have stagnated and currently average less than one ton per hectare. Population pressure has caused both land availability and soil fertility to decline [Cleaver and Schreiber 1994; Henao and Baanante 2005; NEPAD 2003].

This ongoing challenge is predicated on acute vulnerability to hunger that is often triggered by a range of shocks and stressors. Extreme weather events that are increasing in frequency and impact under global climate change, a range of pests, and communicable human and animal diseases undermine fragile livelihoods and pose direct threats to food security. For example, livestock production and pastoral livelihoods in Africa face multiple threats related to trans-boundary disease, water shortages, and climate change related to trade barriers and phytosanitary issues [AUC and NEPAD 2009]. Climate change is likely to create additional challenges and threats to a range of production and livelihood systems in the future, whereas social and political strife and conflict continue to affect food supply in many African countries [AUC and NEPAD 2009].

One of the root causes of food insecurity in developing countries is the inability of people to gain access to food due to poverty because even in good years many households are unable to meet their basic food needs. Mwaniki [n.d.] indicated that more than 70% of the poor in Africa live in the rural areas where food insecurity is prevalent. Poverty is closely related to the lack of a steady flow of income [Rivera and Qamar 2003]. According to the World Bank [2009], the income-based definition of poverty considers the poor to be people whose income levels fall below a certain minimum level to meet basic needs. Philip and Rayhan [2004] elaborated that because of their lack of income and assets the poor are highly vulnerable to, and therefore unable to cope with, uncertainties such as economic down-turns, health hazards, natural catastrophes, and civil conflict. Vulnerability is further increased by the depletion of productive assets and unsustainable livelihoods.

Poor households spend a significant proportion of their household expenditure on food, either by directly purchasing it or by producing it [Staaz 2000]. The amount of money left over for expenditure on basic nonfood items is barely enough to meet needs. Similarly, household expenditure on health and education is sacrificed. Consequently, the standards of living of the poor remain low. This situation is worsened by food price increases that further reduce the purchasing power of poor people's inadequate income. Where households produce their own food, cash and transport constraints limit people's ability to purchase farm inputs and market their produce. Increasing prices of farm inputs associated with supply-side commercial practices in local and subregional markets undermine poor farmers' ability to access the farm inputs for production [European Commission 2001].

Lack of education frequently limits poor households' access to the most lucrative farm and nonfarm employment opportunities. At the same time, a shortage of capital prevents them from investing in transport, mechanical milling, and other high-return farm or nonfarm business opportunities. Instead, the poor depend on low-return unskilled labor activities, such as basket-making, weaving, and casual labor [Haggblade 2007].

Access to and availability of a consistent flow of income is fundamental to the lives of the vulnerable. Therefore, the challenge of emergency programs is to reach the vulnerable, providing the means to access food and prevent the depletion of assets to protect their livelihoods and resilience.

The Nature and Causes of the Global Food Crisis

The recent crisis is different from any previous crisis for two key reasons. First, unlike previous food crises, the recent crisis was not the result of typical covariant shocks affecting food supplies (although irregular weather patterns have led to poor harvests in some areas), but was caused by a complex combination of factors. Second, unlike previous global crises where prices of only a few commodities were affected, the recent price increases affected most food commodities, including cereals, oilseeds, dairy products, and meat. The demand-linked drivers included increased food and fodder demand due to population growth, consumer demand for food commodities (grains, meat, and fish) made increasingly available by market globalization and leading to net importing of nontraditional staples, overconfident market speculation, and the drive for alternative energy sources (biofuels).

Supply factors included the results of long-term underinvestment in the agricultural sector, a reduction of production of maize and wheat in the European Union and United States, reduced global cereal stock levels, and resource constraints related to land availability and productivity and the effects of climate change. These factors were compounded by increased oil prices that pushed prices to higher levels and induced higher transport, input, production, and processing costs, and unprecedented market volatility. Increased fuel costs made transportation of bulky staples particularly unaffordable. This reduced purchasing power and induced panic in international aid circles as budgets were cut or suffered losses and the initial responses of governments led to trade restrictions, bans, and tariff changes that made moving regional stocks improbable. The ensuing global financial crisis added to the panic and misery as development aid dwindled.

What was unusual about the 2008 food crisis was its universality and the fact that governments were taken by surprise despite considerable advances in early warning systems and emergency preparedness. The nature of the crisis presents the most interesting challenges for early warning and emergency response planning as the uniqueness of the crisis lay in the fact that the supply side of the global food system was intact, but demand outstripped global supply. This increased prices quickly, exacerbated by protectionist trade bans implemented by panicking governments, and led to a situation where even having the money to purchase could not secure national stock purchasing to fill consumption gaps. Most contingency and emergency food programs require the purchasing of stocks to fill consumption gaps, but these cannot be implemented when food stocks are not available or when there are no willing sellers. For example, Seychelles was unable to purchase rice from the East as borders were closed, leaving Seychelles with ample cash but no willing sellers.

The collective impact of higher and volatile food, fuel, fodder, and fertilizer prices exacerbate hunger and poverty and affect households differently, determined by their production and consumption patterns [Quinsumbing et al. 2008]. Food shocks harm marginalized and vulnerable groups more deeply than other segments of the population, while increased energy costs put increased pressure on household budgets. Among the very poor are extremely vulnerable and often marginalized segments of society that need careful consideration in designing the package of public investments required to ensure maximum growth and simultaneous poverty reduction in Africa. These groups include women, children, the landless, pastoralists, hired labor, and the increasing numbers of people living with HIV.

The crisis created the realization of the fragile nature of food systems and the changing nature of food security in the wake of increasing proportions of households who are net buyers of food and situations where countries have to rely on imported staple commodities. These factors mean that households are directly affected by any large scale global changes in the food system and policy responses.

Food crises occur when these various hazards and disruptions encounter deeply rooted vulnerability. Food crises can be prevented, or their effects significantly muted, if underlying vulnerabilities are confronted and addressed. However, few African countries possess the required physical, human, institutional, and financial capacities to do so and on-going conflict erodes this capacity further.

Comprehensive Africa Agriculture Development Programme Food Security Agenda

The Comprehensive Africa Agriculture Development Programme (CAADP) has been endorsed by African heads of state and governments as a vision for the restoration of agricultural growth, food security, and rural development in Africa. A specific goal of CAADP is to attain an average annual growth rate of 6% in agriculture. To achieve this goal, CAADP aims to stimulate agriculture-led development that eliminates hunger and reduces poverty and food insecurity. More specifically, the AUC and NEPAD [2009] vision for Africa holds that, by 2015, Africa should achieve the following:

- Attain food security
- Improve agricultural productivity to attain a 6% annual growth rate
- Develop dynamic regional and subregional agricultural markets
- Integrate farmers into a market economy
- Achieve a more equitable distribution of wealth

CAADP is a strategic framework to guide country development efforts and partnerships in the agricultural sector. CAADP directs investment to four mutually reinforcing and interlinked pillars, each with a framework that guides policy alignment and suggests actions for countries to consider in designing their CAADP compacts, policy alignment, program design, investments, and monitoring and evaluation postcompact.

- Pillar I's Framework for Sustainable Land and Water Management seeks to extend the area under sustainable land management and reliable water control systems.
- Pillar II's Framework for Improving Market Access (FIMA) seeks to improve rural infrastructure and trade-related capacities for market access.
- Pillar III's Framework for African Food Security (FAFS) seeks to improve risk management, increase food supply, improve incomes for the poor, and reduce hunger and malnutrition.
- Pillar IV's Framework for African Agricultural Productivity (FAAP) seeks to improve agricultural research, technology dissemination, and adoption through strengthened agricultural knowledge systems to deliver profitable and sustainable technologies that are widely adopted by farmers, resulting in sustained agricultural growth.

The FAFS is a guide and assists stakeholders in Africa simultaneously to meet the objectives of CAADP Pillar III and the broader African development agenda. The food security challenges addressed in the FAFS are four-fold: (1) inadequate management of food crises, (2) inadequate food supply, (3) limited income opportunities for vulnerable people, and (4) widespread and persistent hunger and malnutrition.

These objectives are consistent with the Hyogo Framework for Action on Disaster Reduction (to which many African countries are signatories), which identifies the following strategic goals for disaster risk reduction [International Strategy for Disaster Reduction 2005]:

- More effective integration of disaster risk considerations into sustainable development policies, planning, and programming at all levels, with a special emphasis on disaster prevention, mitigation, preparedness, and vulnerability reduction.
- Development and strengthening of institutions, mechanisms, and capacities at all levels, particularly at the community level, which can systematically contribute to building resilience to hazards.
- Systematic incorporation of risk reduction approaches into the design and implementation of emergency preparedness, response, and recovery programs in the reconstruction of affected communities.

The FAFS aims to provide principles, recommended actions, coordination, peer review, and tools to guide national and regional policies, strategies, investments, partner contributions, and advocacy efforts to overcome these challenges and lead to improved food security risk management, increased food supply of affordable food, increased outcomes for vulnerable people, and reduced hunger and malnutrition. The FAFS represents the first and only internationally concerted attempt to build continent-wide consensus on the challenges and opportunities facing Africa in its efforts to reduce the food insecurity that plagues millions of its citizens.

Effective food security crisis management requires adequate data and systems, emergency plans and strategies, and the resources to respond. The challenges facing African governments with regard to these three essential elements are discussed below through evaluation of the responses of African countries to high food prices. The final section of this chapter presents lessons for improved emergency management, particularly for situations where the cause is not a short-term shock but a structural change affecting large sections of the population, nations, or the globe.

The Impact of the Crisis on African Food Security

Strong upward trends and volatility in global food prices since 2007 led to unprecedented increases in poverty and hunger in Africa between May and September of 2008. The World Bank [2009] estimated that an additional 100 million people (mostly in Africa) were pushed into hunger by the crisis. Between January 2007 and March 2008, food price indices in several countries increased between 11% and 20%. Table 7.1 reports the price changes of key staple foods in African countries (where data were available) during the crisis.

Despite improved global cereal supply and a decline in international prices, food prices are still high in developing countries. Domestic prices of staples have fallen by 78% in 58 surveyed countries, but high prices still affect food access for vulnerable populations [United States Agency for International Development (USAID) 2009]. Twenty African countries were still experiencing food emergencies in the second half of 2009 [USAID 2009]. Although price increases peaked in late 2008 and seemed to have tailed off somewhat in 2009, predications are that a structural adjustment in prices has occurred and the current rates are likely to persist for the next few decades [International Food Policy Research Institute 2008].

The responses of African governments to the crisis were diverse and many were not particularly wise as they compromised long-term development objectives (price controls and reducing input imports) and reduced fiscal reserves (reduced tariffs and taxes). Many chose to close borders to trade, negatively affecting regional relationships with spillover effects on neighbors. Several countries depleted foreign exchange reserves in responding to the immediate crisis or resorted to domestic borrowing, risking higher inflationary pressures and balance of payments [Food and Agriculture Organisation 2008b]. The extent of these measures has been captured in Table 7.2.

The impact of high global food prices was not universally transmitted to all African countries. The extent of the impact is determined by how connected the

Table 7.1 Increases in Food Prices from March 2007 to March 2008

Country	*Percent Increase in Prices (%)*							*Poverty Rate (% of Population Living on Less than US$1 per Day [1990–2005 Data])*
	Beans	*Maize*	*Millet*	*Rice*	*Sorghum*	*Teff*	*Wheat*	
Benin		225		45	79			30.9
Burkina Faso		36	28–46	65	13			27.2*
Burundi	40							81.3
Cote d'Ivoire		20		30				–
Eritrea							115.25	–
Ethiopia		147			133	20	74	39.0
Guinea				14				–
Kenya		30						45.9
Madagascar				11				67.8
Malawi		157.04						73.9
Mali			28–46	43				51.4
Mozambique		85		60				38
Niger		52	28–46	50	17			65.9
Nigeria			116		92			64.4
Rwanda	36				32			56.9
Senegal		37	30	112				33.5
Somalia	317	378			352		287	–
South Africa		–3.93						10.7*
Sudan					75		59.66	–
Tanzania		94						57.8*

(*continued*)

Table 7.1 Increases in Food Prices from March 2007 to March 2008 (Continued)

	Percent Increase in Prices (%)							*Poverty Rate (% of Population Living on Less than US$1 per Day [1990–2005 Data])*
Country	*Beans*	*Maize*	*Millet*	*Rice*	*Sorghum*	*Teff*	*Wheat*	
Uganda	56	12	5					51.5
Zambia		34						63.8

Source: Association for Strengthening Agricultural Research in Eastern and Central Africa, 2008; Food and Agriculture Organisation, 2008b; Munga and Wambugu, 2009; Sanogo, 2009; Save the Children Fund, 2008; United Nations, 2000; United Nations Development Programme, 2008; United Nations Development Programme in Mauritius and Seychelles, 2007.

Table 7.2 Policy Responses Implemented by African Governments in Response to High Food Prices

Policy Response	*No. of Countries*
Lowering of import tariffs	19
Production support	24
Reduce taxes on grain	17
Release stocks at subsidized prices	16
Price controls/subsidies	18
Cash transfers	12
Food stamps/vouchers/rations	10
Export restrictions	9
School feeding	5
Food-for-work	7
Increase supply via imports	3
Lowering of imports for fertilizers/seeds	2

country is with the global economy. However, the wider effects related to reduced global food stocks have had significant impact on countries that are net importers of key global cereals, such as rice, and the countries' responses. Of course, another key determinant of the impact relates to the standing of the country prior to the crisis, how severe food insecurity was at the outset, and what resources governments had on hand to respond (stocks and revenue).

Trade-oriented policy measures were the most common immediate responses of African governments to the crisis in early 2008, including releasing stocks at subsidized prices, restricting exports, and lowering or suspending import tariffs and taxes to protect domestic markets from global price increases and protect the welfare of net food buyers. Lowering import taxes reduces government revenue at a time when emergency and social protection needs increase; in the long-term, this reduces a country's ability to respond to another or protracted crisis and the benefits may not always be passed onto consumers. Although such measures may buffer consumers, they work against regional trade agreements and create disincentives for farmers who desperately needed to reap the long-awaited benefits of higher prices. To counteract this, Niger and Madagascar lowered import taxes on agricultural inputs (fertilizer and machinery) to stimulate agricultural production.

Import taxes usually protect local businesses from unfair competition. Lowering import taxes may harm domestic markets unless domestic supply is able to respond to local demand. In Burkina Faso, it was determined that reducing duties on food imports resulted in a 1.5% loss of expected tax revenue earnings (reducing import duty on fertilizer resulted in an additional 0.6% loss of revenue earnings in the same country).

Export bans were introduced by many countries (Benin, Burkina Faso, Cameroon, Ethiopia, Egypt, Guinea, Malawi, Mali, Nigeria, Tanzania, and Zambia) to protect domestic markets and boost strategic reserves [Association for Strengthening Agricultural Research in Eastern and Central Africa 2008; CGIAR Collective Action News 2009; Economic Communities of West African States (ECOWAS) and Comité Permanent Inter Etats de lutte contre la Sécheresse dans le Sahel (CILLS) 2008; Minde et al. 2008]. This behavior induces greater regional price volatility, affects foreign exchange earnings, leads to regional instability and conflict, and may counteract the impact of reductions in import taxes [Food and Agriculture Organisation 2008c]. Export bans can negatively affect production. Zambia made the mistake of closing its borders just after a bumper harvest, causing prices to fall sharply to the benefit of consumers but farmers responded by planting less maize.

Consumer-oriented policy and program measures include price controls, price subsidies, reduction of taxes (VAT) on staple foods, and food access-based interventions, such as cash or food transfers, handing out targeted food stamps, vouchers, and rations, school feeding programs, and food-for-work programs. Many of these measures were adopted as emergency responses aimed at protecting the purchasing power of consumers. Some governments also increased wages for public service employees (Egypt, Ethiopia, and Niger), a measure with long-term budgetary implications. For example, VAT reduction increases the purchasing power of the poor who spend a large proportion of their income on food but also reduces government revenue.

Price control fixes the price of staple foods to protect consumers but these measures act as a disincentive to farmers. Benin, Burkina Faso, Mali Malawi, and Senegal introduced price controls on key staples (bread and rice). Price fixing at low levels are likely to discourage domestic production and create a thriving black market.

Safety net programs reduce the impact of a crisis and mitigate against hunger and malnutrition among the most vulnerable sectors in both urban and rural areas. Egypt, Ethiopia, Mozambique, Lesotho, and South Africa (with the largest ongoing cash transfer program) used cash transfers to assist vulnerable households in facing the crisis. In West Africa, food or cash-for-work programs were introduced in Sahel countries and Nigeria [ECOWAS and CILLS 2008]. School feeding is an important component of food assistance and income support but was introduced in only a few countries to include Egypt, Senegal, and Sierra Leone. A food ration/stamp system was introduced in Tanzania.

Production-oriented responses included production support, productive safety nets, and fertilizer and seed programs. Some countries introduced or expanded input subsidy programs (mainly seeds and fertilizer) and include Benin, Madagascar, Malawi, Senegal, Tanzania, and Zambia. Benin implemented free distribution of subsidized maize and rice seed and diverted fertilizer, initially earmarked for cotton growers, to staple crop producers. Kenya introduced a targeted input subsidy (fertilizer and seed). Malawi's fertilizer subsidy program has driven the increased production that buffered it from the recent crisis. In 2006–2007, the subsidy program included the sale of 175,000 tons of fertilizer and 4,500 tons of seeds of hybrid maize and open-pollinated varieties to targeted farmers with 72% subsidy (i.e., farmers paid only 28% of the price) [Food and Agriculture Organisation 2008b, 2008c]. The government of Malawi funded 87% of the program, totaling US$91 million. As a result, maize production increased by 26% in the 2006–2007 season. The government continued distributing coupons to poor smallholder farmers to buy fertilizer and seeds at close to 80% and 100% subsidy respectively in 2007 and 2008.

Reflection on the Effectiveness of Africa's Food Security Early Warning Responses

The first respondents to natural and manmade hazards are individuals and households, followed by communities, then local and national governments, and then international actors. Seldom do households and communities have the resources to make decisions that increase their abilities to prevent crises, nor do they have access to early warning systems that integrate traditional knowledge systems with science-based systems. Where early warning systems exist, they tend to concentrate narrowly on food supply and thus rarely extend beyond data collection processes to become people centered mechanisms for identifying vulnerabilities and informing responses. Similarly lacking in most cases are credible contingency plans, backed by solid logistical capacity. Some countries are experimenting with disaster preparedness funds of various types, but the efficacy of these arrangements is unclear. Design and implementation of effective food reserve systems remains elusive. Although most governments and donors have embraced the notions of crisis prevention and mitigation, they have yet to commit significant resources to prevention and mitigation programs. These gaps likely reflect the fact that it is much easier to demonstrate effective crisis response than it is to show that a crisis has been effectively averted.

Strategies for effective disaster risk mitigation are difficult to mobilize when crises are not imminent or, as in many African countries, bilateral budgetary support or food aid are annual expectations. Many African countries are signatories to the Hyogo Framework for Action on Disaster Reduction [International Strategy for Disaster Reduction 2005]. However, few have taken steps as set forth in the CAADP FAFS to ensure that disaster risk reduction is a national and local priority with a strong institutional basis for implementation. As a result, a minority are identifying, assessing, and monitoring disaster risks and enhancing early warning with the result that few are using knowledge, innovation, and education to build a culture of safety and resilience. Thus, a minority of countries are taking steps to reduce underlying risk factors or strengthen disaster preparedness for response at all levels through targeted capacity development. Simultaneously, the international community continues to be less willing to invest in disaster risk reduction, preferring to respond to emergencies. Where food crisis management bodies exist, they tend to be underfunded and poorly integrated with other branches of government [AUC and NEPAD 2009].

Food insecurity, as outlined earlier, is a multifaceted problem affecting the ability of people and nations to access sufficient food in a sustainable way. Effective food security systems require coordination of sectors, sound early warning systems, and the capacity to respond to the crisis (human, financial, logistical, and food). The ability to determine and measure food security is fundamental to early warning systems. However, there are no defined and accepted measures of food security [Hendriks 2005], making the task of monitoring and evaluating the food security situation in countries difficult. Typically, regional economic commissions and development partners and agencies require countries to report crop estimates, commodity prices, poverty estimates (per capita consumption estimated from national stocks), agro-meteorology data, national food balances, and malnutrition rates. Weak baseline data on food insecurity and vulnerability limits capacity to forecast food demand and supply and predict crises. The food crisis illustrates how completely inadequate these measures and systems are at predicting atypical crises emerging from the rapidly globalizing world.

The 2008 crisis would not have been detected by tracking national demand and supply when the crisis was global, affecting prices in the entire global food system. This is because global demand (in China in particular) for food, fuel (from staple food crops in particular), and fodder drove global market prices upward. Monitoring demand would have shown a local decline in demand as purchasing power declined and households reduced expenditures, which should have shown stable food stocks. It is clear these monitoring measures focus on assumptions that African economies are independent of the global food economy, a naive and false assumption but maybe a realization of just how connected the rapidly globalizing world is. Nevertheless, the crisis shows that monitoring national demand and supply is not a reliable early warning measure.

Crop estimate data did not predict the price crisis. Food prices increased rapidly, not allowing lead time for mitigating the crisis. Price increases eroded purchasing power, while incomes were relatively stable, so labor and income surveys did not detect the impending crisis. Agro-meteorology data is effective in monitoring and predicting climate-related trends and shocks, but the 2008 crisis was not caused by these factors. Malnutrition is typically reported as either per capita consumption or stunting rates of children younger than 5 years. Stunting is not easily observable and is also a result of long-term deprivation. Once national surveys detect increased malnutrition, children have already suffered the consequences of inadequate nutrition and hunger that are not easily (if at all) reversible. Besides, reductions in food consumption or rationing, and the resultant increased malnutrition rates, were a result rather than a predictor of the crisis.

In Africa, information sharing across line ministries is typically limited and disorganized, leading to long delays in publication and release of results of key surveys. Data collection and reporting functions are often separated from policymaking processes, and as a result governments and other stakeholders are unable to respond in a timely fashion to information on threats and risks to prevent crises. Food reserves at local, national, regional, and continental levels are often limited or nonexistent, eroding the capacity of a country to respond quickly to increase food supply and stabilize prices by releasing national stocks. The global nature of the crisis and self-protectionist government trade actions mean that neighboring countries are not able or willing to assist, and the international community is unable to respond in moving food to fill consumption gaps. As noted earlier, barriers to crossborder trade are sometimes significant, potentially exacerbating food shortages and worsening crises.

The capacity gaps that constrain adequate crisis prevention and preparedness also limit the depth and breadth of effective action during crises. The principle challenges during food emergencies center on coordination, logistics, and information management at several different levels. National governments are not always able to take the lead in setting the broad framework for emergency response or in prioritizing intervention modalities and locations. International agencies often face difficulties in translating commitments of support into concrete support and action on the ground. NGOs typically operate in tightly defined locations and are thus limited in their abilities to serve as focal points for broad-based coordination, logistics, and information management. Exacerbating this, considerable controversy and debate continue to surround the form in which assistance should be provided to communities in need (i.e., in-kind food versus cash). The appropriate response is necessarily context-specific, based on the nature of vulnerabilities in affected communities. However, reliable information about these differing contexts and forms of vulnerabilities is still lacking. Similarly, the impacts of given interventions on these vulnerabilities is poorly understood. This feeds the controversy and delays emergence of consensus principles and best practices. Weak information systems and poor coordination mean that inappropriate distribution of humanitarian resources is not uncommon.

Efficient responses to food emergencies depend to a large extent on having the ready resources and stocks to mobilize. The management of food reserves in Africa has been problematic; complicated management structures, overlapping responsibilities, poor management practices, and inadequate reserves make food reserves inefficient and ineffective in mitigating emergencies and crises [NEPAD 2003]. Lack of skilled managers and financial resources has crippled food reserve systems in Africa, yet the establishment, stocking, maintenance, and management of strategic reserves (at least a 2- to 3-month supply) is crucial for providing a market for overproduction, mechanisms for mobilizing reserves for rapid responses, and for promoting self-sufficiency and self-reliance.

Lessons for Emergency Responses with Regard to Food Security

As noted, the socioeconomic and political environment is an important determinant of whether sufficient food is available in a society, the degree to which individuals, households, and communities can gain access and effectively use that food, and the level of vulnerability to food insecurity. As such, there is a range of particular challenges to effectively addressing food insecurity and under nutrition at multiple levels. One particularly important area is prevention measures, which are generally less expensive than relief and recovery measures. Given the poor state of financial affairs in African countries, investment in risk-reduction strategies and programs is essential.

Capacity and organizational constraints are important in effective and efficient emergency food situations. However, only a small number of food security and nutrition professionals exist in most African countries with the shortage of trained professionals in nutrition, which is especially acute at local levels. In the absence of national bodies mandated to provide leadership, agencies working on such activities tend to define their own intervention packages. Technical oversight, supervision, and coordination during implementation are often insufficient. Without clear coordination mechanisms to harmonize and integrate approaches and interventions, synergies between activities implemented by different agencies have often proven difficult to realize.

A key challenge regarding coordination of FAFS and Pillar III activities is the lack of policy frameworks and poor definition of mandates and responsibilities for the various sectors and agencies that have a role to play in enhancing food security and nutrition. However, attaining these objectives requires a range of actions that is not neatly circumscribed within a single sector or easily fit into the bureaucratic organization of government. Fragmentation and limited communication across these agencies and the programs that they are implementing further limits the impact. The hierarchical and sector-specific organization of government structures contributes to limited communication and, more importantly, limited sharing of experiences and new technologies in addressing food insecurity and malnutrition within a country or region. As such, high-level political directives and pressures are often needed for effective action to be mounted by the various sectors and agencies concerned to improve food security sustainability and to combat malnutrition effectively.

Although food aid may sometimes be necessary to achieve food security, it is never sufficient. Risk management should be mainstreamed in countries to ensure cost-effective preparedness and capacity for rapid responses that might include food aid as part of a package of possible responses. Comprehensive risk assessments at national, district, and community levels need to be followed by the formulation of risk-reduction strategies at all administrative levels. These systems must be broad enough to include atypical shock information to detect such disasters in a timely manner.

Proactive review and use of alternative instruments is also needed to deal with crises, including food and financial reserves, weather-based insurance, and futures options. Objective criteria for selecting resource transfer modalities focusing on in-kind food and cash transfers are needed to ensure effective targeting. Careful management of strategic reserves is essential for ensuring local food emergency stocks for rapid and economic responses. Integration of local capacities and coping strategies into national and regional crisis preparedness strategies is needed to ensure intersectoral collaboration in crises.

Food security and nutrition are seldom integrated into national development agendas. Responsibilities for these issues within the public sector are typically unclear. Resources for programs to improve food security and overcome malnutrition are often insufficient, with budget constraints resulting in shortfalls in material supplies, trained workers, training and supervision, and monitoring and evaluation. Moreover, there is little coordination of action and use of resources among agencies. With no clear responsibilities established on food security and, especially, nutrition issues, conflict rather than cooperation is likely to characterize the relationships between agencies and sectors of government.

In taking action to improve food security and reduce malnutrition, there is a pressing need to address some of the institutional barriers in the public sector that often make such efforts far less effective than they otherwise might be. The food crisis underscores the need to improve the overall food and nutrition security in countries in a more systematic and sustainable manner, and this means that responses have to look beyond immediate emergency response.

Many countries have now developed comprehensive short to medium term action plans under the Initiative on Soaring Food Prices response missions launched by the AUC and NEPAD's [2009] CAADP, together with the Rome-based food agencies [Food and Agriculture Organisation, World Food Programme, and International Fund for Agricultural Development], and, in some instances, with support from the World Bank under the food price crisis response trust fund. The proposed interventions aim at rapidly increasing food production and availability to alleviate food supply constraints and reduce pressure on food prices. At the same time, the various interventions, policies, and programs aim to improve producer access to inputs, including water, seeds, and fertilizer, and in the interim provide food safety nets to the poorest and most vulnerable people who have been most exposed to price shocks. To ensure coherence and sustainability of the initiative, the country action plans led by the Initiative on Soaring Food Prices need to be integrated into wider national programs, policies, and existing partnerships.

Two main areas have been observed as missing or weak in the short-term to medium-term action plans developed and these areas require attention. First, strategies on information management and how to improve early warning systems are not clearly spelled out. It is not clear how countries will improve emergency preparedness and response mechanisms. Second, responses in the majority of countries are silent on crosscutting issues such as HIV/AIDS and gender. Medium-term to long-term responses should take cognizance of the fact that women constitute the majority of the poor and vulnerable and account for a significant proportion of the rural population and the majority of small scale and subsistence farmers in Africa. The AIDS epidemic is widespread across southern and eastern Africa in particular, and negatively impacts agricultural development.

Improved risk management at the household, community, national, and regional levels will inform decisions that ultimately impact building and protection of assets and investments, strengthen national, regional, and community responses to climatic crises, and mitigate against economic shocks that risk and undermine the coping mechanisms of vulnerable populations. Strategic policy toward agricultural growth will have influence over the design of programs in sectors beyond agriculture, ensuring that all policies protect and further the agricultural growth agenda and its broad-based benefits. Fluid food distribution networks are necessary to channel farm surpluses to deficit households and zones. In addition to widely publicized hunger hot spots, Africa is endowed with a broad range of surplus production zones. Often these food security-enhancing hot spots emerge in regions with good water management or where substitution possibilities among multiple food staples permit flexible supply responses. However, national boundaries, wide rivers, and manmade impediments such as tariffs and export restrictions cut across natural market sheds, impeding the free-flow of food from surplus to deficit zones. Poor infrastructure, lack of appropriate storage, lack of harmonized grades and standards, and cumbersome border procedures further restrain trade. Lack of adequate and efficient food reserves prevents effective and rapid mobilization of food stocks in emergencies.

In the long-term, social protection will be effective when delivered as predictable public support to targeted households through an instrument that is best suited to the local circumstances (i.e., sometimes cash transfers may be preferable over in-kind transfers and vice versa). Small, strategic, and carefully targeted transfers can meet the immediate consumption needs of vulnerable and destitute households, buffering them from asset depletion and adoption of coping strategies that undermine the long-term resilience of households. Although transfers in emergencies and to destitute households aim to smooth consumption, they will not shift households out of poverty. Productive safety nets not only smooth consumption but provide productive investment through conditional transfers that provide linkages and pathways out of poverty through improved agricultural productivity, education, or health care. However, to be successful, social protection programs need to be delivered on a large scale to maximize economies of scale.

Increasing the supply of food through increased production and improved market linkages will increase the food available to households and communities. Strategies to increase the production of staple commodities are also more likely to impact poor small farm holders, increasing their incomes and extending the geographic reach of markets to underserved areas. Increased and improved agricultural productivity are necessary to achieve CAADP's poverty reduction and food output targets, while at the same time reducing production costs and food prices for the poor.

Key Messages and Actions

It is clear that the food price crisis proved to be an unmitigated disaster for many communities in Africa, particularly as the shock exposed deep structural vulnerabilities of food systems. The responses detailed here, many of which initially emerged in the FAFS, resonate with creating a coherent, integrated system that would better withstand such manmade disasters in the future. The key philosophy of facilitating long-term regional collaborations on issues of national disaster risk reduction will go a long way to build resilience.

Careful policy action can alleviate the current crisis, poverty, and hunger and reduce the chances of another such crisis in the future [Von Braun 2008]. Responses to the crisis must therefore accomplish two tasks: they must address the immediate food needs of the poor and correct previous failures in agricultural policy by investing in agriculture and food production [Von Braun 2008].

Long-term food security and the ability of a country to mitigate the effects of future shocks will depend on the adequacy of national food reserves or the means with which to purchase food to stabilize prices and satisfy domestic demand. National food reserves need to be established and incentives created for farmers to supply these reserves. Biofuel policies that prohibit the use of key staple food crops for alternative energy supplies need to be explored to protect food supplies, create better price stability, and restrict the unnecessary and often inefficient use of such crops for fuel.

However, in the short-term, governments will need to provide social protection for the very poor and vulnerable to ensure that they are able to access food, engage in productive activities, and ultimately participate in the agricultural growth agenda in either the farm or nonfarm sector. Investing in social protection will require resources that detract from investment in the longer-term agricultural development agenda but are absolutely necessary to get a country onto the growth path. Carefully designed programs that include productive safety nets (subsidies, provision of inputs, and asset transfers, including livestock where appropriate), specially targeted unconditional cash or food transfers to particularly vulnerable groups (typically young children, women, the elderly, and chronically ill persons), and conditional cash or food transfers that create incentives (such as food-for-public works, school-or-food, and health-for-work programs) have a critical and necessary place in effective responses to the high food price crisis.

The current crisis has highlighted the need for sound information systems that are able to provide timely information and predict future shocks to enable governments to make wise decisions in response to rapidly changing situations and volatile markets and to continually review the impact of any response. National food security information systems need to be strengthened and developed to provide relevant, quality, and timely information.

Acknowledgments

The authors thank Professor Josephine Kimaba, Edna Kalima, and Wezi Chunga for collating the country plans, country responses, and data on Africa's responses to high food prices.

References

African Union Commission (AUC) and New Partnership for Africa's Development (NEPAD). 2009. *The Comprehensive African Agriculture Development Programme (CAADP) Framework for African Food Security.* Pretoria, South Africa: NEPAD.

Association for Strengthening Agricultural Research in Eastern and Central Africa. 2008. *Responding to the food price crisis in Eastern and Southern Africa: Policy options for national and regional action.* Retrieved from http://www.asareca.org/resources/reports/resp2food_pr_main.pdf (accessed on January 20, 2011).

CGIAR Collective Action News. 2009. *The blog of the CGIAR Systemwide Program on Collective Action and Property Rights (CAPRi).* Retrieved from http://www.capri.cgiar.org/caprinews/ (accessed August 4, 2009).

Cleaver, K. M., and Schreiber, G. A. 1994. *Reversing the spiral: The population, agriculture and environment nexus in sub-Saharan Africa.* Washington, DC: The World Bank.

Economic and Social Council. 2005. *President's summary at the special event on "Food Crises in Africa." ECOSOC Chamber Thursday, October 27, 2005.* New York: UN Economic and Social Council.

Economic Communities of West African States (ECOWAS) and Comité Permanent Inter Etats de lutte contre la Sécheresse dans le Sahel (CILLS). 2008. Memorandum on rising food prices: Situation, outlook, strategy and measures advocated. Experts Meeting, Abuja, Nigeria, May, 2008.

European Commission. 2001. Emergency assistance to the vulnerable groups suffering from food shortages in Malawi. Retrieved from http://ec.europa.eu/echo/files/funding/decisions/2005/dec_malawi_01000.pdf (accessed on January 20, 2011).

Food and Agriculture Organisation. 2008a. The state of food insecurity in the world 2008: High food prices and food insecurity—Threats and opportunities. Retrieved from ftp://ftp.fao.org/docrep/fao/011/i0291e/i0291e00.pdf (accessed September 4, 2009).

Food and Agriculture Organisation. 2008b. *Country responses to the food security crisis: Nature and preliminary implications of the policies pursued.* Report prepared for FAO: Rome.

Food and Agriculture Organisation. 2008c. Food and Agriculture Organisation's Initiative on Soaring Food Prices: Guide for immediate country level action. Retrieved from http://www.iaahp.net/iaahnews/detail/en/news/6535/icode/7/?no_cache=1-13k-pdf (accessed February 21, 2009).

Haggblade, S. 2007. Alternative perceptions of the rural nonfarm economy. In *Transforming the rural nonfarm economy: Opportunities and threats in the developing world,* eds. S. Haggblade, P. B. R. Hazell, and T. Reardon, 25–54. Washington, DC: International Food Policy Research Institute.

Henao, J., and Baanante, C. 1999. *Nutrient depletion in the agricultural soils of Africa. 2020 Brief 62.* Washington, DC: International Food Policy Research Institute.

Hendriks, S. L. 2005. The challenges facing empirical estimation of food (in)security in South Africa. *Development Southern Africa* 22:103–123.

International Food Policy Research Institute. 2008. Global food crises: Monitoring and assessing impact to inform policy responses. IFPRI Issues Brief Number 55, October. Retrieved from http://www.ifpri.org/sites/default/files/publications/ib55_0.pdf (accessed September 5, 2009).

International Strategy for Disaster Reduction. 2005. *Hyogo Framework for Action 2005-2015: Building the resilience of nations and communities to disaster.* Geneva: UN Inter-Agency Secretariat of the International Strategy for Disaster Reduction.

Minde, I. J., Chilonda, P., and Sally, H. 2008. *Rising global food prices—Policy challenges and options for southern Africa.* Pretoria, South Africa: ReSAKSS.

Munga, B., and Wambugu, A. 2009. Growth, poverty and income inequality in Kenya: Suggested policy options. Retrieved from http://zunia.org/post/growth-poverty-and-income-inequality-in-kenya-suggested-policy-options-may-2009/?rank=d&cHash=2b63fffb8b922953de40c0ecbe8d619b (accessed on January 20, 2011).

Mwaniki, J. n.d. The impact of informal cross border trade on regional integration on SADC and implications for wealth creation CORN. Retrieved from http://www.streetnet.org.za/wp-content/pdf/CORN.PDF (accessed on January 20, 2011).

New Partnership for Africa's Development. 2003. *Comprehensive Africa Agriculture Development Programme.* MidRand, South Africa: NEPAD.

Philip, D., and Rayham, I. 2004. *Vulnerability and poverty: What are the causes and how are they related?* Term paper for Interdisciplinary Course, International Doctoral Studies Programme at ZEF, Bonn.

Quinsumbing, A., Meizen-Dick, R., Bassett, L., Usnick, M., Pandofelli, L., Morden, C., et al. 2008. *Helping women respond to the global food price crisis,* IFPRI Policy Brief 7. Washington, DC: IFPRI.

Rivera, W. M., and Qamar, M. K. 2003. *Agricultural extension, rural development and the food.* Retrieved from http://www.fao.org/DOCREP/006/Y5061E/y5061e08.htm (accessed September 4, 2009).

Sanogo, I. 2009. The global food prices crisis and household hunger: A review of recent food security assessments. *Humanitarian Exchange Magazine.* Retrieved from www.odihpn.org/report.asp?id=2988 (accessed September 4, 2009).

Save the Children Fund. 2008. *Country profile: Rwanda.* Retrieved from www.savethechildren.org.uk/en/docs/Rwanda_CB_2008.pdf (September 4, 2009).

Staaz, J. 2000. *Strategic pathways and interactions to cutting hunger in half in Africa.* East Lansing, MI: Michigan State University.

United Nations. 2000. Millennium development goals indicators. Retrieved from http://mdgs.un.org/unsd/mdg/SeriesDetail.aspx?srid=580 (accessed September 4, 2009).

United Nations Development Programme. 2008. Human and income poverty: Developing countries. Retrieved from http://hdr.undp.org/en/media/HDR_20072008_Table_3.pdf (accessed on January 20, 2011).

United Nations Development Programme in Mauritius and Seychelles. 2007. *Achieving the MDGs and reducing human poverty.* Retrieved from http://un.intnet.mu/UNDP/html/mdg_1.htm (accessed September 4, 2009).

United States Agency for International Development. 2009. *Global food insecurity and price increase: Situation Report #1, Fiscal Year (FY).* Washington, DC: United States Agency for International Development.

Von Braun, J. 2008. *Responding to the world food crisis: Getting on the right track. IFPRI Annual Report Essay.* Washington, DC: IFPRI.

Von Grebmer, K., Fritschel, H., Nestorova, B., Tolulope, O., Pandya-Lorch, R., and Yohannes, Y. 2008. *The Global Hunger Index: The challenge of hunger.* Retrieved from http://www.ifpri.org (accessed September 4, 2009).

Wik, M., Pingali, P., and Broca, S. 2008. Background paper for the World Development Report 2008: Global agricultural performance—Past trends and future prospects. Report prepared for the World Bank. Retrieved from http://siteresources.worldbank.org/INTWDR2008/Resources/2795087-1191427986785/Pingali-Global_Agricultural_Performance.pdf (accessed on January 20, 2011).

World Bank. 2009. *From agriculture to nutrition: Pathways, synergies, and outcomes.* Washington, DC: World Bank.

World Food Programme. 2009. Global emergency food aid deliveries. Retrieved from http://www.globalpolicy.org/images/pdfs/Z/emergencyregions1996-2007.pdf (accessed on January 20, 2011).

Chapter 8

Faith-Based Humanitarian Assistance in Response to Disasters: A Study of South African Muslim Diaspora

Sultan Khan

Contents

Introduction

Faith-based organizations (FBOs) have historically served an important role in responding to humanitarian needs during disasters and continue to do so. They comprise a considerable share of the nongovernmental organizations (NGOs) sector and are characterized by one or more of the following: affiliation with a religious body, a mission statement with explicit reference to religious values, financial support from religious sources, or a governance structure where selection of board members or staff is based on religious beliefs or affiliation and decision-making processes are based on religious values [Ferris 2005].

FBOs have several characteristics that distinguish them from their secular peers. Elaborate spiritual and moral values underscore their organizational aims and objectives, which result in a significant ability to mobilize adherents who are otherwise estranged from secular development discourse. They are highly networked, both nationally and internationally, and engage politically through processes of governance in both horizontal and vertical terms. FBOs are also less dependent on donor funding and have well-developed capacities, local networks, and expertise in the key areas of development practice. Because of this, they are stakeholders in the development process and have recently drawn significant attention in development policy. Governments have become increasingly aware that the world's largest faith-based NGOs actually provide more aid than do some donor governments, are active in more countries than many governments, and carry greater credibility among taxpayers than do government aid agencies [Ferris 2005, p. 311]. Because development is a normative ideal and moral cause having much in common with the faith discourses from which it has traditionally remained aloof, the importance of FBOs cannot be ignored [Clark 2006].

The capacity of FBOs to mobilize resources is considered greater than the nation state's departments' capacity and provides even more aid during times of disasters than many governments. This greater capacity is due to a great deal of giving inspired by religious beliefs. Different religions emphasize that giving is associated with a deep sense of responsibility, duty, and commitment to realize spiritual salvation. *Zakah*, *tzedalah*, *dana*, and *tithes* are forms of religious compulsory giving for Muslims, Jews, Hindus, and Christians, respectively, that are encouraged by religious leaders of all faiths [Habib et al. 2008]. These beliefs cultivate the habit of charitably giving time and money to faith-based and secular initiatives [Habib et al. 2008].

FBOs are active in most countries that are afflicted by disasters caused by natural hazards, persecution, and war [Redwood-Campbell 2008]. This may be attributed to the end of communism, which enabled the spread of democracy in poor countries, technological change, and economic integration. In short, globalization has created the opportunity for NGOs to increase at a broader level. According to the *Yearbook of International Organizations*, there were approximately 26,000 international NGOs by 2000, compared with 6,000 in 1990 [Ferris 2005]. The increasing number of humanitarian organizations and the wide range of activities that they engage in make any universal typological analysis of its nature and structure difficult. Similarly, it is difficult to draw any typology of FBOs considering the diversity of humanitarian activities that each one pursues [Smith and Sosin 2001].

Although numerous FBOs concentrate their activities on the local level, many tend to make significant contributions within the international humanitarian community. This tradition was instated before International Humanitarian Law formalized the rules of engagement in consolidating relief efforts among different stakeholders in disaster-afflicted environments.

Despite their record in humanitarian aid work, the contributions made by FBOs during times of disaster and development have largely been ignored. However, since the 1980s, the relationship between religion and development has received some scholarly attention [Marshall 2001; Selinger 2004; Ver Beek 2002]. At the turn of the millennium, research interest about the relevance of faith in developmental-related activities increased [Kniss and Campbell 1997]. Notwithstanding the increased scholarly attention that FBOs are receiving, the study of FBOs in development-related contexts can be complex due to the varying nature, size, scope, ideology, and structure, and even mask covert proselytization objectives in the name of humanitarian work. Therefore, the prospect of a universal typology of FBOs emerging through research is an ambitious study. However, the Faith and Technical Education Network [2003] provides a broad typology that classifies FBOs into faith-permeated, faith-centered, faith-affiliated, faith-background, and faith-secular partnership programs to help differentiate the work of different FBOs.

This chapter examines South African Muslim FBOs' humanitarian assistance within the broader international context of the NGO sector, especially during a disaster. A brief history of the FBO's evolution among South African Muslims will be presented and followed by a discussion of trends among Muslim FBOs in response to regional and international disasters. The motivations for such a response, the analyses of the organizational methods used to mobilize resources, the deployment of mercy workers to disaster-afflicted areas, the suitability of resources to disaster victims, the adaptation and provision of aid to disaster conditions, and the interaction between humanitarian FBOs in the region, government institutions, and other international disaster aid organizations will be discussed.

The chapter is informed by several methodological approaches and references, including original sets of data gathered by the author that arose from a national study about the state of philanthropy among South African Muslims from 2004 to 2006.* This data has been supplemented by a study of the profiles of humanitarian

* The data were reanalyzed to inform this chapter. The initial technical report focused on philanthropic resource flows in general within the South African Muslim community. New and more recent sets of data are introduced to strengthen the analysis of Muslim FBOs in humanitarian aid activity.

aid organizations in the country followed by additional interviews with three leading Muslim FBOs engaged in local and international humanitarian aid missions to gain insight on the organizational methods used to mobilize and distribute resources to disaster-afflicted areas, the suitability of resources to disaster victims, adaptation and provision of aid to disaster conditions, interaction between and among FBOs in the region, government institutions, and other international aid organizations.

A Brief Historical Context of Muslim FBOs in South Africa

The early history of South African Muslims may be traced to their enslavement and indenturement under British colonialism. The arrival of Muslims in South Africa may be traced to Muslim soldiers in 1652 in the army of the Dutch East India Company known as Mardyckers [Nadvi 1988; Naude 1992; Randeree 1997]. However, these Muslims were prevented from overtly practicing Islam and never had a long-term presence in the country. A more permanent presence of Muslims in South Africa began in 1667 in the Cape with the arrival of slaves from Malaysia and the East Indies at the behest of the Eastern Batavian Dutch Empire [Islamic Council of South Africa 1984; Nadvi 1988; Randeree 1997]. After the abolishment of slavery in 1833, Islam became a flourishing religion and by 1840 had gained approximately 6,435 adherents, equivalent to one-third of the total colony population [Nadvi 1988]. A separate stream of Muslims from the subcontinent of India began to arrive in South Africa in 1860 to work on the sugar plantations of the Natal colony as indentured laborers [Meer 1980; Ramphal 1985]. These Muslims came from Malabar, located on the western coast of South India and Hyderabad, located in the south. Around this time, a third group of passenger Muslims—mainly Sunni Vhoras from Surat and Memons from Kathiwad and Kutch—followed their indentured counterparts. Their history and origin varies from the indentured Muslims. The passenger Muslims consisted of the merchant class and were referred to by the colonialists as Arabs because of their distinct code of dress.

In 1873, a fourth group, Zanzibaris,* also supplemented the Muslims of Indian origin in Natal. Zanzibaris were descendents of Muslim slaves and were brought to Natal between 1873 and 1880 to alleviate the labor problem besieging the colony. They originated largely from northern Mozambique, Portuguese East Africa, Malawi, and Zanzibar, and mainly settled in Kingsrest on the Bluff in Durban [Randeree 1997]. From the historical evolution of Muslims in South Africa, its

* Muslims of African origin were referred to as Zanzibaris, another racial definition to isolate African Muslims from the broader community.

social composition is heterogeneous in nature and reflects the same diversity that characterizes the rest of the South African population.

Despite differences in the historical evolution of the Muslim community in South Africa, the colonialist and subsequent apartheid negatively impacted the social well-being of Muslims, but more severely affected Muslims with a history of indentured labor and slavery [Swilling and Russell 2004]. During their early evolution, poverty and related social problems necessitated a philanthropic response from the more economically privileged sectors of the community, which resulted in the provision of community infrastructure, including places of worship, orphanages, old age homes, schools, social welfare services, cemeteries, and religious institutions. The socioeconomic hardships of the earlier times prompted the emergence of FBOs that acted as safety nets to preserve the different facets of the Islamic way of life.

One of the most important foundations that FBOs established in the country was the institution of charity, which is most revered among adherents of the Islamic faith. Regularly contributing to charity is an affirmation of faith for Muslims and is institutionalized through rigid religious rules and practices. The fourth article of faith prescribes that all individuals professing to be Muslims are religiously obliged to contribute 2.5%* of their annual surplus earnings toward the betterment of their community and those afflicted by social and economic misfortunes. The institution of charity among Muslims is considered an act of building community through faith and, concomitantly, building of faith through the deepening of community. Therefore, Muslims, in their conception of faith and community, subscribe to the notion that humans are linked to one another through their obligations to God [Alterman and Hunter 2004] and the act of giving to humanitarian causes is central to their religious belief.

As South African Muslims advanced socially and economically, their community resource investment increased into different forms of social capital. The growth of the population not only made the availability of different forms of charity possible, but it also helped recapitalize its social capital through investments in community infrastructure. This investment resulted in the creation of further resources to advance different aspects of humanitarian work. For example, community centers are used to house refugees and disaster victims as a repository for the collection and distribution of food and clothing to afflicted areas and a space for humanitarian aid volunteers to meet and organize their mercy missions. The period characterized by colonialism and, subsequently, apartheid may be defined as one of social capital accumulation pursued through FBOs to meet certain structural challenges to

* It must be noted that *Zakah* is a compulsory act of charity and is the fourth pillar of Islam. Apart from *Zakah*, there are other optional charities (i.e., *Lillah*, *Fitra*, and *Sadaqa*) that become due for specific religious purposes and situations. However, *Zakah* constitutes the primary act of charity.

Table 8.1 Distribution of Muslim Faith-Based Organizations by Places of Worship and Social and Educational Institutions by Province

Number	*Province*	*Places of Worship*	*Social Institutions*[a]	*Community Organizations*	*Total*
1	Free State	10	6	8	24
2	Gauteng	122	128	108	358
3	Mpumalanga	35	8	22	65
4	North West	25	6	19	50
5	Northern Cape	4	2	3	9
6	Western Cape	111	172	108	391
7	Eastern Cape	21	12	20	53
8	Northern Province	11	6	11	28
9	KwaZulu-Natal	116	125	109	350
TOTAL		455	465	408	1328

Source: Data from Davids, M., *Directory of Muslim Institutions and Mosques in South Africa*, Johannesburg, South Africa, Islimu Publications, 1997; Khan, S., and Ebrahim, A. F. M., in *Religious Pluralism in the Diaspora*, vol. 4, ed. P. T. Kumar, Neatherlands: Brill, 2006.

[a] A *social institution* is defined as organizations comprising madressas, theological training centers, Islamically oriented preschools, crèches, schools, and community centers.

reproduce itself as a community. Table 8.1 provides a breakdown of FBOs in South Africa 3 years after democracy was established.

When one compares the number of FBOs serving the Muslim community in different sectors and parts of the country against the population,* the ratio of FBOs serving the Muslim population is 1:489, which suggests that, as a minority

* The actual population of Muslims in South Africa cannot be accurately ascertained. Crude estimates suggest that the population could have ranged from 500,000 adherents to more than 1,000,000 during the 1990s [Naude 1992; Randeree 1997; Tayob 1998]. The lack of accurate data about Muslims in the country during the apartheid era resulted from a complicated and cumbersome race classification system, which inaccurately identified individuals' religious beliefs. In recent years, this has been partly addressed with the two postapartheid censuses occurring in 1996 and 2001 [Census Data 1996, 2001], which recorded the religious affiliation of South Africans across all race groups. It put the Muslim population at 553,717 and 654,063, respectively. This serves as an important official baseline for any demographic analysis of Muslims in South Africa. The third census is scheduled to take place in 2011.

community, it is more than adequately serviced by a variety of social institutions necessary for its continued existence and reproduction.

The extent of giving among Muslims is difficult to quantify based on the different types of mandatory and nonmandatory giving. Attempting a crude estimate of the community's charitable monetary contributions is statistically daunting considering the limitation imposed by the lack of reliable information about the community's size and organizations' reluctance to provide vital information about finances and the personal nature of giving to humanitarian causes. Nonetheless, authoritative sources within the community estimate that the population ranges between 600,000 and 1,000,000. Guided by calculations formulated by the South African National Zakaat Fund, a leading FBO in the country, some quantitative

Table 8.2 Forms and Frequency of Humanitarian Assistance

	Frequency of Charitable Contributions (%)			
Form of Charity	*Regularly*	*On Auspicious Religious Occasions*	*Special Projects and Events*	*Whenever Personal Circumstances Permit*
Cash	**63.0**	7.3	2.9	26.3
Feeding schemes	**46.2**	9.6	23.1	21.1
Household goods	**34.5**	5.0	13.8	46.6
Clothing/blankets	**33.8**	5.0	26.3	35.0
Food hampers	**31.0**	28.7	16.1	24.1
Medicine	**31.3**	6.3	15.6	46.9
Voluntary service requiring no specific skill	16.1	1.3	8.7	11.7
Books, school uniforms, and stationery	12.7	1.3	4.7	13.3
Professional service	11.3	1.3	6.0	11.3
Technical expertise	5.3	0	3.3	8.7
Raw material/ hardware	2.7	0	4.7	11.3

Note: Data in boldface indicates the most frequent forms and frequency of humanitarian assistance by the study sample.

insight into the monetary value of mandatory charity (*Zakah*) can be ascertained [SANZAF Biennial National Conference 2004]. The projected cash value of this form of charity is estimated at R300 million per annum. This estimate is based on the assumption that there are 1 million Muslims in the country and an average four members per household. Therefore, the total number of Muslim households is estimated at 250,000. Of this number, 50,000 are households that are assumed to be in possession of cash assets less than R100,000 (one-fifth of the household population size). When these population and income estimates are factored into calculations using the 2.5% mandatory charity formula (*Zakah*), a figure of R300 million for mandatory forms of charity is extrapolated. It must be noted that this estimate is conservative and its calculation is primarily based on one level of household income and excludes other religious charities, those that accrue from Muslim businesses, and the corporate sector.

The forms and frequency with which South African Muslims give to humanitarian causes are represented in Table 8.2.* Cash forms of giving are most popular (63%), with approximately one-third of charity contributions comprised of basic consumables (e.g., food, household goods, medicines, and clothing).

Trends among Muslim Philanthropic Organizations Responding to Regional and International Disasters

The political and economic isolation of South Africa during apartheid meant that different diasporic communities were marginalized from their ancestral links and networks internationally. Similarly, Muslims and FBOs that represented their social, political, and economic interest developed a narrow intracommunity development orientation considering the antidevelopment atrocities perpetrated by colonialism and apartheid on the country's entire Black population. Muslims focused on strengthening community life through various social and economic investments and forging links with other diasporic communities on common issue of social coexistence.† Consequently, little attention was devoted to humanitarian activities outside the countries' borders. The political restrictions placed by apartheid were a further impediment to any humanitarian engagement. However, the Russian inva-

---

* This table is drawn from the original study conducted by the author on philanthropy among South African Muslims between 2004 and 2006. A random sample of 150 respondents comprising 50 male and female respondents from the three major South African cities (Durban, Gauteng, and Cape Town) from which a database of 1,200 regular donors was drawn.

† South African Muslims have a history of building common social infrastructure (e.g., schools, community halls, and social welfare agencies) for the common good of their non-Muslim Indian and colored counterparts. The term *colored* was the official South African racial classification for the country's mixed descent inhabitants during the apartheid era.

sion of Afghanistan in the late 1970s increased Muslim FBOs' interest in disaster and development-related international work [Osman 2003].

Beyond the Afghan conflict, other global forces served as catalysts for the increase in Muslim FBOs' international humanitarian aid focus. For example, the decline of communism and the end of the Cold War fuelled the rise of politics centered on new blends of ethnic, cultural, and religious identity. Identity politics bolstered another form of development: the emergence or revitalization of public religion, which resulted in religious leaders and organizations becoming more willing to enter the public stage and highlighting the moral and spiritual importance of public policy [Clark 2006]. Therefore, humanitarian aid work provided a morally acceptable platform through which FBOs could publicly pursue their religious objectives.

Another wave of Muslim NGO activity took place from 1992 to 1993, during the war in Bosnia. This trend was also noted with South African FBOs. During this time, apartheid rule was showing signs of unsustainability and the prospect of a new democracy was becoming eminent. At the same time, communism was being dismantled and hopes of South Africa's reentry into the global stage were closing in. This period of political change was exemplified by the substantial humanitarian aid provided by South African Muslims at an international level. Under the newly established humanitarian aid group, Gift of the Givers Foundation, South African Muslims provided aid to Bosnia, Herzogovena, and Chechniya.

Over the years, South African Muslim's aid for international disasters has increased and their solidarity with nation states in times of distress has been noted. A national study that focused on the state of philanthropy among South African Muslims was conducted by the author and revealed that 24% of South African Muslims provided monetary resources during natural disasters and humanitarian crises outside of the immediate community. This finding suggests a trend among South African Muslims to direct humanitarian resources to causes outside of their immediate community.

Another trend with South African Muslims and their response to international and regional disasters is the emergence of FBOs that specialize in humanitarian work. Muslim FBOs that have emerged with a humanitarian focus since the country's years of democracy included Crescent of Hope (founded in 1992), Gift of the Givers Foundation (founded in 1992), and Islamic Relief South Africa (founded in 1994). Older and more established FBOs also entered the field of international and regional humanitarian aid work during the same period. The Africa Muslims Agency (founded in 1982) has a continental focus and is known to undertake development-related work in 30 African countries. Similarly, the Islamic Medical Association (founded in 1980) provides volunteer medical professional services during mercy missions in disaster-afflicted regions. The Al-Imdaad Foundation (founded in 1980) has a well-established focus on Africa and provides long-term development assistance. Table 8.3 provides a comparison of three South African Muslim FBOs' responses to international and regional disasters since the inception

Table 8.3 Trends among South African Muslim Faith-Based Organizations in Responding to International and Regional Disasters

International and Regional Disaster Areas	*Crescent of Hope*[a]	*Gift of the Givers*[a]	*Islamic Relief South Africa*[b]
Afghanistan	X	X	X
Africa	X	0	X
Bangladesh	X	0	X
Bosnia	X	X	X
Goma	0	X	0
India	X	X	X
Indonesia	X	0	0
Iran	0	X	0
Iraq	0	X	X
Kashmir	X	0	0
Kosovo	0	X	X
Lebanon	0	X	X
Mozambique	X	X	0
Niger	0	X	X
Pakistan	X	X	X
Palestine	X	0	X
Sri Lanka	0	X	X
Somalia	X	X	X
Sudan	0	X	X
Turkey	X	0	0

Note: Information in the table was adapted from the organizations' Web sites. Trends among South African Muslim faith-based organizations in responding to international and regional disasters are highlighted with the symbol X, which denotes a positive response. A 0 denotes no response.

[a] Established in 1992.
[b] Established in 1994.

of democracy in these countries. It should be noted that both the Crescent of Hope and Gift of the Givers are South-African-based, while Islamic Relief South Africa is a chapter of Islamic Relief Worldwide, an international humanitarian aid organization with a strong focus on Albania, Herzegovinia, Chechniya, China, Yemen, Jordan, Indonesia, and Africa (Chad, Ethiopia, Kenya, Malawi, and Mali).

Although these are specialized regional and international relief giving organizations, the mobilization of resources during times of disasters is a community activity that involves the support of other smaller FBOs in different areas of the country. Given the extent and size of Muslim FBOs in the country and the administrative and volunteer infrastructure at the disposal of more established FBOs, networking and mobilizing resources during disasters is executed within hours of a disaster occurring. Therefore, as a community, South African Muslims' response to disasters is spontaneous and this readiness may be attributed to the credibility of the networks established in the community by regional and international humanitarian aid organizations.

Motivation for Responding

Because a large portion of charities solicited by Muslim FBOs are religiously motivated, stringent rules and regulations guide the manner in which the receiving organization can expend these resources. For example, it is mandatory that the receiving organizations of charities in the form of *Zakah* expend this within the Islamic year for which it becomes due. In addition, custodians of religiously mandated charity are obliged to honor the giver's wishes should any precondition be attached to the donation. Therefore, charities received with a particular religious prescription are administered accordingly by humanitarian aid organizations. The application of these religious prescriptions on the administration of the different forms of Islamic charity is universal and adherence to it is essentially unambiguous and uniform. It is through conformity to these religious prescriptions that the credentials of charitable organizations are established, which is a strong source of motivation for Muslims to support a particular humanitarian cause. This is supported by the findings depicted in Table 8.4,* which confirm that 45.5% of South African Muslims give to organizations that have a credible record for meeting important social needs among Muslims.

A second motivation for South African Muslims to support a particular humanitarian cause relates to the level of identification that they have with the FBOs' goals. From Table 8.4, it may be gauged that a significant number (30.6%) of those

* This table is drawn from the original study conducted by the author on philanthropy among South African Muslims between 2004 and 2006 on the State of Giving undertaken in collaboration with the Centre for Civil Society, University of KwaZulu-Natal, Durban, South Africa. It will be noted that although much of the charitable resources is religiously mandatory, only 6% of respondents were motivated by strong religious reasons.

Table 8.4 Motivation for Supporting a Particular Humanitarian Cause

Factors Motivating Support for a Charitable Cause	%
Strong personal identification with the goals of the organization or institution	30.6
The organization has a credible record in meeting important social needs in the Ummah	45.5
Historical and/or traditional association with the organizations' activities	0.7
The organization is within my locality and deserves my support	16.4
The important status, competence, and reputation that these organizations have established over the years in promoting sociopolitical and economic influence on matters concerning spiritual and religious priorities in the community	0.7
No particular reason—purely for the love of God	6.0
Total	100

that support humanitarian causes do so due to strong personal identification with the goals of the organization.

The Africa Muslims Agency is one of the more established regional humanitarian aid organizations in South Africa and is known to attract a large portion of the communities' charitable resources. Its aims and objectives are characterized by a strong sense of religiosity. The organization considers itself "the pacesetters of the Islamic Renaissance in the Continent of Africa," and claims credit for the "revival of Islamic, social, moral and intellectual learning and development" manifested through "the existence of the range of Institutions and activities initiated by the Agency across the length and breadth of the Continent of Africa" [http://www.africamuslimsagency.co.za]. The display of such a strong religious ethos may be viewed as technique to religiously solicit mandatory charity from individuals within the community who firmly adhere to the principle of serving humanity through charity.

Some Muslim humanitarian aid organizations' mission statements are defined by the religious prescriptions that broadly appeal to individuals who subscribe to the principles of aiding humanity in general. For example, although having a strong support base within the community, the Gift of the Givers Foundation seeks to network, engage, and collaborate with individuals of different cultures and religions via goodwill, harmonious coexistence, tolerance, and mutual respect. This practice aligns with the divine injunction embodied in the Holy Qur'an that states:

"O Mankind! We created you from a single (pair) of a male and a female, and made you into nations and tribes, that ye may know each other (not that ye may despise each other). Verily the most honored of you in the sight of Almighty is (he who is) the most righteous of you" [Qur'an 49:13, www.giftofthegivers.org]. With such an Islamic perspective, the Gift of the Givers Foundation's mission aims to appeal to a wide range of donor communities within the country. Similarly, both al-Imdaad Foundation and Islamic Relief South Africa are motivated by the Qur'anic verse that states "If you save the life of one human being then it is as if you have saved the lives of all human beings" [Surat Al-Mā'idah 5:32].

Organizational Method to Mobilize Resources

The organized nature of charity among South African Muslims has resulted in the establishment of many formal offices among FBOs with full-time personnel dedicated to meet the organizations' objectives. This trend is most prevalent in larger, more established FBOs. Smaller organizations with little or no formal infrastructure tend to largely depend on volunteer support to meet their organizational goals. Volunteer participation within organizations is heightened during the implementation of special projects, which includes disaster aid.

However, as FBOs mature and establish their humanitarian accomplishments for the community, there appear to be a greater likelihood for them to secure a stable set of donors over time. One example is the case of one FBO, South African National Zakáh Fund (SANZAF), which helps to mobilize resources for humanitarian aid organizations. Originating in KwaZulu-Natal more than 30 years ago with a little less than R20,000 during their first year of operation, SANZAF has grown considerably.* Today, the organization has 19 offices nationally and for the 2008 financial year, its annual charitable income was greater than R40 million. This organization is coordinated nationally and is served by full-time national and regional directorates who employ national and provincial fundraisers. SANZAF has a well-developed marketing and publicity network and draws on a wide range of volunteer expertise to mobilize resources on its behalf [Khan and Ebrahim 2006]. Similarly, humanitarian aid organizations such as the Gift of the Givers Foundation, Al-Imdaad Foundation, Africa Muslims Agency, and Islamic Relief South Africa have formal offices in major parts of the country. Over the years, many of these organizations' activities have grown, resulting in administrative expansion programs. For example, the Gift of the Givers Foundation initially started with one office located in Pietermaritzburg, the capital of KwaZulu-Natal. Currently, it has four other offices in the major centers of the country, with one based in Malawi. The Al-Imdaad Foundation also has registered offices in Jakarta, Indonesia, and in Amman, Jordan, where it is engaged in long-term development work.

Many established organizations use mailing lists to reach potential donors and sponsors. For example, one FBO reaches out to 14,000 households annually through

* According to the audited financial statement as of February, 1977, the total income for this period was R19 049.25.

the mail using cash deposit and debit order systems to solicit funds. The average income derived from this method amounts to a million rands per annum based on a response rate of approximately 800 donors. During the month of Ramadaan when most of the Islamic charity becomes due, it is not uncommon for South African Muslim's mailboxes to be flooded with letters of appeal for donations. During this holy month, most FBOs make a concerted effort to mobilize cash resources to support their organizational aims and objectives [Khan and Ebrahim 2006].

The use of debit orders by larger organizations to sustain regular donations comprising compulsory forms of charity spread over a period of time is also a widely used fundraising strategy. In 2004, the SANZAF Western Cape branch had 1,527 donors using the debit order system and generated R153,040. The organization reported a 24% increase compared to its 2002 financial year [SANZAF Biennial Report 2004]. This method of fundraising provides donors with an opportunity to spread their contributions over monthly installments and helps organizations secure stable resources for an operational year instead of being dependent on charities due during religious occasions.

Many larger FBOs, including the Gift of the Givers Foundation, Al-Imdaad Foundation, Islamic Relief South Africa, and Africa Muslims Agency, have well-established Internet Web sites that encourage giving through electronic banking and dedicated Web sites. In their general study of South African patterns of giving, Everatt and Solanski [2005] confirm that 1% of South Africans make their charitable contributions through this medium. This trend among the South African Muslims is also becoming a popular way of mobilizing resources. Personalized e-mail appeals are another form widely used by humanitarian aid organizations to request humanitarian aid resources. Mobile phones and text messages are also used to coordinate the call for resource sponsorships, delivery, and collection. For example, the Gift of the Givers Foundation sends approximately 4,000 text messages to their supporters to immediately mobilize resources when a disaster occurs.

The Muslim business sector also plays a role in the provision of tangible resources on short notice. Many Muslim businesses that produce and manufacture supplies needed in disaster conditions are able to adapt their production plants to meet the delivery schedules of humanitarian aid organizations. In many instances, the turnaround time is within 24 hours of a disaster occurrence. Often, these are produced and packaged according to the specification of the requesting aid organization and delivered by water, air, or road transport. The additional logistics and costs incurred in the production and distribution of such goods and services are often discounted.

The local written, audio, and visual media is widely used to promote the humanitarian activities of Muslim FBOs. Regular newsletters and letters of appeal are made on special development-related projects such as sponsorship of water boreholes, provision of schools and clinics, and feeding programs. The theological fraternity plays an active role in making a public appeal to support special humanitarian projects. Often these appeals are followed by mosque fund collections from the local congregation after weekly Friday prayers. Collection points for tangible donations are also arranged via schools and community-based organizations. Community radio stations profile the work of humanitarian aid organizations and often dedicate regular on-air time slots for such purposes.

The extent of resources collected for any given disaster relief situation in monetary terms is difficult to quantify because the nature and extent of disasters and aid resources needed vary. Further, given the number of Muslim FBOs involved in the mobilization and distribution of resources to disaster-stricken areas, a consolidated account of the total contribution made by South African Muslims requires extensive research. Based on the quantitative figures of resources provided by the Gift of the Givers Foundation alone, there is some indication about the quantitative dimension of resources provided by the South African Muslims in times of national and regional disasters. The following are some examples of the disaster relief provided by South African Muslims that illustrate the quantitative dimension of resources:

- Mozambique floods, 2000. Food, clothing, medicine, and shelter valued at R1.500,000 were air lifted to victims.
- India earthquake, 2001. R1,100,000 was spent on the reconstruction of schools, houses, clinics, and hospitals, and R100,000 in cash was made available.
- Goma volcanic eruptions, 2002. R1,000,000 was spent on food, medicine, blankets, and water purification tablets within three days of the occurrence of this disaster.
- Somalia, 2004. R3,000,000 of aid supplies including food, medicine, water purification tablets, and tents were provided.
- Tsunami Disaster, 2004. R12,000,000 of aid was provided to the different areas afflicted by this disaster. In Sri Lanka, a Gift of the Givers Foundation housing village comprising 500 homes at a cost of R6000 per unit was established [http://www.giftofthegivers.co.za].

Figure 8.1 (a) Relief aid to Goma during the 2002 volcanic eruption. (b) Disaster aid in volcano ravished Goma, 2002. (Courtesy Gift of the Givers Foundation.)

According to the Gift of the Givers official Web site, it estimates that it has provided R260,000,000 of aid in a 17-year period (since its establishment) to 23 countries, including South Africa.

Deployment of Mercy Workers to Disaster-Afflicted Areas

The deployment, recruitment, and training of mercy workers to disaster-afflicted areas vary among South African Muslim humanitarian aid organizations. For example, in the Gift of the Givers Foundation, mercy workers are classified into three categories. In the first category, volunteers are mobilized to pack resources. In the second category, volunteers are trained to undertake trauma counseling. In the third category, professionals are deployed on mercy missions. It has a stable cohort of 15 to 30 mercy mission workers consisting mainly of medical practitioners who venture into disaster hot spot areas. The foundation organizes special recruitment drives to find volunteers with specific skills when needed. Engineers, specialist medical practitioners, and paramedics are some of the specific volunteer skills sought by the organization. All mercy workers are trained to work under disaster conditions and are prepared to work under high-risk conditions (interview, Dr Imtiaz Sooliman).

Islamic Relief South Africa does not use mercy workers in its humanitarian aid activities. It relies heavily on its world-wide staff, the communities that is has trained in countries prone to disasters, and, occasionally, highly specialized professionals who complete specific tasks on condition that there will be knowledge and skills transferred at the local level. In addition, the organization places much attention on training in the different aspects of disaster management. Over the past 4 years, Islamic Relief South Africa has trained 50 local people in South Africa, especially individuals from FBOs. The content of such training is based on international best practice for emergency aid response, such as preparing disaster management plans, managing logistics in emergencies, and understanding the Red Cross Code of Conduct and the Humanitarian Charter and Minimum Standards in Disaster Response (Sphere Guidelines).

The Suitability of Resources for Victims and Its Adaptation to Disaster Conditions

South African Muslim FBOs involved in regional and national disaster work are mindful of local disaster conditions and the need to provide resources that are suitable for disaster victims. Beyond the provision of basic necessities such as food and clothing, South African Muslim FBOs have displayed innovative ways to meet the needs of disaster victims and the local conditions prevalent at disaster sites. In 1994, the Islamic Medical Association, comprising more than 2000 members from different fields in the medical and health-care profession, took the lead in making an innovative contribution to disaster conditions. This organization is known to provide medical care and support to remote parts of the country that have little or no health-care services. It has set up several mobile clinics that largely focus on primary and preventative health-care programs on a regular basis. The organization is noted for its innovative design and outfitting of sophisticated medical facilities in partnership with the Gift of the Givers Foundation. Created from shipping containers and equipped with state of the art medical equipment at a cost of R6,000,000 using South African technology, it is described as comparable to any of the best equipped medical facilities in Europe and is unmatched to this day. In recognition of the facility's assistance with the health-care needs of local victims, the Gift of the Givers Foundation was presented with a special Bosnian award for humanitarian aid work in 1994.

In 2004, the Gift of the Givers Foundation made another major international breakthrough by introducing a high energy and protein food supplement called Sibusiso. Sibusiso is known to play a leading role within the broader national and international health and disaster programs, with particular focus on malnutrition, tuberculosis, and HIV and AIDS. Currently, the product is used by 230 health facilities in South Africa, Malawi, Lesotho, and Botswana [http://www.sibusiso.net]. Given the wide prevalence of the HIV/AIDS pandemic and the extent of malnutrition in the southern African region, this locally produced product has been endorsed by the South African medical fraternity for its enormous health supplementary benefits. Because it is considered easy to use, requires no preparation or refrigeration, and can be eaten from the container, spread on bread, or mixed with porridge, it has become a highly recommended product for use by disaster victims.

After examining the resources provided by South African Muslim humanitarian aid organizations, it becomes evident that they increasingly respond beyond the provision of food and clothing. Especially during times of natural disasters (earthquakes, floods, hurricanes, and volcanic eruptions), the tangible aid packages contain items that not only bring immediate relief to the victims, but that also prevent secondary disasters (i.e., outbreak of cholera, waterborne diseases, and malnutrition) and items that can help reconstruct victims' social functioning after a disaster. These are classified in the sidebar, which notes whether the types of resources provided are practical, relevant, and necessary.

Humanitarian aid organizations vary in the type of aid provided to disaster situations. Some provide support only in the acute phase of the disaster by meeting the immediate needs of the victims, whereas others engage in the preventative and postdisaster phases. The Gift of the Givers Foundation concentrates primarily on providing aid in the acute phase of the disaster, whereas other FBOs provide aid to prevent the spread of human catastrophe through diseases. Depending on resources available, some continue working in disaster-afflicted areas for longer durations, helping to build infrastructures and community facilities destroyed by natural disasters.

Sidebar. Types of Aid Provided for the Different Phases of Disaster

Immediate Need	*Preventative*	*Postdisaster*
Food	Water purification tablets	Building materials
Baby food	High-energy protein supplements	Building tools
Candles		Cooking utensils
Soap		Fertilizer and seeds
Medicine		Agricultural implements
Medical equipment		
Tents		
Blankets		
Mattresses		
Fuel		
Transport		

Making a local assessment of disaster situations is considered an important aspect in providing appropriate resources. One humanitarian aid organization commented:

> Islamic Relief's assessment capacity and field experience over 25 years have taught us many lessons. Inappropriate resources are mistakes rarely made now. Having staff who know local conditions if there was never an Islamic Relief presence in the disaster affected country/region before is a major asset. These vary, from shelter provision, to water points, to hygiene kits, to food and clothing especially. All [of] these take the

Figure 8.2 (a) Foundation scale of relief aid to war torn Lebanon in 2006. (b) Demonstration of water purification tablets used during times of disaster. (Courtesy Gift of the Givers Foundation.)

> cultural, gender, and power relations into account. Special resources would be for interventions of a sensitive nature. Gaza demanded specialized hospital equipment before the 27 December 2008 attacks. [Islamic Relief]'s strongest partner in the region is Qatar Charity, who managed to bring in large amounts of equipment ordered by [Islamic Relief] Palestine and included four ambulances during the war [personal communication, C. Khan, April 3, 2009].

Although there is little control of the type of tangible resources donated during times of disaster, humanitarian organizations increasingly make special appeals to the South African Muslim public for specific sponsorships to meet local disaster conditions. For example, Islamic Relief South Africa, by way of policy, discourages the collection and distribution of second-hand clothing to disaster victims because the people believe that they need to have the very best under such trying social, psychological, and physical circumstances. This is based on the organizations' policy to restore dignity to victims of disasters [personal communication, C. Khan, April 3, 2009]. Similarly, the Gift of the Givers Foundation provides aid in the acute phase of disasters and would only mobilize resources needed in this phase.

Interaction of FBOs in the Region and Beyond

Broadly, South African Muslims are diverse in their social composition, influenced by sociohistorical factors. Hence, it is not atypical that class, ethnic, ideological, and theological differentiation* permeates the structures of FBOs. Muslim FBOs vary in their theological orientation and this has a bearing on their organizational composition, the ability to interact within and between FBOs, the level of participation it attracts from within and outside of the community in its activities, and the ability to mobilize resources. Because most resources mobilized by humanitarian aid organizations are intracommunity, strong religious prescriptions define how, who, and under what circumstance these resources should be used. To have access to resources within the community, humanitarian aid organizations are required to abide by these religious definitions. Independent regulatory bodies sanction all fundraising campaigns within the community and require FBOs to abide stringently to a code of conduct or risk losing their credentials to fundraise [Khan and Ebrahim 2006]. However, not all FBOs adhere to the rules of the independent

* Muslims in South Africa are theologically, spiritually, and ideologically divided and are composed of different groups (i.e., Sunnis, Soofies, Tabligh, Shafi, Hanafi, and Shite). In addition, these groupings are subdivided into the modernists who would like the Islamic world to become part of global modernity (they want to reform Islam to bring it in line with the contemporary age in every respect) and postmodernists (those who view Islam manifesting in the form of religious resurgence that rejects modernist institutions, such as secularism and nationalism, and instead advocates a different moral and political ethic).

regulatory bodies and, in some instances, are not called to account for their financial activities.

South African Muslims have three broad religious traditions that impact their social organization and cohesiveness. The *Sunni*, *Shafee*, and *Deobandi** groupings have historically contested interest in FBOs not only because of the resources at its disposal, but also as a vehicle to drive their theological agenda in development-related aid work. Guarding one's humanitarian aid turf and spaces is emulated either overtly or subtly within FBOs. There is also a tendency for business entrepreneurs, professionals, and politically influential individuals to take up important positions within these organizations. Ethnic and racial exclusivity in the organizations' membership exists at varying levels. Organizations with a history of supporting a particular theological or ideological thought are less open to membership from opposing groups. Broadly, African Muslims feature less prominently in the membership profile of FBOs, although Islam is considered to be the fastest growing religion within the country's indigenous Black African population. It is these differences within the South African Muslim context that determine how FBOs interact, coexist, and respond to each other within and outside of the community in times of disasters.

Given that the humanitarian aid sector is often not characterized by a sense of unity and cooperation among South African Muslims FBOs but conflict defined by religious and cultural barriers, a bold initiative was initiated in 2006 by the World Humanitarian Forum to help Muslim FBOs network and partner with each other. Approximately 150 FBOs attended this day-long workshop, and the following challenges were identified in the manner in which Muslim FBOs operate:

- Programs are not planned to be sustainable in the long term.
- International donors are not aware of constraints faced by local NGOs.
- Poverty in South Africa causes severe isolation and lack of community.
- Although standards are set by the South African government, many NGOs fail to adhere to them.
- NGOs do not feel that they are in partnership with government or that they are afforded enough authority.

Since this symposium, the International Humanitarian Forum has initiated a series of training programs among local Muslim FBOs on different aspects of international disaster-relief work. Similarly, the Al-Imdaad Foundation provides life orientation skills on disaster management at local schools in the country to make young South Africans aware of disaster situations and ways to respond to them.

* Muslims of indentured origin largely follow the Sunni tradition, as compared to Muslims of slave heritage who settled largely in the Cape and follow the Shafee school of thought. Merchant class Muslims are known to follow the Deobandi School of thought and have enormous influence in public institutions, both in Gauteng and KwaZulu-Natal.

The extent of political linkages established with the state is an important determinant on the ease with which humanitarian aid organizations are able to offer relief in foreign countries. It is not uncommon for some humanitarian aid organizations to partner with the South African government on certain missions. Both the Al-Imdaad and the Gift of the Givers Foundation have strong partnerships with the South African government. They are both registered as Proudly South African NGOs, which provides them with a particular status in the country and abroad as mercy ambassadors. The Gift of the Givers Foundation has established a strong partnership with the South African government to facilitate its mercy mission to disaster-afflicted areas outside of the country. Based on this partnership, the South African government awarded R60 million in 2002 to distribute food parcels in the provinces of KwaZulu-Natal and Eastern Cape, which had the highest incidence of poverty.

The entry of international relief organizations into the country has influenced the way South African Muslim FBOs undertake aid work in foreign countries. The Islamic Relief South Africa helps as an implementing organization for local Muslim FBOs. Major purposes are to decrease travel costs and the logistics of resource distribution and to ensure accountability in how resources are collected and distributed. The Director of Islamic Relief South Africa commented as follows on the principle of accountability and collaboration with governments.

> South African aid organizations are known for their parachuting in of aid and doing high profile media handovers to government officials of aid. When they leave, the officials usually distribute the aid to their cronies and sell the rest. Humanitarian aid principles require aid agencies to be independent and international humanitarian law protects that independence. Independence does not mean no coordination with local authorities. But coordination is not collaboration. To save [South African] aid agencies of this embarrassment Islamic Relief provides relevant information and contacts as well. Aid cannot be donor driven. [South African] aid agencies who engage in media stunts claim that the [South African] Muslim donors and the [South African] government want to see the media coverage. The South African aid agencies do not educate donors about humanitarian principles [personal communication, C. Khan, April 3, 2009].

Clearly, from the above account, one may conclude that not all South African Muslim humanitarian aid organizations are fully in agreement on certain key principles. Hence, the level of interaction between and among humanitarian aid organizations is more than likely characterized by organizational barriers. In an interview with Dr. Imtiaz Sooliman of the Gift of the Givers Foundation, he confirmed that his organization interfaces with the South African government and the disaster-afflicted country only. He further added that "my organization does not interface with other organizations in the disaster country or within the country," which suggests that this organization, by virtue of enjoying state support by way of diplomatic privileges, can afford not to network with organizations with similar objectives.

However, among humanitarian organizations that interface with each other, one finds them interacting on the logistics of mobilizing resources within the community, the type of resources needed, and storage, packaging, and movement of such resources. Some do so through established networks in foreign countries and local networks with organizations to reach out to the victims of disasters [http://islamicfocusarticles.blogspot.com]. The following example derived from Islamic Relief South Africa illustrates the point being made:

> A good example is when the Mormon Church approached Islamic Relief to provide aid to refugees in Abeche, Chad, but [Islamic Relief] did not have a strong enough presence there yet [and] informed them that CAFOD (Catholics) [Islamic Relief]'s partner has. [Islamic Relief] received the aid at its strong base in El Geneina West Darfur and facilitated the transfer to Abeche 200 km across the Chadian border. Islamic Relief received 200,000 USD for the floods in Guatemala and handed it to CAFOD. CAFOD raises funds for many disasters in [mostly] Muslim countries such as Chechnya, Iraq, recently Gaza, [and] Afghanistan, and uses [Islamic Relief] as its implementing partner [personal communication, C. Khan, April 3, 2009].

Outside the country and especially on mercy missions, South African Muslim humanitarian aid organizations are known to interact and partner with other aid organizations. Over the years, interaction and collaboration with Islamic Relief Worldwide, Red Cross Relief Agency, Mercy Malaysia agency, and Direct Aid International has increased. Further, a trend among South African Muslim humanitarian aid organizations to become part of the international aid community and subscribe to universal disaster relief codes can be noted. In 2008, the Al-Imdaad Foundation became signatory to the Code of Conduct for the International Red Cross and Red Crescent Movement and NGOs in Disaster Relief. It was one of 462 international organizations that signed this code.

Conclusion

This chapter has highlighted the nature and extent of Muslim FBOs involvement in regional and international humanitarian aid activity and the extent to which Muslim FBOs endeavor to adapt resources to local disaster conditions based on the needs of victims and the extent and nature of interaction among and within humanitarian aid organizations at a regional and international level. For a minority community, the response of South African Muslims FBOs to regional and international disasters since its democratization has been responsive, significant, and consistent, and has shown appreciable signs of growth. They have taken innovative measures to respond to disaster conditions and the needs of victims and are increasingly recognized by disaster-stricken nations for their humanitarian gestures.

However, certain gaps in the way in which South African Muslim humanitarian aid organizations operate have become evident and need to be considered. For example, some do not work independently of government, which has both merits and demerits. Working with governments ensures that diplomatic protocols are observed and provides access to disaster-afflicted regions with ease, but disadvantages include the restrictions it places on long-term political advocacy and the support needed to reconstruct disaster victims' livelihoods beyond the acute phase of disaster. Because post-disaster reconstruction is a long-term event and an unanticipated cost to governments, a longer-term presence will have a greater impact on sustaining disaster recovery. Nonetheless, some humanitarian aid organizations have taken the challenge to sustain a long-term presence in disaster-afflicted countries and truly emulate the spirit of identity that accompanies NGOs of being independent of state influence.

The nature and extent of interaction at a local, regional, and international level appears to be just starting. Although some humanitarian aid organizations interface with other stakeholders in disaster-afflicted areas, others choose to work independently. Interfacing with other disaster stakeholders has an advantage in ensuring that disaster interventions are not fragmented, but instead are consolidated to provide the maximum relief to victims. It appears that consolidating Muslim FBOs capacities is only beginning to take place in South Africa. This is evident through the network initiatives and emerging training programs formulated for FBOs in the country.

Volunteer participation appears to be restricted to either paid personnel of humanitarian aid organizations or professional sectors within the Muslim community. The technical sectors of the community have hardly been tapped as a potential source of expertise for postdisaster reconstruction programs.

Insofar as conformity to international disaster relief codes of conduct are concerned, there is some evidence to support that Muslim FBOs indeed do. Given that a Muslim presence in political hotspots in the world is characterized by Islamophobic reactions, it is important for Muslim FBOs engaged in regional and international humanitarian activities to seek the protection of such agencies for their volunteers and personnel. Recently, South African Muslim FBOs volunteers engaged in humanitarian aid work have been intimidated and deported from politically tense areas because of being suspected of association with some form of terrorist activity.

Considering that South African Muslim FBOs engagement in humanitarian aid activity is only a recent phenomenon due to decades of political isolation from the rest of the world, they have made a major impact on regional and international humanitarian aid activity in a short time. It is evident that they have made enormous strides despite gaps and limitations and there are sufficient indicators to attest that their involvement in regional and international disasters is more than likely to increase.

References

Alterman, J. B., and Hunter, S. 2004. *The idea of philanthropy in Muslim contexts.* Washington, DC: The Center for Strategic and International Studies, USAID.

Clark, G. 2006. Faith matters: Faith based organizations, civil society and international development. *Journal of International Development* 18:835–848.

Statistics Council. 1996. Census. Pretoria, South Africa: Statistics South Africa.

Statistics Council. 2001. Census. Pretoria, South Africa: Statistics South Africa.

Davids, M. 1997. *Directory of Muslim institutions and mosques in South Africa.* Johannesburg, South Africa: Islimu Publications.

Everatt, D., and Solanki, G. 2005. A nation of givers? Social giving among South Africans: The state of social giving in South Africa. Research Report No.1. Durban, South Africa: Centre for Civil Society, University of KwaZulu-Natal.

Faith and Technical Education Network. 2003. *Philanthropy and faith: An introduction.* New York: National Crime Prevention Council.

Ferris, E. 2005. Faith-based and secular humanitarian organizations. *International Review of Red Cross* 87:311–325.

Habib, A., Maharaj, B., and Nyar, A. 2008. Giving, development and poverty alleviation. In *Giving and solidarity: Resource flows for poverty alleviation and development in South Africa,* eds. A. Habib and B. Maharaj. Cape Town, South Africa: HSRC Press.

Islamic Council of South Africa. 1984. *Meet the Muslims in South Africa.* Durban, South Africa: Liberty House.

Khan, S., and Ebrahim, A. F. M. 2006. The state of philanthropy amongst the Muslim diaspora in South Africa. In *Religious pluralism in the diaspora,* Vol. 4, ed. P. T. Kumar. Netherlands: Brill Publishers.

Kniss, F., and Campbell, T. D. 1997. The effect of religious orientation on international relief and development organizations. *Journal for the Scientific Study of Religion* 36:93–103.

Marshall, K. 2001. Development and religion: A different lens on development. *Debates Peabody Journal of Education* 76:339–375.

Meer, Y. S. 1980. *Documents of indentured labour. Natal:1851–1917.* South Africa: Durban Institute of Black Research.

Nadvi, S. H. H. 1988. Muslim minorities in South Africa. *Annual Journal of Arabic Studies X11*:144–149.

Naude, J. A. 1992. South Africa: The role of a Muslim minority in a situation of change. *Journal of the Institute of Muslim Minority Affairs XIII*:17–32.

Osman, M. 2003. Muslim NGOs can help bridge culture gap. Retrived from http://www.globalpolicy.org/component/content/article/176/31237.html (accessed October 16, 2008).

Ramphal, R. 1985. Marital conflict among Hindus in the metropolitan area of Durban. Unpublished thesis. Durban: University of Durban-Westville.

Randeree, Z. B. 1997. Muslim minorities with special reference to South Africa. *Journal of the Centre for Research in Islamic Studies* 17:63–82.

Redwood-Campbell, L. 2008. Editorial comments—Faith based organizations and sustainable operations in Mississippi after Katrina: Implications for informal network utilization. *Pre-hospital and Disaster Medicine Editorial Review* 1:55–56.

SANZAF Biennial Report. 2004. Western Cape Region, South Africa: SANZAF.

Selinger, L. 2004. The forgotten factor: The uneasy relationship between religion and development. *Social Compass* 51:523–545.

Smith, S. R., and Sosin, M. R. 2001. The varieties of faith-related agencies. *Public Administration Review* 61:651–670.

Swilling, M., and Russell, B. 2004. *The size and scope of the non-profit sector in South Africa.* Durban, South Africa: Centre for Civil Society, University of Natal.

Tayob, A. K. 1998. Muslim political space in South Africa: Imagining a local *Ummah. Annual Review of Islam in South Africa* 1:6–9.

Ver Beek, K. A. 2002. Spirituality: A development taboo. In *Development and culture.* Oxford: Oxfam.

Chapter 9

Analysis of the Interorganizational Cooperation in the Jewish and the Arab Sectors in Israel

Alex Altshuler

Contents

Interorganizational Cooperation in the Context of Emergency Preparedness

The emergency preparedness process involves a wide range of organizations and groups. Clearly, efficient coordination is necessary between them [Peek and Sutton 2003], and the quality and extent of interorganizational cooperation and coordination are highly influential factors among those concerned with the level of emergency preparedness [Drabek 2007]. Intersectoral and interorganizational cooperation throughout the emergency preparedness process on the local level was one of the main conclusions drawn after Hurricanes Rita and Katrina [Simo and Bies 2007]. Similar lessons regarding the significance of intersectoral cooperation on the local and national level were learned in Israel following the war between Israel and Lebanon in 2006 [Katz et al. 2006]. Those lessons were put in practice in Israel through various projects aimed to increase interorganizational coordination and cooperation on the local as well as on the national level.

The importance of cooperation was highlighted not only during the preparedness stage, but also in the framework of emergency response activities of the local communities in the southern regions of Israel in the time of the warfare between Israel and Gaza in the winter of 2008–2009. Those local communities tried to coordinate the efforts aimed at addressing multiple and constantly changing needs of the civilian population who suffered from missile attacks with other stakeholders. The governmental agencies also tried to provide a high level of coordination by conducting daily meetings with local representatives, learning their needs and trying to supply a coordinated interdisciplinary governmental response to the actual civilian needs.

It should be mentioned that in southern Israel both Jewish and Bedoiun Arab populations suffered from missile attacks. The Bedouins in this region live in small towns, nomad settlements, and in Rahat, the largest Bedouin town in the world, where 43,700 people live [Central Bureau of Statistics 2008]. This mixture of Jews and Arabs in Israel and its empirical implications in the field of emergency preparedness will be extensively presented and analyzed in this chapter.

It should also be mentioned that a great variety of stakeholders are present at the local arena: governmental agencies, third-sector organizations, business companies, nonformal civil society groups, and residents. A recent study that was conducted in Israel [Altshuler 2008] found empirically that joint activities of local emergency managers with governmental and nongovernmental stakeholders were among the most significant predictors of the level of preparedness of local authorities for war-caused disaster. Although the level of cooperation with governmental structures was higher than with nongovernmental entities, the two factors had similar levels of prediction. This fact highlights the theoretical foundations and practical implications regarding the importance of joint activities as a measure to promote the level of emergency preparedness.

Connection between Interorganizational Cooperation and Ethnic Composition of a Country

Ethnic minorities are more vulnerable in the face of disasters [Peacock et al. 1997; Rodriguez and Barnshaw 2006] and should be treated accordingly as a group at risk [Laor et al. 2005]. The reasons for this situation are related to economic and political factors [Bullard 2000], such as economic inequity and minorities' limited political power. These factors may lower the preparedness level and increase the magnitude of disaster damage in communities whose residents mostly belong to ethnic minorities [Peacock et al. 1997]. In terms of social responsibility, more attention is clearly needed and more financial resources must be invested in emergency preparedness of the communities in which most people belong to ethnic minorities. Unfortunately, this is frequently not the case because of the political and economic factors mentioned above, and that is why we should expect lower emergency preparedness levels in minority communities in most cases. In Israel, for example, it was revealed during the war between Israel and Lebanon in the summer of 2006 that local communities inhabited mostly by Arabs lacked public shelters, sirens to warn of impending danger, and guidance material in Arabic [Bavly et al. 2007; Ben-Harush 2007; Sharkansky 2007].

The Israeli local authorities are mostly monoethnic; namely, they are inhabited either by Jews or by Arabs. Usually in Jewish towns and cities there is an extremely low percentage of Arab citizens, and in the vast majority of Arab towns there are no Jews at all. The only exceptions to that phenomenon are some mixed cities (such as Haifa, Lod, and Ramla), in which the percentage of Arab population is higher, but it is also less than a quarter of the total population. Based on that reality, it is possible to use the terms Jewish sector and Arab sector in Israel and analyze different aspects that are related to emergency preparedness in those sectors.

Until recently, there was no scientific empirical research regarding various aspects related to the emergency preparedness of local authorities in Israel. This lacuna was filled by the first nationwide research on emergency preparedness of local authorities in Israel that I conducted in the summer of 2008 [Altshuler 2008]. This examination is currently the only empirical source for thorough analysis of emergency issues in Israel. One factor that was examined in that research was the joint activities of local emergency managers with various stakeholders: governmental and nongovernmental. This factor provides us with information regarding interorganizational cooperation on the local level through the view of emergency managers. The extent of the activities was examined in both the Jewish and Arab sectors. On the basis of that research, it is possible to conduct a specific analysis and comparison of the data regarding both sectors (116 Jewish local authorities and 25 Arab ones).

Generally speaking, it may be said that the extent of the joint activities with the majority of the stakeholders are similar in the Arab and the Jewish sectors. However, there are joint activities between some stakeholders and the local emergency managers that were found to be significantly higher in the Jewish sector: police, one of the emergency departments of the Ministry of Interior, other local authorities, neighborhood committees, community centers, and national and local philanthropic foundations. Regarding neighborhood committees, community centers, national and local philanthropic foundations, it seems that the reasons for the fewer joint activities with them in the Arab sector are connected with broader social phenomena than pure or specific emergency preparedness issues. All of the above institutions and structures are generally less widespread and developed in the Arab sector—a fact that influences, among others, the field of emergency preparedness. Regarding the three additional stakeholders—police, one of the emergency departments of the Ministry of Interior and other local authorities—it is possible that those stakeholders are perceived as less relevant by Arab emergency managers in comparison with their Jewish colleagues. Future research is needed to identify the possible reasons for this phenomenon.

As was mentioned above regarding most of the stakeholders, the extent of joint activities between them and local emergency managers was found to be similar in both the Jewish and the Arab sector. Those stakeholders are the National Emergency Economy Authority and its local branches, Fire and Rescue Services, Emergency Medical Services ("Magen David Adom in Israel"), Home Front Command, governmental ministries, mayor and vice-mayors, top managers of the local authority, local and national assistance and volunteer organizations, professional unions, health funds, local and national businessmen, professional consultants who are not staff members of a local authority, and residents. The above fact regarding the similar extent of joint activities regarding the majority of the stakeholders in both sectors is of great importance. It means that the level of interorganizational cooperation in communities inhabited mostly by minorities and those inhabited mostly by the majority group in Israel are close. These findings may be seen as surprising given the theoretical expectation for lower levels of different aspects that are related to emergency preparedness in minorities' communities. Unfortunately, this general expectation is usually true all over the world. However, specific processes and circumstances may change that reality. It seems that certain characteristics influenced the above reality of similar levels of interorganizational cooperation (with the majority of the stakeholders) in both the Arab and the Jewish sector.

The research examination of the level of interorganizational cooperation in both sectors was conducted in the summer of 2008, 2 years after the war between Israel and Lebanon. In this war, the Arab civilian population in Israel was hit by missile attacks. Before that, most of the local leadership in Arab towns in Israel believed that their residents would not be exposed to attacks by the neighboring Arab countries. Emergency preparedness in Arab villages and towns was not seen as a relevant issue before the 2006 war, and most of the local communities were not interested in any drills or other preparedness activities. Another obstacle was connected with the fact that the central institutions in the field of emergency preparedness (Emergency Economy Authority and Home Front Command) are related to the Ministry of Defense, despite the fact that they deal with civilian emergency preparedness issues. This symbolic connection with the Ministry of Defense made the cooperation with those stakeholders complicated for the Arab local officials because of the feelings of solidarity with the Arabs in other countries and unwillingness to be somehow related to the Israeli Defense structures. However, the missile attacks on Arab villages and towns increased awareness dramatically regarding the need for and willingness of Arab local authorities to deal extensively with emergency preparedness for war-caused disaster and to cooperate with the relevant stakeholders to achieve that goal.

Following the 2006 war between Israel and Lebanon, the central conclusion was connected with the need to improve greatly interorganizational cooperation and coordination in the field of emergency preparedness for war-caused disaster [Altshuler 2008; Gidron et al. 2007; Katz et al. 2006]. Rivera and Miller [2007] related that disasters have a potential to reflect to the authorities the dysfunctions of the current disaster policy. It seems that this potential was practically realized in that case. The above lesson from the 2006 war was implemented in various governmental and nongovernmental initiatives aimed specifically to improve and intensify the interorganizational cooperation. Local authorities were related as a key factor in the framework of those efforts because of the understanding of the great importance of the local level. Those initiatives were administered in both the Jewish and the Arab sector. In addition, there was initiative established by the Arab nongovernmental organizations that aimed to improve the cooperation between the Arab and the Jewish stakeholders in the field of emergency preparedness, as well as to enhance it on the local level in the Arab sector between all relevant stakeholders.

Those changes following the 2006 war increased greatly the interorganizational cooperation in the Arab sector and created the reality of almost similarity in the extent of the interorganizational cooperation on the local level in the Jewish and the Arab sector in 2008.

General Policy Implications

Interorganizational cooperation is a key factor that may predict the level of emergency preparedness in both the Arab and the Jewish sectors in Israel. It is recommended to continuously increase the level of cooperation with the existing stakeholders and to identify other possible partners as an important way to improve the actual level of preparedness. In addition to cooperation on the local level, it is proposed to establish formal regional frameworks for continuous cooperation in which representatives of both the Arab and the Jewish local authorities will be able to cooperate in a more effective way, while reciprocally assisting each other. Establishment of a framework for regional cooperation may also improve the level of cooperation with national emergency authorities and achieve higher level of preparedness for the residents by comprehensive representation of the needs of the whole region and all its ethnicities. This regional framework should include not only local officials from different town and villages, but also representatives of business and social organizations that operate in those regions.

The above lessons seem to be relevant for various types of disasters and for different regions of the world. Earthquakes, floods, hurricanes, and other types of disasters do not distinguish between the conflicting ethnic groups or fighting neighbor countries. Those contingencies may similarly affect the confronting sides in the same region through creation of severe humanitarian consequences. The main question is whether disasters will be addressed using joint efforts of various relevant stakeholders and regional comprehensive response or in separate, autonomous ways where each organization or country affected by a disaster acts in a clearly independent, competing, and sometimes even hostile way toward other stakeholders. Certainly, there are different in-between modes of action, but the main opposing directions are those presented above—cooperation versus segregation. Current research illustrates that cooperation is a real option that is relevant all over the world. It enables organizations to work jointly on emergency preparedness issues despite deep political controversies to create another sphere in which the main goal is to save people's lives, to prepare the local communities, and not to let the prejudices control the reality. This is a challenging road, but it proposes an extraordinary reward (decrease of loss and pain in face of disasters) to those who are ready to step on, it creates an extremely important opportunity to build bridges and promote a dialogue between the conflicting sides. The multiple stakeholders who may be involved in that process promote both an increase in the level of emergency preparedness and a reduction in the intergroup or intercountry tensions.

In the local context, the consequences of the war between Israel and Lebanon in the summer of 2006 and the suffering of Arab civilian population in Israel because of the missile attacks increased greatly the awareness regarding the relevancy of the civilian emergency preparedness for war-caused disaster for the Arab sector in Israel. In general, that process reveals the influence of the actual exposure to disaster on the extent and intensiveness of emergency preparedness, including interorganizational cooperation. It seems that thorough analysis of future risks may cause greater emergency preparedness and is essential. Unfortunately, there is a difference between the actual exposure to disaster and the possible exposure in terms of the potential to enhance the level of preparedness. Actual exposure shows everyone that the danger is real, whereas possible danger is regarded many times by the officials as something only theoretical and, hence, deserving of less effort and investment.

War is clearly a manmade disaster. It is caused by political, social, and other factors that may be changed. It seems that the greatest and the most important challenge is to prevent wars. However, in terms of saving lives, emergency preparedness activities should be conducted to minimize the potential damage if a war occurs. In conflict regions, emergency managers and their colleagues should constantly deal with preparedness for war-caused disaster, whereas the political leadership should take maximal measures to prevent wars and find other solutions for disagreements. The above dialectics of a need to be prepared and, in parallel, the duty to prevent war may seem complicated, but it is believed to be the most effective way to save lives in conflict regions. Political situation and ethnic divisions greatly affect the technical process of emergency preparedness. It should also be mentioned that achieving a sustainable peace is many times more effective than the most effective emergency management. However, when armed conflicts occur in different parts of the world, civilian emergency preparedness for war-caused disaster can't disappear. It must be effective and integrated. Hopefully, this type of emergency preparedness will become nonrelevant in the future, when sustainable and continuous peace will replace wars and great suffering.

One of the main proposed global generic policy implications is that there is an urgent need to increase the emergency preparedness level for various types of disasters, both in areas of military conflict and in the regions where lasting peace already exists, via effective interorganizational cooperation based on forums, task-groups, community initiatives, think tanks, and other organizational frameworks aimed to facilitate and increase the joint activities of multiple stakeholders from different sectors of society, such as governmental institutions, nongovernmental organizations, business enterprises, and volunteer citizens. The exact form of those activities should be defined locally because they depend on the norms, culture, traditions, constraints, and other social characteristics of the specific community, but the general claim regarding the contribution of interorganizational cooperation to the level of emergency preparedness (which means saving people's lives) is indeed global.

References

Altshuler, A. 2008. Emergency preparedness of local authorities for war-caused disaster: The Israeli case [in Hebrew]. MA dissertation. Haifa: University of Haifa.

Bavly, M., Raday, F., Dabush, A., and Eytan, H. 2007. *Civilians on the frontline: Citizens' views of home front failures during the second Lebanon war* [in Hebrew]. Jerusalem: Shatil Organization Press.

Ben-Harush, E. 2007. The responsibility of the state with regard to a weakened citizen in the second Lebanon war: The elderly and invalids as a case study [in Hebrew]. Graduation paper. Israel: College of National Defense.

Bullard, R. D. 2000. *Dumping in Dixie: Race, class and environmental quality*, 3rd ed. Boulder, CO: Westview.

Central Bureau of Statistics. 2008. *Tables of population in communities* [in Hebrew]. Jerusalem: Central Bureau of Statistics Press.

Drabek, T. 2007. Community processes: Coordination. In *Handbook of disaster research*, eds. H. Rodriguez, E. L. Quarantelli, and R. R. Dynes, 217–234. New York: Springer.

Gidron, B., Yannai, A., Liel, R., Limor, N., Sokolov, Y., Amichai, S., et al. 2007. *Report of the committee for examination of the third sector roles and functioning in the Second Lebanon War* [in Hebrew]. Tel-Aviv: Civic Leadership Association Press.

Katz, H., Raviv, E., Yogev, H., Yaacobi, M., Levinson, A., Alon, Y., et al. 2006. *The Israeli civil society in Second Lebanon War* [in Hebrew]. Beer-Sheva: Ben-Gurion University of the Negev, Israeli Center for Third-Sector Research.

Laor, N., Wiener, Z., Spirman, S., and Wolmer, L. 2005. Community mental health in emergencies and mass disasters: The Tel-Aviv model. In *The trauma of terrorism: Sharing knowledge and shared care*, eds. Y. Danieli, D. Brom, and J. Sills, 681–694. New York: The Haworth Press.

Peacock, W. G., Morrow, B. H., and Gladwin, H. 1997. *Hurricane Andrew: Ethnicity, gender and the sociology of disasters.* New York: Routledge.

Peek, L. A., and Sutton, J. N. 2003. An explanatory comparison of disasters, riots and terrorist acts. *Disasters* 27:319–335.

Rivera, J. D., and Miller, D. S. 2007. Continually neglected: Situating natural disasters in the African American experience. *Journal of Black Studies* 37:502–522.

Rodriguez, H., and Barnshaw, J. 2006. The social construction of disasters: From heat waves to worst-case scenarios. *Contemporary Sociology* 35:218–223.

Sharkansky, I. 2007. Local autonomy, non-governmental service providers and emergency management: An Israeli case. *Journal of Homeland Security and Emergency Management* 4:1–8.

Simo, G., and Bies, A. L. 2007. The role of nonprofits in disaster response: An expanded model of cross-sector collaboration. *Public Administration Review* 67:125–141.

Chapter 10

Role of Schools in Rebuilding Communities after Disaster

Beryl Cheal

Contents

The world's weather is changing and it has resulted in more storms, more human tragedies, and more physical devastation. However, schools can help communities become more disaster-resistant and recover more successfully by preplanning. First, the community needs a useable, site-specific preparedness plan, which could be developed by schools in cooperation with the wider community. The school plan, developed jointly by school personnel and students, should specify risks that have been mitigated and insure that participants know what they will do and where they will go when disaster strikes. Second, schools must be prepared to help students and their families recover from the traumatic experiences that often accompany disaster.

With these two components in place, in cooperation with other community entities, schools can be of great service to their students and families and thereby play a major role in rebuilding communities after disaster.

The Importance of Schools

The role of schools in recovery is similar to that stated by Ishaq [2008] in describing an International Child Art Foundation's Healing Arts program. Ishaq clearly describes the importance of schools in community recovery when he states that a focus on schools accomplishes the following:

A. Forces the community to focus on its future needs, rather than merely restoring the past that is lost
B. Aides and accelerates the children's emotional recovery, which in turn channels and guides the community's mental recovery
C. Leads to a focus on family as the most important social unit and the school as the community center for postrecovery
D. Deepens the children's emotional bonds to the place where they grew up, thus offsetting the unbearable memories of the disaster, which may cause them to leave and restart their lives somewhere else
E. Revitalizes the community around the strength, hope, and optimism of its children [Ishaq 2008, pp. 28–29]

Schools have an obligation, a responsibility, and an opportunity to play a major role in helping their students recover from disaster.

Importance of Preplanning

When we think of disasters, we often discuss a phenomenon called a natural disaster. Actually, Romano, risk management specialist at the Humboldt Center, tells us:

> We cannot say that disasters are natural. Some physical events are natural, but the effects of the events are not natural; they are built on social practices. Disasters are socionatural events. . . . A closer look at how events like earthquakes and volcanoes, droughts and floods and tsunamis become tragedies reveals that these events of nature, while powerful, only trigger disasters if vulnerable communities are exposed to their destructive force [Romano 2006, p. 1].

For instance, several thousand earthquakes occur in Alaska in any given year, but we do not consider them disasters or tragedies because no vulnerable communities are affected. They become a disaster when nature conflicts with human life and living.

Erikson, a renowned American sociologist and an authority on the social consequences of catastrophic events, makes this important point in his foreword to *The Sociology of Katrina: Perspectives on a Modern Catastrophe*:

> Katrina cannot be visualized usefully as a natural disaster that blew across the horizon. . . . The openings Katrina took advantage of to reach New Orleans and other vulnerable parts of the coast were of human construction. The susceptibilities of the persons victimized by Katrina were of human construction. And virtually everything that followed was a product of human hands or human imaginations [Brunsma et al. 2007, p. 1].

Additionally, disasters may appear to affect rich and poor alike, but that is not the case. For instance, earthquakes produce landslides on steep hillsides and the edges of rivers or ravines, areas that are frequently populated by the poorest residents. Our own striking examples of the tragedies resulting from Katrina confirm this. Those who lost the most, either in a material way or by a loss of life, were the poorest residents. They lived most directly in harm's way. Whereas extreme acts of nature affect us all, it is those with the fewest resources who suffer the most. Romano [2006] again reminds us that:

> Those with the fewest resources can seldom afford to build earthquake-resistant features into their houses. And the poor lack insurance to provide for them if an earthquake destroys their homes or their means of making a living [p. 2].

Because the effects of nature are due to human activity, it is possible to make changes to lessen impact. A nongovernmental organization that has been involved in many emergency or disaster mitigation or preparedness plans throughout the world, Oxfam America, reports that communities that have implemented risk reduction (or preparedness) plans sustain less damage during disasters than communities without such planning. Much of the negative impact of disasters, they contend, can be significantly reduced with proper planning and training. Residents recover more rapidly, fewer lives are lost, and children and the community are more resilient [Thompson and Izaskun 2004].

Cuba, a sterling example of successful risk reduction and preparedness, has experienced a dramatic reduction in human costs and physical devastation because they have made major efforts in reducing risks and developing a culture of safety in the past few years. As a small, poor country located in the mouth of the Caribbean, Cuba experiences frequent destructive storms. Because of the human and physical toll that hurricanes have taken over the years, residents have developed

comprehensive risk reduction strategies and they have had positive resu
of deaths have been attributed to hurricanes in Cuba prior to 1996
hurricanes hit Cuba between 1996 and 2002 and a total of only 16

In contrast, when hurricane Isabel hit the United States in
that one storm was directly responsible for 22 deaths and indire
6 additional deaths [Thompson and Izaskun 2004]. The areas hit
had inadequate or nonexistent preparedness plans and mitigatio

There are more success stories worldwide because people k
Amanda Ripley, the award-winning *Time Magazine* reporter and
the fate of two cities in Southeast Asia during the 2004 tsunami.
high percentage of casualties, while casualties were minimal in the other. Both villages were located near the epicenter of the earthquake that triggered the tsunami. Jantang was a coastal village on the northern coast of Sumatra. Residents felt the ground shake and approximately 20 minutes later, a wall of water 45 to 60 feet high engulfed the area. All of the village's structures were destroyed and more than 50% of the people were killed.

Langi, on the island of Simeulue, was even closer to the epicenter. Residents had only 8 minutes after they felt the shaking, the shortest interval between the earthquake and tsunami anywhere. Waves reached 30 to 45 feet, slightly less high than Jantang but still deadly, and all of the village's buildings were destroyed as they were in Jantang. However, unlike Jantang, where residents did not leave the village, the 800-person population in Langi fled to higher ground and stayed there. Very close to 100% of the population survived. When asked why they all ran, the answer was that is what they do in that village. In the early part of the 1900s, the island had experienced a devastating tsunami that killed approximately 70% of the villagers. From that time on, the culture of the village changed. Whenever an earthquake struck, everyone ran to higher ground. That was the tradition, no matter what [Ripley 2008]. They had a plan and they had developed a culture of safety.

Stevens [2006] gave an example of another community that had a plan with good results: the village of San Jose Costa Rica, El Salvador. In January 2001, a powerful earthquake shook the village. Houses collapsed, many residents suffered broken bones, and a 4-year-old girl was killed. In addition, the main road, a vital link to the outside world, was destroyed, so for a time the community was cut off from outside help and village people had to rely on their own limited skills and resources. Determined not to let future earthquakes, hurricanes, and other natural events from becoming full-scale disasters, the people of San Jose Costa Rica worked to develop a risk reduction plan. They developed emergency communications, trained community members in first aid and other skills essential for first responders, and held drills to simulate emergencies. They developed what Oxfam America describes as a culture of safety. Four years later, in October 2005, Hurricane Stan pounded El Salvador. San Jose Costa Rica's preparations resulted in no deaths or serious injuries and vital links to main roadways were reopened in a short time. In the words of Claudia Dalila Sanchez, a resident of San Jose Costa Rica, "The difference with Stan was that we were organized. After the earthquake, people didn't have the consciousness to help each other, but after Stan the community was united. . ." [Stevens 2006, p. 2].

What are the key factors in helping a community be ready for disaster? Oxfam America argues that strengthening community capacity, strong coordination of local actors, and investing in social capital are determinate factors for successful community risk reduction and preparedness [Thompson and Izaskun 2004]. Not only must there be physical preparedness—something tangible—but the intangibles are also important.

In discussing a culture of safety and thinking of intangibles, Oxfam America describes an interview with an international aid worker living in Havana in 1996, when Hurricane Georges was about to strike Cuba. A steady stream of neighbors was going in and out of the aid worker's apartment, helping to prepare, giving advice on what needed to be done, and answering questions. As the aid worker describes it:

> Everyone in the apartment building was out helping to tape up the windows in the entryway. The neighborhood representative from the Federation of Cuban Women was out checking on her "vulnerable population" to see how they were doing. The community doctor was checking on patients in the neighborhood, determining if anyone needed to be moved to a hospital as a precautionary measure. . . . Everyone, even the children, knew what to do [Thompson and Gaviria 2004, p. 36].

In writing her book, Ripley [2008] interviewed numerous people who had experienced disaster in many forms. Some of those survived because of good preplanning, and not only did they have a higher chance of survival, but in her analysis their recovery was also made easier. She stated that:

> People who knew where the stairwells were in the World Trade Center were less likely to get injured or have long-term health problems. That's partly because they had the training they needed to take action under extreme stress and later they could take comfort in their own competence. The same is true with police officers or firefighters. If they have the skills they need, they not only have a higher chance of survival; they fare better psychologically after the crisis. They saved themselves once; they can do it again [Ripley 2008, p. 71].

In addition to national and state emergency planning, which is important, local community organizing is an essential and effective part of risk reduction. Among other community-based disaster risk reduction programs [Rich and Kelman 2007], Community Emergency Response Training (CERT) is frequently offered in U.S. local communities. The CERT program, run by the Federal Emergency Management Agency (FEMA), trains participants in recognizing disaster specifics, gives training in knowing what to do before first responders arrive, provides basic first aid skills, and trains participants in developing family and community emergency plans. Those who are trained develop confidence in their abilities to withstand an emergency and get prepared to minimize loss and destruction.

Based on CERT, a similar program has been developed for youth, Teen School Emergency Response Training (SERT), particularly between the ages of 15 and 18 years, although younger students have also been successfully trained. Rich and Kelman [2007] describe SERT as a program that "ensures the intersection between disaster-related education, school safety and community disaster response. . ." [pp. 2–3]. Through experiential learning, students increase their knowledge and skills in disaster-related areas of planning, thinking, and decision-making: recognizing hazards, risk taking and individual responsibility, and mitigating hazards and vulnerabilities through basic information and skill development [Rich and Kelman 2007].

SERT is part of the health curriculum and is required of all incoming freshmen at Pueblo West High School in Pueblo, Colorado. It has proven to be effective for students immediately and also in preparing them for giving major assistance when disasters occur in their community. As a result of first aid training students received in SERT, at least 10 Pueblo citizens' lives have been saved. There is also an example of a SERT-trained student with Down syndrome being instrumental in saving the life of a younger child [personal communication, Haley Rich 2009].

Learning SERT skills and becoming interested and involved in one's community, students take their new learning home to family and learn leadership skills and community participation by teaching neighbors and friends about what they have learned [Rich and Kelman 2007]. These new-found feelings of competence frequently carry over into student's academic studies and into other parts of their lives where they often begin to take leadership roles never before considered [personal communication, Haley Rich 2009]. This is community-based because each SERT program must reflect the issues of the area in which it is taking place and the needs of the students being trained. It is a community capacity-building program for the short- and long-term. Students can be ready to assist when an emergency arises at any time and they will carry with them wherever they go the skills they have practiced, the interest they have developed, and the feelings of competence and confidence. They are developing a culture of safety (Figure 10.1).

There are other examples of community organizations and emergency preparedness efforts. Various states have school laws and regulations for training in emergency preparedness and developing emergency management plans. One of California's most challenging hazards is the devastating earthquakes that occur periodically in that state. However, since the Oakland fires in the early 1990s, California schools have made a concerted effort to develop emergency plans and

Figure 10.1 Teen School Emergency Response Training participants practice their skills in simulated exercises. (Photo reproduced from the Teen SERT brochure.)

train school personnel throughout the state. Training in school crisis response has become an ongoing activity in California and is accomplished through county offices of the superintendent of schools [personal communication, Linda Sargent 2009].

Many areas of the country no doubt have examples of communities being prepared, but one such example is located in the Pacific Northwest. During the January, 2009 flooding in the Seattle, Washington area, the U.S. Corps of Engineers determined that rivers in the Puyallup Valley had a high probability of overflowing their banks and causing extensive flooding. In warning affected areas, personnel contacted the two nearby small communities of Fife and Orting.

As reported in the *Seattle Times* [2009], on hearing of possible flooding, the City of Fife struggled to get people to assist with the anticipated emergency. On receiving the warning, Fife city employees quickly sent people door-to-door into newer housing developments to advise residents of the expected event, asked the county to send out evacuation advisories to 5,400 homes in the city, began looking to find shelter for evacuees, and began to accumulate food, water, and other essentials.

In contrast to the struggles and haste in Fife, the City of Orting was calm and efficient. When the mayor of Orting was called, she knew that all of the preparedness planning had been done in her city and the community had already logged many hours of practice. By her account, she didn't even have to consult the city's emergency plan. "It's one thing to have drills, but to go through it time and time again, you get to know what to do," Mayor Cheryl Temple explained. "I didn't even look at the book this time" [*Seattle Times* 2009].

The city of Orting has had an emergency response plan for many years. That plan was put together and is maintained by city officials and interested community individuals, including the mayor, chief of police, local organizational representatives, and representatives from the school district. It gets reviewed at least yearly and

Figure 10.2 Mt. Rainier, a 14,411-foot volcano. Photo taken from Orting, Washington. (Photo: Beryl Cheal. With permission.)

updated when needed. However, Mayor Temple does not depend on that plan; she has her own detailed personal plan that she carries with her everywhere she goes, every day [personal communication, Mayor Temple 2009]. Drills are scheduled frequently, so that everyone is familiar with the plan and its details (Figure 10.2).

Situated in a valley at the base of Mt. Rainier, which volcanologists believe could erupt again at any time, the community of Orting takes their hazards very seriously. Not only could Mt. Rainier erupt, with lahars flowing swiftly down the

Figure 10.3 Street signs for volcano evacuation route in Orting, Washington. (Photo: Beryl Cheal. With permission.)

Figure 10.4 Community emergency preparedness fair advertised on the Orting High School reader board, Orting, Washington. (Photo: Beryl Cheal. With permission.)

narrow valley, but the two rivers on either side of the valley could flood in devastating fashion. In addition, there are earthquakes—always a possibility on the coast in the Pacific Northwest, frightening winter storms, wildfires, and other catastrophes. The emergency response plan will be adjusted for use in any emergency or any disaster, according to Mayor Temple. But Orting is prepared in case any of these events should occur. Not only does the community have a working emergency response plan, but every effort is made to get the community involved in thinking about safety and in developing a culture of safety (Figure 10.3 and 10.4).

The city recently organized a community emergency preparedness fair held at the local high school with a good response, especially for the first year. Plans have already begun for a bigger and more extensive fair next year. The community has established street signs to designate volcano evacuation routes, and concerned PTA parents have developed a nonprofit organization that has designed a walking bridge over the river nearest the schools to facilitate student's evacuation as they walk to safety. The "Bridge for Kids" is well on its way to fulfilling the dream of involved and concerned parents [personal communication, Orting community members, January 2009].

Schools in Orting consistently work with the mayor's office and other city officials in ensuring the safety of students. They have a responsibility to keep students, staff, and visitors safe. But schools are also where children are the most vulnerable when disaster strikes. Salvano Briceno, Director of the United Nations, International Strategy for Disaster Reduction, in announcing the Disaster Risk Reduction Begins at School campaign writes:

> When a natural hazard strikes, children are among the most vulnerable population groups, especially those attending school in times of disaster. Disasters such as the October, 2005 earthquake in Pakistan, where

> over 16,000 children died in schools that collapsed or the. . . mudslide on Leyte Island in the Philippines, where more than 200 school children were buried alive, are just a few tragic examples of why more needs to be done to protect our children before disasters strike [Briceno 2005, p. 1].

Of the estimated 70,000 people killed in the May, 2008 earthquake in China, more than 11,000 of those were teachers and students trapped inside some 7,000 schools that collapsed [O'Donnell 2008].

However, schools do not have to be traps for children. Whereas many countries do not have adequate construction regulations for schools and others are missing trained construction inspectors, there are examples of school directors taking the responsibility themselves to provide safer school structures. Principal Ye Zhiping of Sangzao Middle School in An County, Sichuan Province, China, took 3 years to renovate his school that had been built in 1983. An County was one of the hardest hit areas of the May, 2008 earthquake. In the renovation, toilets had been rebuilt in Sangzao Middle School, balconies and stairwells were reinforced, and monthly emergency evacuation drills had been introduced. None of the 2,323 students in the middle school were injured during the earthquake, but a nearby school lost all of its pupils. Principal Ye Zhiping sadly informed reporters that, in the 1980s, some schools had been built with great speed, but with low architectural standards and often with flimsy materials used for walls and windows. In a more hopeful tone, he said that things changed in the 1990s. Among other things, standards were upgraded and it became a basic requirement to have two exits in schools [O'Donnell 2008]. Principal Zhiping found that it saved lives to be prepared.

Child-Led Risk Reduction

In looking at responses to disastrous events worldwide, the safety and well-being of children are of paramount importance because children are one of the most vulnerable populations. The United States can learn a great deal from their counterparts in the developing world with respect to caring for and ensuring the safety of children in a disaster. In comparing responses to Hurricane Katrina in Louisiana with those of the tsunami in Aceh, Philippines, Lauten and Leitz [2008] see an apparent disregard for the well-being of children and their families in the aftermath of Katrina, and well-planned, caring responses on the part of government figures in Aceh. Sadly, they have concluded that whereas "internationally accepted standards and best practices in protection was the basis for the child protection response to the tsunami, they were entirely absent from the response to Katrina" [Lauten and Leitz 2008, p. 159].

Central to the philosophy of a culture of safety is keeping children safe and caring for them well during and after disasters. One way to develop this focus on child and family well-being is to develop plans prior to disaster that will keep children safe when an event occurs. Because children are among the most vulnerable population groups and because children are also the hope for any community, the United Nations International Strategy for Disaster Reduction (UN/ISDR) developed an international campaign in 2005: Disaster Risk Reduction Begins at School. The focus was on disaster risk reduction education and safer school facilities. With this effort, the UN/ISDR gathered 168 governments to develop and adopt a 10-year plan to make the world safer from natural disasters. This plan, the Hyogo Framework, was accepted by all attendees at the meeting and specifies guiding principles, priorities for action, and practical means for achieving disaster resilience in vulnerable communities.

Each of the participating governments reported 1 year later on their achievements in meeting the goals of the Hyogo Framework and the Disaster Risk Reduction Begins at School campaign. The good practices reported by each country were organized by the UN/ISDR for reporting purposes in three sections: raising awareness within school communities; building a culture of prevention; and making school buildings safer. Projects involved school children, teachers and nonacademic staff, and in some cases even the surrounding communities.

Several countries concentrated on raising awareness within school communities. Cape Verde, for instance, institutionalized disaster risk reduction in schools; Mozambique trained teachers to help mainstream disaster risk reduction into school curricula; and the United Kingdom implemented a project through the Internet to educate students on how to prepare for and respond to several hazards [Valency and Lazarte 2007].

Other countries concentrated on building a culture of prevention: Costa Rica developed a national education plan for disaster risk reduction; Iran instituted annual earthquake and safety drills in all schools across the country; France taught students prevention and protective measures against major risks; and the United States institutionalized the Red Cross Disaster Education Program into a nationwide curriculum [Valency and Lazarte 2007].

The third emphasis, that of making school buildings safer, was the focus of the last group of countries: Nepal implemented a community-based program where masons were trained in earthquake-resistant construction; Peru developed a simple practical way to reduce school buildings' seismic vulnerability; India developed a project to help reduce vulnerability in schools in mountain areas; and Japan developed national guidelines on school retrofitting [Valency and Lazarte 2007].

Not only have individual governments worked toward disaster risk reduction, but other organizations have also. Save the Children Sweden, a nongovernmental organization, became a major actor in involving children in disaster risk reduction based on goals and objectives of the Hyogo Framework. Knowing that disaster risk reduction involves not only mitigating physical hazards, but also training individuals in developing a disaster risk reduction mindset in addition to learning new skills, Save the Children Sweden's intent was to empower children and youth to become actively involved in developing safer communities and cultures of safety.

Save the Children Sweden's organizational commitment, experience, and history of international projects led them to develop a unique and effective child-led disaster risk reduction project to meet the goals of the Hyogo Framework. They worked with children in several Asian countries that had histories of disaster devastation. The philosophy of Save the Children Sweden in directly involving children in risk reduction activities is supported by the American Academy of Pediatrics. The Academy has found that, as frightening as disasters are for adults, they can be even more traumatic for children. Feelings of anxiety, sadness, confusion, and fear are all normal reactions but have even more devastating emotional effects if experienced over long periods of time. It is their contention, however, that children can develop skills and resiliency needed to overcome these effects and possibly even grow from traumatic experiences. Children, the Academy says, can regain their sense of power and security if they feel that they can help in some way. This can be done by encouraging children to participate in or organize community-response efforts [Markenson and Reynolds 2006, p. e347].

Save the Children Sweden has found that children who participate in child-led disaster risk reduction projects have "a greater capacity to cope with disasters; their sense of security is increased; their knowledge of the risks is developed; and their sense of control and survival potential is enhanced by knowing how to respond to disasters" [Benson and Bugge 2007, p. 2]. Each community is different worldwide and each requires assessment and a community-specific response to their unique needs. The following are Save the Children Sweden's descriptions of children's risk reduction activities in their individual communities in the aforementioned six Asian countries.

In Thailand, children interviewed community members, mapped their communities' designated hazards and resources, assessed risk and vulnerabilities, educated the community about those risks, advocated for government understanding of children's needs in disasters, were instrumental in leading the community in developing action plans to mitigate risks that were found, and conducted program assessments to measure the success of their work. A child spokesperson involved in the effort commented that, "At first we were worried that adults would not help us when we visited the community to gather information. But it turned out that they were very cooperative. Many of them say that the map is very helpful" [Benson and Bugge 2007, p. 3].

In Sri Lanka, children participated in developing their communities' preparedness plans. Schools included risk reduction material in the curriculum, children helped in the reconstruction of schools to make sure they were child-friendly, children developed hazard evacuation maps, and official plans included issues affecting children and how to make sure their needs were met. After participating in these activities, children acknowledged that, "Of course the risks were there before, but we did not have the knowledge to identify them" [Benson and Bugge 2007, p. 3].

In Vietnam, children participated in community meetings to assess risk and hazards in their area, they produced risk maps and action plans for schools and advocated for mitigation of hazards that were found. Children taught community members about preparedness, talked with decision-makers about the impact of disasters on children, and educated their communities about child rights and child protection. At the end of the project, the children reflected on their learning: "In school we learnt about disasters, and we participate in disaster risk reduction in our commune with the adults" [Benson and Bugge 2007, p. 3].

In India, children participated in designing house construction, formed committees in school to identify and mitigate risks and hazards, used cultural performances to raise awareness about disasters and related risks, and created a method to use drums as early warning systems. "We want to inform all the community members on knowledge and skills related to disaster preparedness," they stated after participating in the activities [Benson and Bugge 2007, p. 3].

In the Philippines, schools formed children's emergency response teams who trained in risk assessment, participated in advocacy and awareness-raising within their communities, held regular safety drills for earthquakes and tsunamis, and got involved in peer-to-peer education for other children. Comments from participating children reflected competency and capability, with one youth declaring, "I'm not scared of any disasters anymore, because I already know what to do" [Benson and Bugge 2007, p. 4].

In Indonesia, children participated in local government disaster planning, developed hazard maps, and contributed to contingency plans, including learning about evacuation, and developed educational materials to share with their communities. Children reflected on their participation through what might be some of the hardest tasks for children, that of negotiating and advocating for their ideas to government officials and other adults. But with determination, they declared, "We can teach the community and government about the issues that children face during disasters—we must share our experiences with them" [Benson and Bugge 2007, p. 4].

Developing emergency management plans and disaster management itself has traditionally been done from the top down in the United States and internationally. This assumes that children and youth are passive victims with little or no interest or expertise in communicating risks or preventing or responding to disasters. Adults have been the ones who have taken it on themselves to develop and implement emergency or disaster plans, almost as if they alone are the ones with knowledge, expertise, and ideas. However, as the frequency and intensity of disasters has increased worldwide, the value of talking and listening to children and getting them directly involved in disaster risk reduction has taken on new meaning. Resources are scarce and children are an untapped resource. Although this is still a new concept and lacks formal research on its effectiveness, practical findings suggest that children and youth can be helpful in describing disaster risks, disseminating information, sharing their experiences of disaster, learning survival skills, helping to develop risk reduction plans, and developing a formal plan [Mitchell et al. 2008].

In developing their initiative of child-led disaster risk reduction, Save the Children Sweden demonstrated the value of involving children. Knowing that children need adult guidance in such a mission, they developed community partners. Some of those partners were schools; others were nonschool community organizations. Working together, goals were accomplished. Children, however, took the lead.

> Save the Children believes that children should be centrally engaged in reducing the risks of disaster in order that the negative impacts on communities are significantly reduced. Children have the capacity to contribute, bring a unique perspective to disaster risk reduction preparations and have the right to play a part in making themselves and their communities safer [Benson and Bugge 2007, p. 2].

Students need the guidance and support of adults in the community. To make disaster risk reduction a real community-supported event and to assist students in their work, it is absolutely necessary for adults to be involved because they are the ones who give leadership, continuity, and stability to the project. It is the adults who guide the direction and know the process of developing a preparedness plan. However, the key to the integrity of child-led disaster risk reduction is that children must be major players. The understanding, skills, and experience of adults can be used in further developing children's skills and the knowledge, skills, and experience of children as children are vital to the process and the outcome.

Students Participation in Emergency Planning

How can schools help in developing a preparedness plan? The work of the UN/ISDR and its campaign Disaster Risk Reduction Begins at School give some clues, as does Save the Children Sweden. Because students live, study, and play for so many hours during the week at school and because schools are in such a unique position to provide students with structure and continuity, they are in a perfect leadership position in developing risk reduction/preparedness plans.

All ages of students can help in risk reduction. Disaster risk reduction involves not only mitigating physical hazards, but also developing an awareness of safety and learning about how and why things happen. Younger children can talk about safety, learn about living safely (e.g., crossing streets when the light is green, keeping walkways clear of bicycles and clothing and not playing with fire), and help teach others about safe practices. Older students' risk reduction activities will be more sophisticated (observing safety hazards at school and telling school personnel about them, helping develop a family emergency plan, or getting to know their neighbors so that neighborhood communities will become more inclined to help each other).

At school, there are many resources to use for students. Learning gives greater understanding about what makes disasters happen, which can often mitigate fear and be a foundation for designing safe actions to take when an event occurs. Teachers will want to use various resources to help children learn, such as regular science curricula, textbooks, and possibly some of the following online resources:

- FEMA [www.fema.gov]
- FEMA for Kids [http://www.fema.gov/kids/]
- *Investigating Natural Disasters Through Children's Literature: An Integrated Approach* [Fredericks 2001]
- Live Science [http://www.livescience.com/]
- *Masters of Disaster*, developed by the American Red Cross [2007]
- Plan International [http://www.planusa.org/contentmgr/showdetails.php/id/560953]
- Risk Reduction Education for Disasters [http://www.riskred.org/]
- Save the Children U.S. [http://www.savethechildren.org/]
- Save the Children Sweden [http://www.savethechildren.se/]
- Save the Children UK [http://www.savethechildren.org.uk/]
- The Weather Channel Online [http://www.weather.com/]

The following is not an exhaustive listing of possible activities, but it gives an idea of some of the things students can do to become knowledgeable about safety and what to do in case of an emergency (Table 10.1) [American Red Cross 2007; Fredericks 2001].

There are several levels of child-led disaster risk reduction. Reducing the risk of disaster is one phase of emergency planning—mitigation. Other phases of emergency planning include writing down the actual plan, practicing it, and evaluating it. The following chart describes five levels of students' participation in emergency planning (Table 10. 2). Most schools are at least at Level 1, but many have incorporated parts of all levels. Few, however, have achieved Level 5. Each level of involvement is based on the levels preceding it. The goal is to include students at every level of involvement because students feel empowered when they are given opportunities to participate and contribute to their own safety and well-being.

Schools' Role in Recovery

A key element in students' recovery is their ability to be resilient (to be able to overcome the effects of the disaster). Actually, psychological effects of war, violence, and natural disasters can be more damaging than the physical nature of an event. For students, such incidents can have a devastating impact because the developing brain of the young child is especially vulnerable [Karr-Morse and Wiley 1997]. All experiences, both good and bad, change the brain. The brain is designed to respond to stimulation, and the stimulation associated with fear and trauma can be particularly devastating [Perry 2001]. Schools can provide opportunities and support for resiliency in children.

Table 10.1 Sample Activities for Child-Led Disaster Risk Reduction

Age Group	*Sample Activities*
Preschool kindergarten, early elementary	• Learn how to be safe at school and in the community. • Learn where to go in an emergency. • Decide what to do while waiting to get back to classrooms if evacuated. • Help find hazards inside and outside the school. Talk with teacher about findings and what could be done. • Learn about disasters that might happen in area. • Decide on ways to share with family and friends about what is being learned. • Practice disaster response. • Talk with family about safety at home.
Middle school	• Help decide where to go in case of emergency. • Help decide what supplies will be needed for 72-hour waiting period. • Find out about disasters that might strike community. • Help decide what should be done when event occurs. • Map school inside and out to determine hazards.

	• Discuss findings with school personnel. • Learn basic first aid for appropriate age. • Develop ways to educate friends, family, and community. • Practice disaster response. • Talk with family about developing family safety/emergency plan.
Older children	• Calculate amount of supplies needed for 72-hour period. • Interview community experts on possible disasters. • Discuss what specific actions need to be taken in responding to disaster. • Study and recommend response personnel. • Assist in writing school disaster plan. • Help in devising and implementing periodic practice of school disaster response plan. • Practice disaster response. • Talk with community and political leaders regarding community disaster plan and integrating school plan into that of the community. • Give input to community leaders in developing community disaster plan. • Talk with family about developing family safety/emergency plan.

Table 10.2 Levels of Student's Participation in Emergency Planning

Level 1
School has an emergency plan, trains students in its use, and schedules periodic practice. Examples include fire drills, earthquake drills, and hurricane drills
Emergency plan developed by adults
Students follow directions
Level 2
School has an emergency plan, trains students in its use, and schedules periodic practice. Students learn about hazards in the area and why things happen
Emergency plan developed by adults
Desk learning
Information about disasters included in various school classes—reading, discussions, lectures, and media
Level 3
School has an emergency plan, trains students in its use, and schedules periodic practice. Students learn about disasters/hazards through projects/participation. Students help their families develop family plans
Emergency plan developed by adults
Experiential learning
Information about disasters included in various school classes—reading, discussions, and lectures; students use classroom
Information in projects at school and the community
Level 4
Information gathered by students is incorporated into school plan. Students are involved in the development, implementation, and evaluation of the plan
School recognizes need for integrated emergency plan; emergency plan developed together by adults and students
School and community working separately
Experiential learning through participation, negotiation, and discussion
Level 5
School's emergency plan integrated with community plan. Students participate in community disaster planning and development
Emergency plan developed by adults and students
School and community working together
Experiential learning through participation, negotiation, and discussion

Although we want students to be resilient, they also need adult help to support their efforts and struggles. Schools are not only where students spend many of their waking hours, but equally as important, schools have a socializing and normalizing influence on children's lives. After the family, schools have the greatest capacity for providing children with a range of experiences that protect them and serve in supportive capacities [Johnson 1998].

Schools can play key roles in recovery because, as Judith Herman makes so clear in her extensive work with traumatized people, disempowerment and disconnection from others are the core experiences in psychological trauma. She says that recovery is based on relationships, and those can not occur in isolation [Herman 1997]. From the bus picking up the students in the morning, to inside and outside activities, nutrition, and health and social services, students are together at school and can find strength and comfort in that setting. Relationships can be developed and the school staff has opportunities to be sensitive to student's needs.

In extensive research with children in war zones, Garbarino et al. [1992] confirm Herman's thesis with their discussion of the vital importance of the relationship teachers have with their students—their caring, their understanding, their ability to listen without judgment, and their emotional availability for their students. It is the relationship between the teacher and students that is established that develops sound mental health for all students—kindergarten through high school [Axline 1969].

Barrett et al. [2008] described the positive experiences of Katrina-evacuated youth and the importance of positive relationships developed with their teachers at a new school. Findings indicate that many of the evacuated adolescents turned to school personnel and other significant adults to help them successfully cope with the tragedies they had experienced [Barrett et al. 2008]. Although individual therapy should be done by mental health professionals, teachers can support those mental health efforts through a therapeutic atmosphere and by providing therapeutic programming and activities.

School staff need to help students learn to trust again and feel physically and emotionally safe and they need to create a stable and consistent schedule. Students' behaviors demonstrate how they are feeling, and these behaviors may change after experiencing a traumatic event. Recognizing these changes gives adults opportunities to provide assistance. Disaster may cause students to be unusually quiet or unusually aggressive, appear detached, become easily upset or angry, show fear of new situations, revert back to younger behaviors, or have symptoms of illness. These are normal reactions to abnormal circumstances.

In response, schools may want to provide special staffing patterns for a time. Limiting the number of adults with whom a student relates can assist in developing trust and attachment again. For younger children, schools may want to develop family groupings to assist this process; older children might benefit from longer periods with their homeroom teachers.

Good schools and programs for students already provide many activities that are helpful for those who have survived disaster, but some things could be done with more intention. For example, in a preschool classroom, the sand or water table that is often placed against the back wall could be brought out into the room so students will notice it and be drawn to its soothing activities. The housekeeping corner could include more props that will enhance student's playing out scenes and feelings from the event they have experienced and have opportunities to talk about them, if they want to do that [Cheal 2002; Koplow 1996, 2002; Oehlberg 1996] (Figure 10.5).

Many opportunities should be available for students of all ages to tell their stories and express their fears, sadness, uncertainty, and relief [Johnson 1998]. This can be done with younger children through activities such as talking, storytelling, art, music, dramatic play, or playing with blocks. Teachers will want to give older students many opportunities to tell their stories through class discussions, role play, writing, music, drama productions, art activities, scientific research, and sharing information with other students, family, and the community. Reports or creative writing on topics that have to do with the event frequently are useful [Fredericks 2001].

Because students may feel that their lives are out of control, activities should be planned and materials should be made available that will help them feel in control again. Rituals should be developed for they give something for students to hang onto and help them feel secure. Children want to be needed, so giving appropriate classroom responsibilities can give them a sense of contribution [Koplow 2002]. Teachers may also find ways for students to visit a children's hospital, write messages with good wishes for older people in the neighborhood, or make some other contribution in their community to help someone else's life and well-being (Figure 10.6).

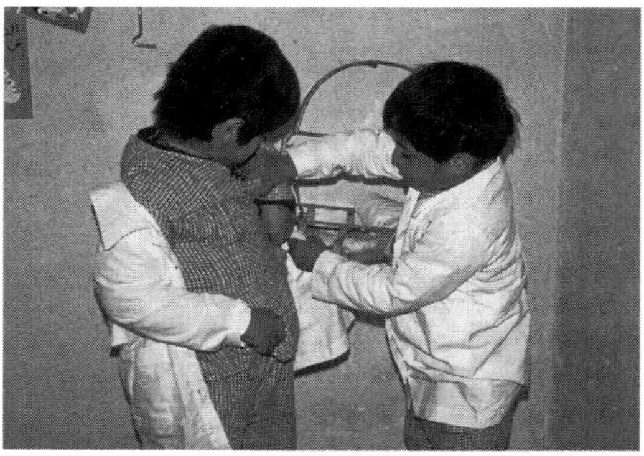

Figure 10.5 For younger children, teachers may want to give opportunities to play out some of the events that have occurred, as in the above picture of two children trying on doctor coats in a refugee camp kindergarten in the Middle East. (Photo: Beryl Cheal. With permission.)

Figure 10.6 Playing games that have rules and structure can be helpful for children who need boundaries. These children are playing Duck, Duck, Goose, Goose in a refugee camp kindergarten in the Middle East. (Photo: Beryl Cheal. With permission.)

It is crucial that schools working with students consider culture in planning activities. This is equally important for events involving violence as well as natural disasters. In an attempt to provide for psychosocial needs of students affected by experiences of war and displacement, Save the Children Sweden initiated a school in Sudan. Their intention was to equip the teachers with the knowledge and skills required to enable them to lead group discussions and have supportive conversations with individual students. However, for various reasons, that training was never accomplished. Instead, the teachers developed their own methodology based on cultural norms of not directly discussing painful issues. They developed a range of extracurricular activities for children, some involving parents. Through singing, drama, storytelling, sports, and games, teachers encouraged students to express feelings and verbal communication in a culturally appropriate manner and in the context of a supportive and caring community [Tolfree 1996].

From her work with communities surviving violence and war in many countries, psychologist Nancy Baron chose the centuries-old art of storytelling to help children who have been affected by war, violence, and loss. Her two books for elementary age children, *A Little Elephant Finds His Courage* and *The Little Elephant Helps to Find Peace*, both set in the forests of Sri Lanka, are "directed towards promoting resilience, building courage, strengthening family and community bonds and preventing future violence" [Baron 2005]. Both stories have been translated into several languages and are used with many groups of children. They can be used by teachers, psychologists, social workers, parents, religious, and traditional or community leaders—anyone who works with children.

In her discussion guides, Dr. Baron suggests that reading the books to children and then talking about what they have heard in the story and seen in the illustrations can stimulate understanding and empathy about dire circumstances to innocent victims [Baron 1994, 2005]. Talking about what happens to the animals and how they feel can be easier, and sometimes more culturally appropriate, than talking about one's own experiences and feelings. Listening and talking about the stories may help some children, however, to identify their own feelings and then be able to verbally express them.

Good nutrition can also be a challenge for students surviving a disaster. After a traumatic event, students may not feel like eating, families may not be able to buy or prepare nutritious food, and for some other reason, students may come to school hungry; this is possible for older and younger children. School staff should be prepared by having extra quantities of nutritious food available that students especially enjoy. Make it available at many times during the day, fixed in attractive ways, in small pieces that students can eat easily. Remember that likes and dislikes may revert back to earlier tastes during these times [Cheal 2002, 2007].

Student health needs sometimes go unobserved during times of crisis because children may have been injured during the event and it might not have been noticed or families may not have had the opportunity to take their children for regular health checkups. School personnel should be particularly sensitive to those possibilities and design protocols where those needs will be met.

School staff should be particularly aware of traumatic reminders for older as well as younger students. If, for instance, the devastation of a flood has been experienced, the student may express particular stress when it begins to rain. Students may be helped to see water as a resource rather than as a destructive force. Wind might trigger memories of a frightening hurricane. Secondary stressors can cause continuing trauma. If a child's home has been lost and the family is required to find other housing, the child may adapt well to the first move, but by the time the family has moved several times from the emergency shelter to a relative's home and then to a more permanent location, adapting capabilities may become strained. That becomes a secondary stressor and can be experienced by children of all ages.

Some students appear to return to normal within a few days, whereas others take weeks or months. Even after a year or two, some children's behavior may show that they are still dealing with the effects of the disaster. If a student's fears, worries, or behavior problems continue or increase, staff should suggest that parents consult a mental health counselor, social worker, religious person, or community leader who can give assistance. Some of these resources might also be used at school to discuss with staff and parents about how to best work with their children.

A plan regarding how to work with students after disaster is very important, and training the staff is equally as important. Training on what to do and how to respond to students should include all staff and volunteers who interact with students and their families—bus drivers, cooks, janitors, administrators, office workers, health, social service personnel, counselors, and teaching staff. Staff must be prepared and trained so that they know how to apply their usual skills to trauma situations [Johnson 1998]. They may also want to develop additional skills in dealing with what they hear and see as well as be able to alert others to important observations.

Because the family is the most important social unit in a child's life, schools will want to find ways to strengthen those relationships. Parents frequently do not have opportunities to talk about the event with other adults and thereby defuse some of their own fear and anxiety. Schools could provide space for parents to relax and talk together. Parent meetings with outside experts in mental health or other topics of interest would allow them opportunities to see how other families are dealing with the crisis and gather specific information that might help them in getting their lives and those of their families back together. Schools might provide opportunities for parents to meet with community leaders, local healers, or others important in various ethnic communities served by the school. They might want to know how to work with their children after the experiences of disaster and teachers could share some of the healing activities that occur in classrooms or in out-of-class activities and give parents ideas of things they might do with their children.

Parents can organize projects for the community, which can be a way to contribute to someone else's well-being, socialize with others, and help them focus on something outside their own challenges. Family get-togethers for fun are also good opportunities for sharing experiences. Schools may want to organize fun nights or have a series of creative nights where families could get together at school and have opportunities for art activities, develop puppet shows, write and produce dramatic presentations, participate in creative dance events, or find creative opportunities to express their emotions in other ways. Some cultures mourn traumatic events by specifically not participating in dances and songs, but rather work through their grief quietly. Schools providing space and opportunities for quiet contemplation and grieving could be a community service in such circumstances.

Activities for parents and families can continue for a short time after a disastrous event or can continue for longer periods of time depending on the needs and desires of participants. Theater Arts Against Political Violence [http://www.itspnyc.org/theater_arts_against.html/], a community theater group organized in lower Manhattan immediately after 9/11, continues giving public presentations throughout the area more than 8 years later.

Staff can work best with students if they have been given opportunities to do their own personal healing. They should have the opportunity to talk about their experiences, share stories, cry and laugh, and feel sad, frustrated, and relieved. Opportunities and spaces for staff to talk together or be alone, if needed, should be provided. Schools might also give opportunities for staff to express themselves through discussions, art, music, writing, drama, or other activities. Several specifically scheduled optional staff meetings can give occasions to work through some of these issues, formalize the healing process, and let staff know they are being supported. Depending on the circumstances, an outside person may be brought in to conduct these sessions. Staff should continue to be supported as long as it is needed. If they need personal mental health assistance, encourage them to make those arrangements. The school may want to give special assistance for this to occur in the form of grants, loans, time off, or other supportive actions.

Conclusion

Students are some of the most vulnerable in times of disaster, but they can also play a major role in saving lives. On a beach in Thailand in December of 2004, 10-year-old British schoolgirl Tilly Smith showed the world that students are not passive victims. Enjoying the beach that sunny afternoon, she noticed the sea retreating and remembered recent classroom discussions back home in Britain. She quickly told her mother she thought a tsunami was coming and together they were instrumental in saving the lives of everyone on the beach [Ripley 2008].

Schools can help greatly in rebuilding communities after disaster by developing their own school-specific preparedness plans, bringing children into the development of that plan as well as cooperating and coordinating with the community as they develop their community-wide disaster preparedness plans. Vitally important, too, in rebuilding communities is for schools to be prepared to help students, their families, and staff recover from disaster. Using our creativity and imagination, we can forge new alliances, develop new partnerships, and find new ways to use available resources to save lives and make communities dynamic again.

References

American Red Cross. 2007. *Masters of disaster: Quick-start guide—Educator's kit*. Retrieved from http://www.redcross.org/disaster/masters/ (accessed January 26, 2011).

Axline, V. M. 1969. *Play therapy*. New York: Ballantine Books.

Baron, N. 1994. *A little elephant finds his courage*. Retrieved from http://www.littleelephant.org/(accessed January 30, 2009).

Baron, N. 2005. *The little elephant helps to find peace*. Retrieved from http://www.littleelephant.org/ (accessed January 30, 2009).

Barrett, E. J., Martinez-Cosio, M., and Barron Ausbroks, Y. 2008. The school as a source of support for Katrina-evacuated youth. *Children, Youth and Environments* 18:202–235.

Benson, L., and Bugge, J. 2007. *Child-led disaster risk reduction: A practical guide*. Retrieved from http://www.rb.se (accessed January 26, 2011).

Briceno, S. 2007. *World disaster reduction campaign: Disaster risk reduction begins at school.* Retrieved from http://www.unisdr.org/eng/public_aware/world_camp/2006-2007/pdf/WDRC-2006-2007-English-fullversion.pdf (accessed March 29, 2009).

Brunsma, D. L., Overfelt, D., and Picou, J. S., eds. 2007. *The sociology of Katrina: Perspectives on a modern catastrophe*. Lanham, MD: Rowman & Littlefield Publishers, Inc.

Cheal, B. 2002. *Helping children cope with frightening events, what you can do!* Retrieved from http://www.disastertraining.org (accessed January 26, 2011).
Cheal, B. 2007. *Developing programs for young children that help in healing after disaster, a curriculum.* Unpublished manuscript, International Trauma Studies Program, Columbia University.
Fredericks, A. D. 2001. *Investigating natural disasters through children's literature: An integrated approach.* Englewood, CO: Teacher Ideas Press.
Garbarino, J., Dubrow, N., Kostelny, K., and Pardo, C. 1992. *Children in danger.* San Francisco, CA: Jossey-Bass Publishers.
Herman, J. 1997. *Trauma and recovery.* New York, NY: Basic Books.
Ishaq, A. 2008. *Rebuilding after disaster: A child-centered approach.* Proceedings of the International Conference on Rebuilding Sustainable Communities for Children and Their Families after Disasters, November 16–19, 2008. Boston, MA: University of Massachusetts.
Johnson, K. 1998. *Trauma in the lives of children.* Alameda, CA: Hunter House.
Karr-Morse, R., and Wiley, M. 1997. *Ghosts from the nursery.* New York, NY: Atlantic Monthly Press.
Koplow, L. 1996. *Unsmiling faces: How preschools can heal.* New York, NY: Teachers College Press, Columbia University.
Koplow, L. 2002. *Creating schools that heal: Real-life solutions.* New York, NY: Teachers College Press, Columbia University.
Lauten, A. W., and Lietz, K. 2008. A look at the standards gap: Comparing child protection responses in the aftermath of Hurricane Katrina in the Indian Ocean Tsunami. *Children, Youth and Environments* 18:159–201.
Markenson, D., and Reynolds, S. 2006. The pediatrician and disaster preparedness. *The American Academy of Pediatrics Journal* 117(2):560–565.
Mitchell, T., Haynes, K., Hall, N., Choong, W., and Oven, K. 2008. The roles of children and youth in communicating disaster risk. *Children Youth and Environments* 18:254–279.
O'Donnell, L. 2008. At China quake schools, life and death in stark contrast. Retrieved from http://www.yahoonews.com (accessed May 26, 2008).
Oehlberg, B. 1996. *Making it better: Activities for children living in a stressful world.* St. Paul, MN: Red Leaf Press.
Perry, B. D. 2001. Violence and childhood: How persisting fear can alter the developing child's brain. Retrieved from www.childtrauma.org (accessed March 29, 2009).
Rich, H., and Kelman, I. 2007. Teens learn to save lives—and do. *Disaster and Social Crisis Research Network Electronic Newsletter* 32:6–7.
Ripley, A. 2008. *The unthinkable: Who survives when disaster strikes—and why.* New York, NY: Crown Publishers.
Romano, L. 2006. *In harm's way: Rethinking "natural disasters."* Retrieved from http://www.oxfamamerica.org (accessed February 16, 2009).
Seattle Times. 2009. Tale of two disaster plans: Orting, Fife, p. A9.
Stevens, E. 2006. Signs point to success: Reducing disaster risks in El Salvador, Oxfam America. Retrieved from http://www.oxfamamerica.org (accessed 2006).
Thompson, M., and Izaskun, G. 2004. Cuba, weathering the storm: Lessons in risk reduction from Cuba. Report for Oxfam America. Retrieved from www.oxfamamerica.org/newsandpublications/publications (accessed March 29, 2009).

Tolfree, D. 1996. *Restoring playfulness: Different approaches to assisting children who are psychologically affected by war or displacement,* Radda Barnen (Swedish Save the Children). S-107 (88). Stockholm, Sweden. Retrieved from http://www.rb.se.

Valency, R. A., and Lazarte, J., eds. 2007. *Towards a culture of prevention: Disaster risk reduction begins at school, good practices and lessons learned.* Report for the United Nations International Strategy for Disaster Reduction. Retrieved from http://www.unisdr.org/eng/about_isdr/isdr-publications/11-education-good-practices/education-good-practices.pdf (accessed March 29, 2009).

EUROPE III

Chapter 11

Citizen Seismology: How to Involve the Public in Earthquake Response

Rémy Bossu, Sébastien Gilles,
Gilles Mazet-Roux, and Fréderic Roussel

Contents

Introduction

Documentation of the effects of earthquakes has always been, and remains, essential in seismology. Understanding tectonic setting and seismic hazard studies requires knowledge of seismicity over long periods of time, well before the instrumental era. These effects can be reported in historical documents, recorded on historical monuments, or observed in the landscape. For more recent earthquakes, field surveys are performed to better understand the distribution and nature of the damage, to calibrate historical observations, and ultimately to improve risk assessment and enhance our capacity to predict damage scenarios in future earthquakes.

The Internet opened a new way to collate this information directly from witnesses' accounts. Online macroseismic questionnaires (i.e., questionnaires that described earthquakes' effects) were first developed in 1999 at the U.S. Geological Survey [Wald et al. 1999] before being implemented in seismological observatories around the world. It is not only an inexpensive and easy way to assemble data of scientific interest, but its rapidity also makes it useful for response purposes by providing initial, field-based indications of the actual effects of an earthquake. Such in situ clues can be decisive in identifying the likely damage scenarios, particularly for moderate earthquakes where uncertainties in location and magnitude can dramatically expand the ranges of possible damage [Bossu et al. 2009].

Online macroseismic questionnaires are common examples of community involvement in data collection. Denizens are significant providers of information (e.g., testimonies, photographs, and videos) to institutionalized media such as television, radio, and newspapers. Information on disasters is more often spontaneously posted on social networks and discussion forums [Winerman 2009]. The first picture of the emergency landing of an airplane into the Hudson River on January 16, 2009, appeared on a social networking site, Twitter, only 10 minutes after take-off [Deards 2009].

We do not intend to aggregate and assimilate useful seismological information available on various Web sites. Instead, we take advantage of the presence of witnesses, looking for earthquake information on our Web site [http://www.emsc-csem.org], to rapidly collect data based on their observations. We will discuss the European-Mediterranean Seismological Centre (EMSC), the tools we developed, and their results. We will then illustrate what can be learned regarding the reaction of the public when subjected to earthquake shaking and how this understanding may help optimize data collection. We will also explore the implications of the relationship between those directly affected by earthquake shaking and the scientific community.

In science, projects that involve volunteers in observation, measurements, or computation are described as citizen science. These projects range from bird censuses to distributed computing on volunteers' computers. Thus, we name our current work citizen seismology.

The EMSC: Roles and Web Site

The EMSC is a scientific nonprofit, nongovernmental organization created in 1975 by the seismological community as an instrument to facilitate international data exchange and to rapidly locate earthquakes. Currently, 84 seismological institutes from 55 different countries are members of the EMSC, spanning the whole European–Mediterranean (Euro–Med) region from Iceland to Dubai and from Morocco to Russia.

The EMSC provides current earthquake information on its Web site by collating seismological observations from approximately 70 different monitoring networks and then offering a composite service at a broader geographical scale than its individual contributors [Mazet et al. 2009]. Earthquake information is coordinated with national seismological institutes in the affected region for a variety of reasons. It provides redundancy in case of technical problems. Its availability improves location accuracy in border regions and offshore earthquakes. Making this information available also maintains the information flow outside the affected areas in case of a severe earthquake and offers unique data for international organizations like European authorities.

Until 2004, our Web site was exclusively devoted to seismologists and the information it provided was prone to mistakes. Despite disclaimers, misuses and misunderstanding grew as the audience of the Web site expanded. A new Web site was developed to circumvent these problems. It offers answers to the most common questions the public may have after an earthquake. It also provides earthquake characteristics, information, and contextual maps (e.g., geographical maps, localities and population in the vicinity, local time, and background seismicity). Traffic has constantly increased since the redesign; in 2008, the daily average was 300,000 loaded pages compared to 220,000 in 2007 and 120,000 in 2005, with visitors originating mainly from the Euro–Med region and the United States. Brutal surges of traffic following large or felt earthquakes necessitated a new Web infrastructure in early 2009. Difficulties aside, our site offers numerous scientific benefits.

Tools that Collate Earthquake Information from the Public

Earthquakes that make headlines also generate large increases in the number of visits on Web sites offering earthquake information. In the case of a felt earthquake, observed surges of Web site traffic are significant, usually beginning within the first few minutes after the earthquake occurrence (Figure 11.1). We surmise that people who feel the ground shaking rush to the Internet to investigate cause of the shaking, which causes our traffic to increase. In other words, it is earthquake witnesses who cause the Web site traffic to suddenly increase.

One can then map where an earthquake was felt simply by locating Internet Protocol (IP) addresses during traffic surges. In practice, the level of traffic is monitored in real time to detect significant surges. Once detected, the observed traffic at each IP location is compared with the average traffic observed at the same place, during the same days of the week, and at the same hour of the day during

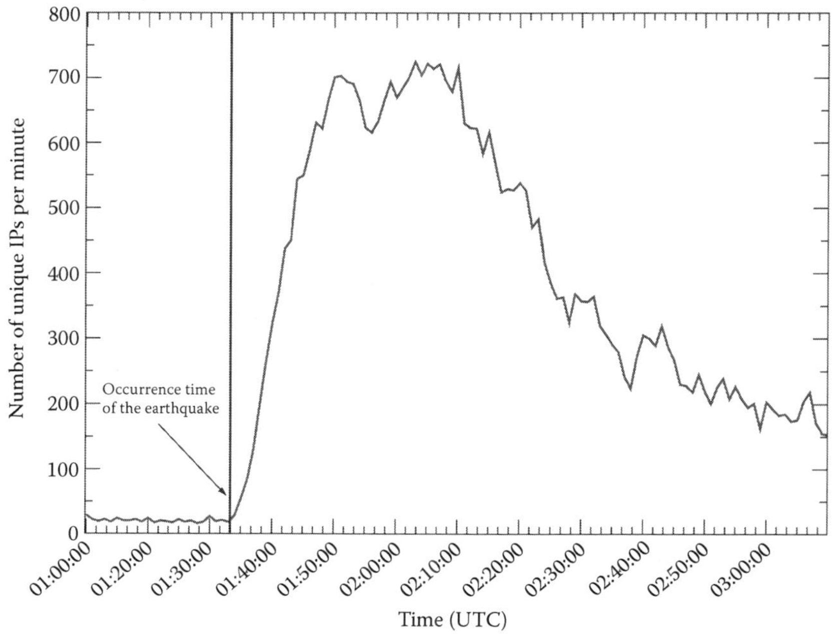

Figure 11.1 Time variations of traffic observed at the EMSC Web site during an April 6, 2009, magnitude 6.3 earthquake in the Abruzzi region of Italy. Traffic variations are represented in relation to origin time of the earthquake.

the previous four weeks. The area where the earthquake was felt is then delineated by the geographical distribution of the IP locations exhibiting significant traffic increases (Figure 11.2a). Areas that might have been hit hard should also be identifiable based on a significant lack or complete absence of visitors due to damaged infrastructure. This original product is called a felt map [Bossu et al. 2008].

A list of ignored IP locations is maintained to increase accuracy of felt maps. The list includes identified indexing robots and seismological institutes. The EMSC uses Digital Element's* IP Intelligence software to identify the geographic locations of site visitors, a technology that was initially developed to fight fraud on the Internet by matching IP location of a client with billing and shipping addresses for Web statistics and for targeted online advertising.

This method of tracking IP locations to map felt earthquakes was first applied in 2004 by Bossu et al. [2007]. To the best of our knowledge, it was the first developed application that automatically and rapidly characterized an external event through the reaction it generated in the public via the Internet. Similar applications are likely to continue to appear in various fields where involved citizens converge

* Digital Element delivers the de facto standard in IP intelligence and geolocation technology. http://www.digital-element.net.

on specific Web sites [Bossu and Walker 2009]. Google Flu Trends is an example. It uses aggregated Google search queries to identify weekly flu activity at a state level in the United States with a reporting lag of 1 day [Ginsberg et al. 2009]. The main difference between our method and the method employed by Google Flu Trends is reflected in the typical time scale of the considered external event; a quasi-instantaneous phenomenon for an earthquake occurrence in our case and a months-long epidemic period in the case of influenza.

An example of the application of the felt map algorithm is presented in Figure 11.2a for the magnitude 5.3 (M5.3) earthquake that occurred in Vrancea, Romania on April 25, 2009. IP location and frequency data clearly illustrated that the event had been widely felt in Moldova (to the north), eastern Romania, and Bulgaria (to

(a)

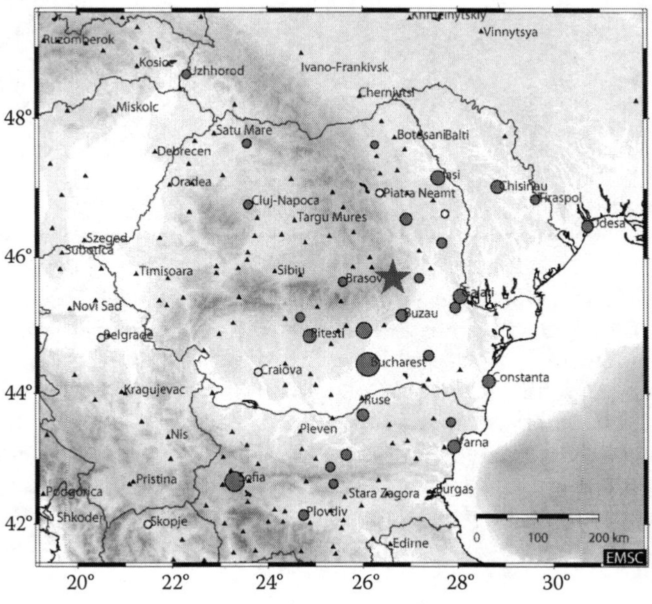

Figure 11.2 (a) Felt maps created during the April 25, 2009, M5.3 earthquake in Vrancea, Romania, within 5 minutes (duration) of earthquake occurrence (start time). Gray circles represent geographical origins of statistically significant increased traffic observed during the 5-minute time window. White circles represent traffic with no significant variations. Black triangles represent the observed geographic origins of visitors in past 12 months and characterize the regional audience of the EMSC as well as the maximum spatial resolution. Circle size is a function of the difference between expected and observed numbers of unique IPs, with the smallest circle equal to 0 and the largest greater than or equal to 100. The first epicentral location (represented by a star) for this earthquake was reported on the EMSC Web site 13 minutes after its occurrence. IP locations are determined by Digital Element.

(b)

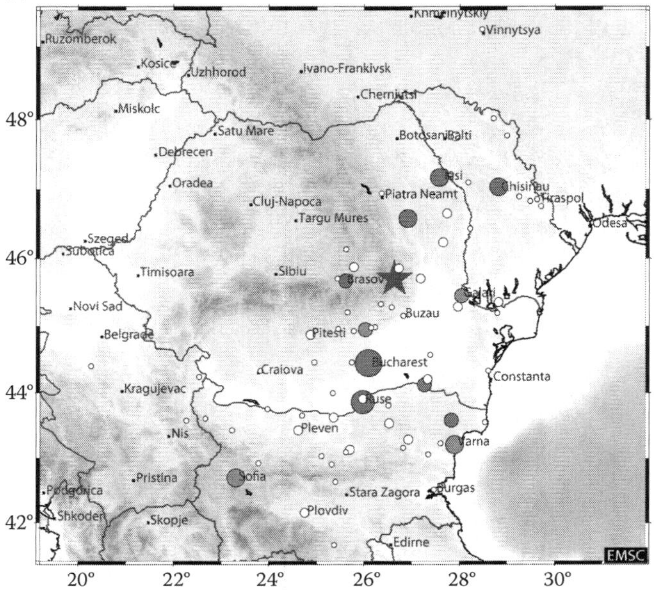

Figure 11.2 (Continued) (b) Macroseismic map based on collected online questionnaires. The area where the earthquake was reportedly felt is comparable with the map in Figure 11.2a. There were only two levels of intensity reported. Light gray circles indicate EMS-98 level III reports of weak seismic activity, and darker gray indicates a slightly stronger but largely observed EMS-98 level IV intensity. The white circles indicate reports that the quake was felt, but gave no indication of intensity. There were 300 reports received from 77 areas, and no reports of strong or damaging intensity. Circle size indicates number of reports received from a given location, ranging from as few as 1 report to as high as 30 or more reports received. Intensity was calculated in those communities from which 5 or more reports were received.

the south) within 5 minutes of the earthquake's occurrence (Figure 11.2a). The IP location pattern is consistent within these three countries. Isolated traffic surges from western Romania (Figure 11.2a) might indicate that the event had also been felt in this part of the country, but no conclusion can be drawn because isolated variations of Web site traffic could be purely accidental. More important, the existence of traffic originating from the locations closest to the epicenter (i.e., where the ground shaking is maximum) (Figure 11.2a) indicates the stability of the local Internet infrastructure and, in turn, the absence of widespread damage.

The second tool implemented at the EMSC Web site is an online questionnaire available in 32 languages (which cover the vast majority of languages in use in seismically active countries of the Euro–Med region). Witnesses are requested to locate their observations using a global database of cities. Then, through closed questions they characterize the level of shaking they were subjected to and the damage they may have observed. Collected questionnaires are automatically converted in macroseismic intensities (i.e., values describing the effects of earthquakes along the predefined scientific European Macroseismic Scale) [Grünthal et al. 1998]. The resulting macroseismic maps are immediately put online and regularly updated as new questionnaires are collected. They are labeled "Not Validated" until a seismologist has visually checked them. If fewer than five questionnaires are returned from the same location, reports of intensity are not considered reliable enough to assign an intensity value and simply appear as "felt" on the map.

The macroseismic maps (Figure 11.2b) for the same Vrancea earthquake and the felt map (Figure 11.2a) yield similar results in terms of area where the earthquake was felt. No questionnaire was collected from western Romania, which strongly suggests that the event was not felt in this area and supporting the validity of the felt map. The overall northeast–southwest extension with a limited attenuation of the intensity is typical of an intermediate depth earthquake (its hypocentral depth was estimated to 96 km) in this region [Mandrescu et al. 1999]. It is important to note that, although a felt map can be created within 5 minutes, it generally takes 90 minutes to a few hours to obtain a reliable macroseismic map and the last questionnaires are often collected several days after the earthquake.

According to the International Telecommunication Union [ITU 2008], there are 4 billion cellular subscribers worldwide. Recent mobile device models generally include a camera, making cell phone users the largest possible network of potential reporters. Citizens can share their photographs or videos on the effects of earthquakes with the EMSC by email, by uploading them onto the Web site, or through a dedicated phone line. All photographs are georeferenced. Before sending the photographs, users indicate the causative earthquake and the location of their observation using the same database of cities used for macroseismic questionnaires. A temporary code integrating this information, which must be attached to the multimedia messaging system (MMS), is generated online for witnesses who want to send photographs directly from their cell phones. Photographs and videos are only relevant for damaging earthquakes or for earthquakes large enough to generate long-duration shaking. Photographs are systematically checked before publication for relevance, sorted by location and causative earthquakes, and archived to build a scientific database.

The collected photographs are mainly related to damage, but some include other earthquake-related observations. In November 2007, a large earthquake (M7.7) shook Chile. A witness based in Tocopilla (24 km from the epicenter) sent us photographs taken during the shaking that showed a dust cloud above the mountains. The cloud was caused by rocks rolling along the mountains' slopes (Figure 11.3), a transient phenomena not easy to document. This type of data is a good example of the peripheral benefits of the structure of our Web site.

Data collection using IP location and submitted questionnaires or photographs has been in operation at the EMSC for several years. These methods are still subject

Figure 11.3 Picture taken by a citizen of the city of Tocopilla, Chile, and submitted to the EMSC. This photograph was taken 24 km east of the epicenter of a M7.7 earthquake that occurred in November 2007. A dust cloud over the mountain caused by displaced rocks rolling along slopes is observed during the shaking.

to fine tuning and improvements based on experience feedback. Since 2006, we have also been developing a way to collect ground motion data, recorded by laptops, to provide additional input on earthquake effects. Apple laptops and several personal computer models include a three-component accelerometer that detects free falls and parks the head of the hard disk before a shock to avoid the head damaging the disk. Being able to rapidly and automatically collect this data on a large scale following an earthquake will potentially provide dense data from cities (i.e., where the density of laptops is high and where the damage can be the most significant). It will also help with mapping the spatial variations in actual ground acceleration, a parameter well known to significantly correlate with the level of damage [Wald et al. 1999]. Our feasibility study for this type of data collection, performed in 2007, was successful and our version developed in collaboration with the National e-Science Centre in Edinburgh is to be publically released in 2011. A similar initiative in the United States, called Quake Catcher Network, confirmed this feasibility by collecting a few waveforms for a M5.1 earthquake recorded on laptops at a few tens of kilometers [Cochran et al. 2009].

Lessons Learned

Information collected from the public becomes more reliable and accurate as it is crosschecked against a large number of witness accounts. In general, no conclusion can be drawn from a unique witness account. This is certainly the case on the Internet where some users may simply want to test the available tools without having experienced an earthquake. The key to the success in citizen seismology via the EMSC Web site, then, is to attract as many witnesses as quickly as possible after an earthquake's occurrence. This goal can only be achieved if we understand witness reactions and respond to their needs.

In this part of the study, we present and discuss our experience: the necessary trade-off between preciseness—which is essential for scientific usage of collected data—and a user-friendly interface, without which no significant participation can be expected. We will also explain how our analysis of the use of tools on our Web site sheds light on public reaction following an earthquake.

Lessons Learned from Felt Map

A felt map focuses on the reaction of a population, rather than on physical seismological data from an earthquake itself. This focus differs from that of earthquake monitoring, where the magnitude of an earthquake is the main (and often the unique) criterion used to identify significant events. For the public, the important earthquakes are not simply the larger ones, but those that were felt, regardless of magnitude. Following a felt event, the evolution of the hit rate* illustrates obvious similarities between earthquakes from different countries and magnitude, ranging from M3.6 to M6.7 (Figure 11.4). An increase in the hit rate begins in the first minutes after an event. A typical rise time (i.e., the time to reach the maximum amount of hits) of approximately 20 minutes can then be observed. Rise is then followed by exponential decay. Significant variations are limited after 90 minutes from earthquake origin time (Figure 11.4). Felt-map time characteristics will later be compared with those from the macroseismic questionnaires.

Hit-rate data confirms that Internet users are expecting real-time information as they rush to the EMSC Web site to investigate the cause of the shaking they just experienced. The amplitude of the hit-rate increase can be considered a quantitative measurement of the size of an audience requesting information from our Web site. Monitoring hit-rate variations promotes understanding of the expectations of the public. It also highlights elements leading to short- and long-term modifications of our audience and of their Internet usage.

The five widely felt earthquakes that have occurred in the Vrancea region of Romania since 2004 illustrate changes in the reaction time of the Romanian population regarding our Web site. (This data can probably be extrapolated to other European regions.) Traffic increased with each successive earthquake, regardless of magnitude. This likely reflects an ever increasing Internet penetration rate among the public (Figure 11.5a). Reaction time (i.e., the time lapse between earthquake origin time and hit-rate increase) shrunk significantly (Figure 11.5b). If defined as the time required to double the level of traffic, reaction time was 8 minutes in 2004, 5 minutes in 2005, 3 minutes in 2006, 2 minutes in 2007, and less than 2 minutes in 2009 (Figure 11.5b). Reaction times, which can be measured in minutes, are

* The hit rate is an audience metric for a Web site. Higher hit rates characterize increased Web site traffic.

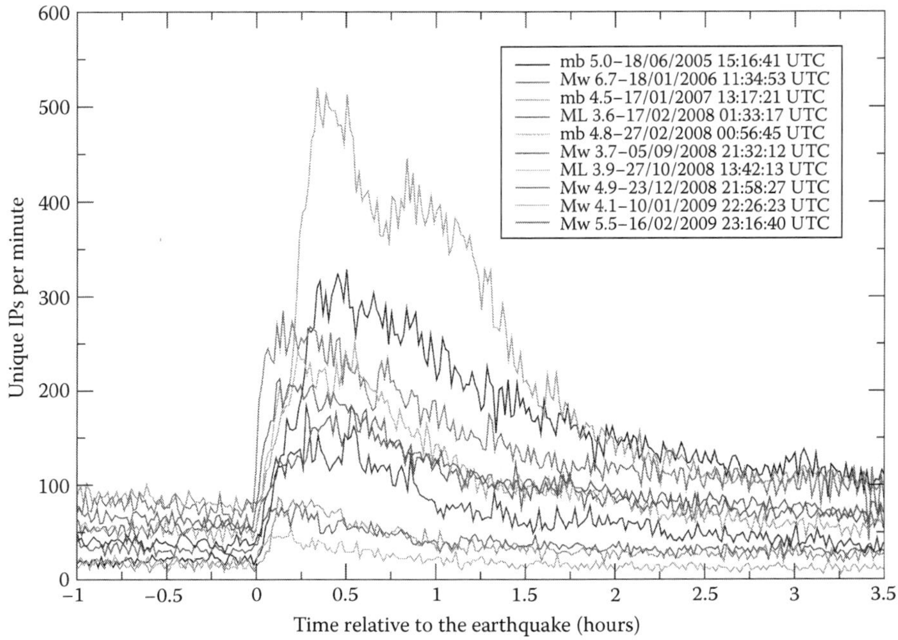

Figure 11.4 Observed temporal variations of hit rates on the EMSC Web site following a series of Euro–Med felt earthquakes, ranging from M3.6 to M6.7, that occurred between June 2006 and February 2009. Hit rate is measured as the number of unique IP addresses.

typical for earthquakes widely felt by our audience in 2009, even for events occurring in the middle of the night (e.g., the l'Aquila earthquake; Figure 11.1). On the EMSC Web site, the existence of a hit-rate surge is the first indication of the existence of a felt earthquake. Initial location and magnitude estimates are typically posted on our site 8 to 12 minutes after an earthquake occurrence.

A possible explanation for the reduction of reaction time in the past 5 years is the change in public Internet access, specifically the shift from dial-up to broadband access. A dial-up connection requires the full use of a telephone line, causing it to provide intermittent Internet access in many cases. In addition, establishing access using a dial-up connection takes several minutes. Broadband access allows a user to remain permanently connected and to immediately access a Web site at any time in a matter of seconds without affecting telephone lines. The gradual reduction of reaction time on the EMSC Web site could be a sign of a progressive change of access technology rather than a modification of the behavior of the Internet users.

Web-site audience can also change rapidly. Three earthquakes were felt widely in Sofia, the capital of Bulgaria, over a three-day period in November 2008. No

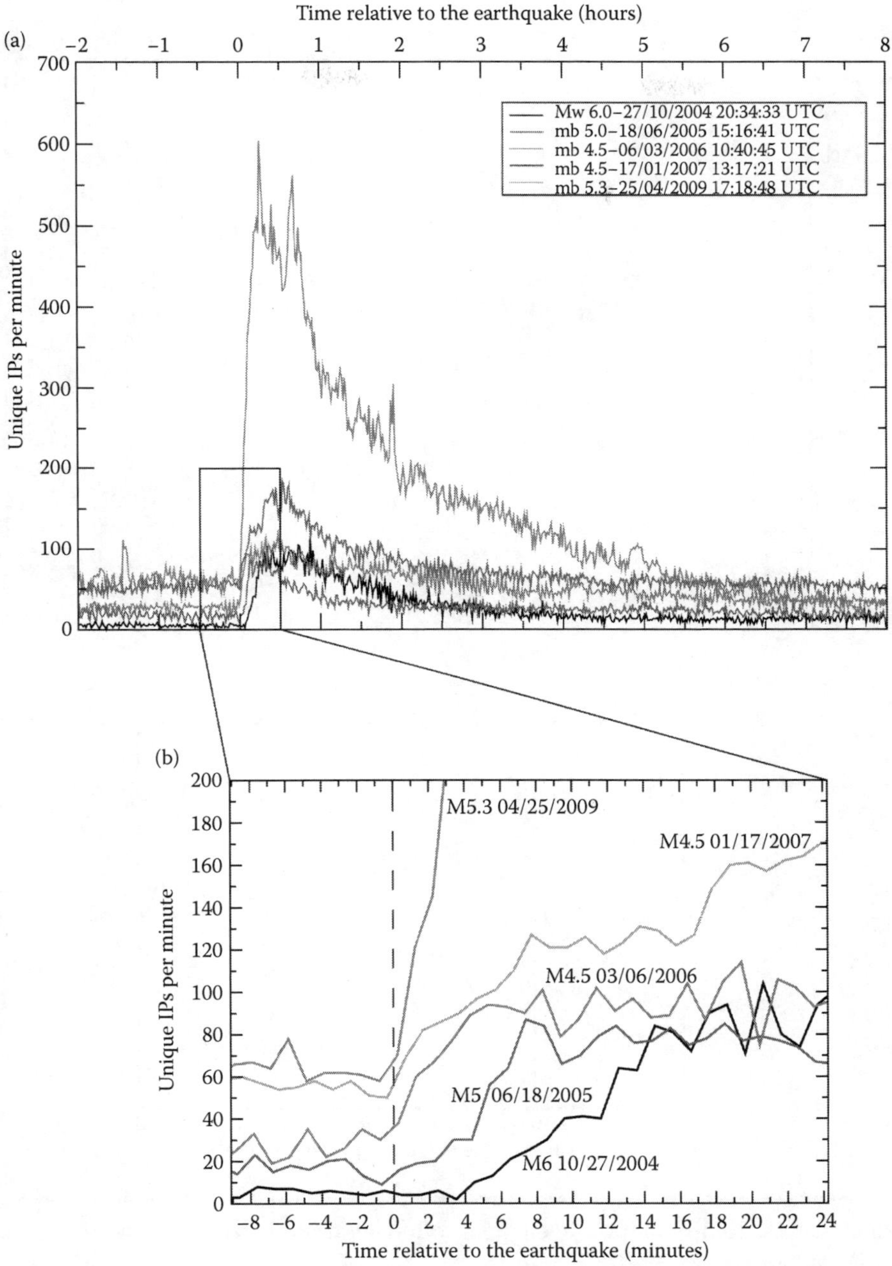

Figure 11.5 Comparison of hit rate temporal change, measured in unique IP locations per minute, for five earthquakes that shook the same region of Vrancea, Romania, between 2004 and 2009.

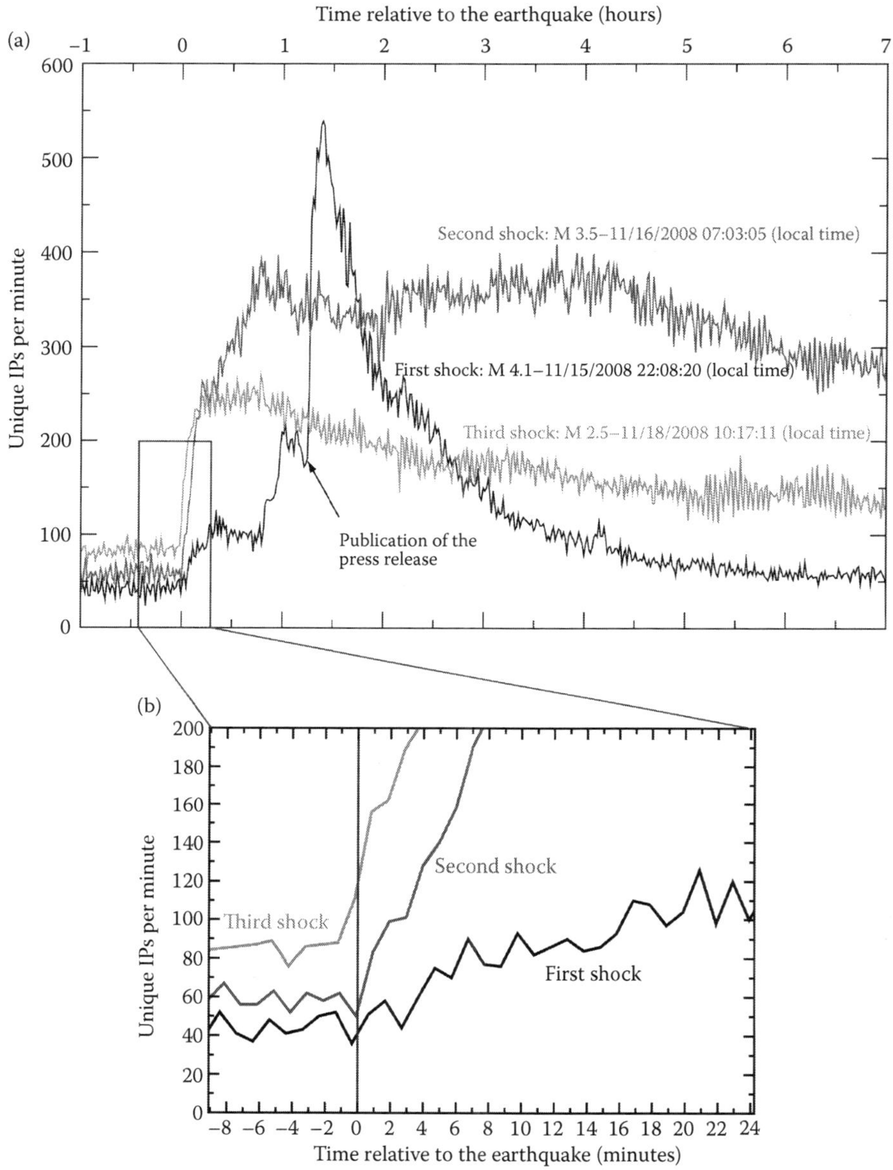

Figure 11.6 Temporal hit rate changes following three earthquakes felt in Sofia, Bulgaria, in November 2008.

significant surge of traffic was observed (Figure 11.6a) immediately after the first main shock (M4.1), which occurred at 10:08 P.M. local time, indicating that the EMSC Web site was not well known by Sofia's inhabitants at that time. Although the amount of traffic did not increase sharply, the felt map computed *a posteriori* (i.e., after the location and origin time of the earthquake were determined), identified a marginal increase of the number of visitors from Sofia (Figure 11.7a). Traffic not only from Sofia but also from Bulgaria as a whole began to increase significantly after links to the EMSC Web site were posted on several discussion forums 40 to 60 minutes after the earthquake occurrence. The main surge, however, was caused by a national press release that mentioned the EMSC Web site, published approximately 90 minutes after the event (Figure 11.6a). Actual geographical audience of this press release can be mapped by simply computing a felt map from this surge (Figure 11.7b). This map shows that the press release drove visitors from the whole of Bulgaria to the EMSC Web site.

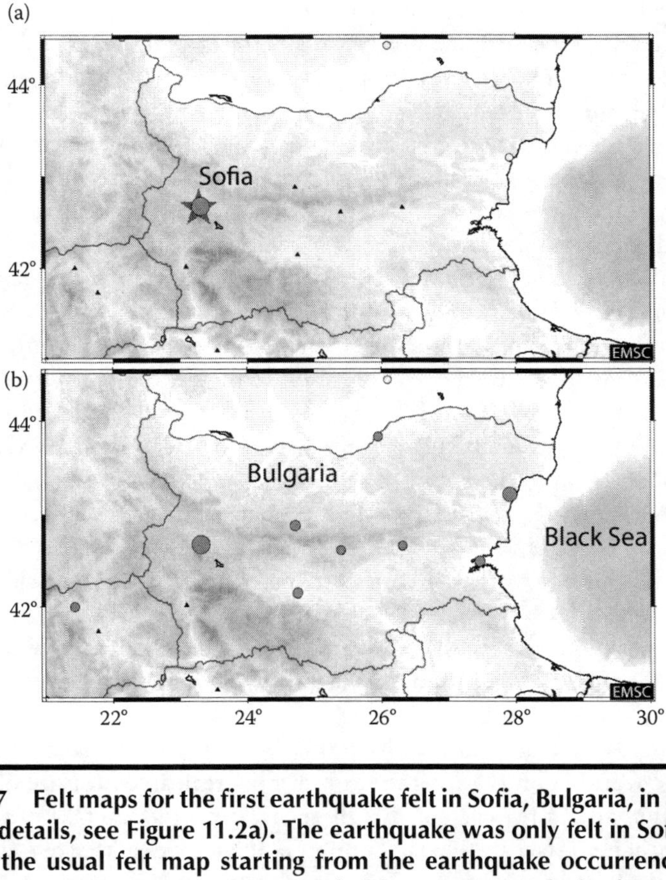

Figure 11.7 Felt maps for the first earthquake felt in Sofia, Bulgaria, in November 2008 (for details, see Figure 11.2a). The earthquake was only felt in Sofia as illustrated by the usual felt map starting from the earthquake occurrence (a). The press release mentioning EMSC attracted visitors to our Web site from the whole of Bulgaria (b).

The press release enlarged the EMSC audience in Bulgaria in the few hours following the first shock. Consequently, and contrary to the first shock, a hit-rate surge was clearly detected for the next two shocks despite their smaller magnitudes (M3.5 and M2.5), and the reaction time decreased to a couple of minutes (Figure 11.6). Referrals from sites such as forums, where the information about the occurrence of an earthquake can rapidly be shared, can theoretically pollute the result of the felt map by attracting to EMSC's site visitors who have not felt the event. Tracking down referrers to exclude visitors linking from them cannot entirely solve the problem, however, because telephone calls and television or radio programs can spread information about an earthquake occurrence without being traced using the access log of the Web site. In practice, the best approach is to shorten the time window used to compose the felt map. The time window is currently 10 minutes; in the case of massive surges, the felt map is computed with a window of no more than 5 minutes (Figure 11.2a). We are confident these limited windows will be short enough to avoid the problem of information spreading beyond the felt area because we have never observed such a phenomena occurring in the first 20 minutes. Conversely, 5 minutes is probably the shortest possible window that will allow enough time for seismic waves to propagate at long distances and for witnesses to visit our Web site. Five minutes is the value obtained by summing the observed reaction time (several minutes) with the propagation time of the violent S seismic waves at long distances (typically 500 km in the same time interval). This 5-minute processing window can be reached for any widely felt earthquake occurring in the Euro–Med region thanks to a constantly growing audience.

The only noticeable traffic surge that was not associated with any earthquake occurred on May 7, 2009, where the number of visitors exhibited a six-fold increase in 1 minute, jumping from 81 to 482 individual IP identities per minute (Figure 11.8a). This surge was caused by the mention of the EMSC Web site address during a television program about earthquake and earthquake prediction in Romania [Ionescu 2009, personal communication]. Similar to what was observed in Bulgaria after the publication of the press release (Figure 11.7b) in 2008, the May 7, 2009 felt map shows that the television program's sphere extended all over the country (Figure 11.8b). The only difference with the earthquake in Romania was that 72% of the traffic within the first 5 minutes originated from Google search "Romania" in comparison with 26% for the Vrancea earthquake on April 25, 2009. We surmise that the majority of visitors related to the TV-generated traffic were unable to remember the Web site address mentioned in the program and had to use a search engine to reach our Web site.

Mapping data using a felt map is an innovative approach that takes advantage of social networking on the Internet to provide a real-time measure of a population's reaction after a felt earthquake. By focusing on the reaction of the population rather than on the earthquake, our program identifies all the felt earthquakes, which are the only ones of significance for the citizens, irrespective of magnitude. From a seismological point of view, a M2.5 earthquake, such as the one near Sofia

(a)

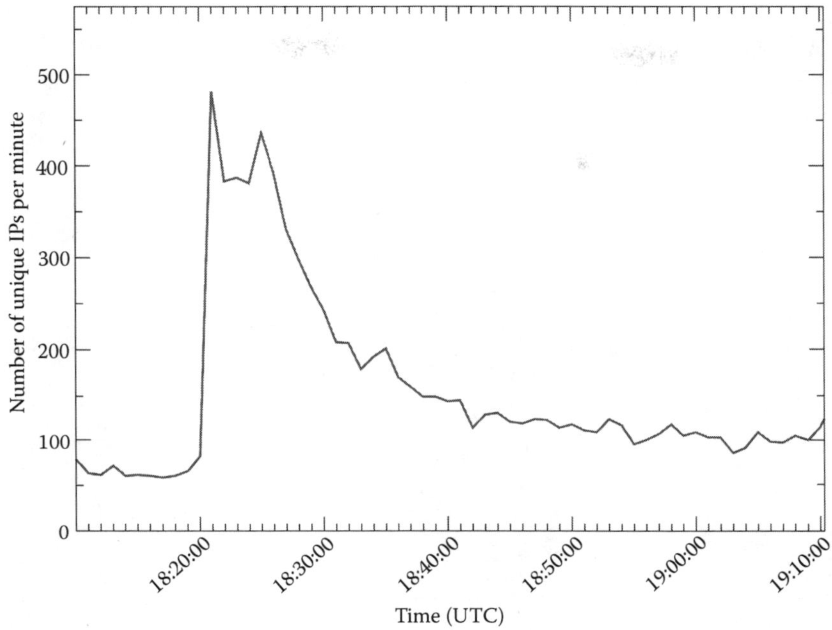

(b)

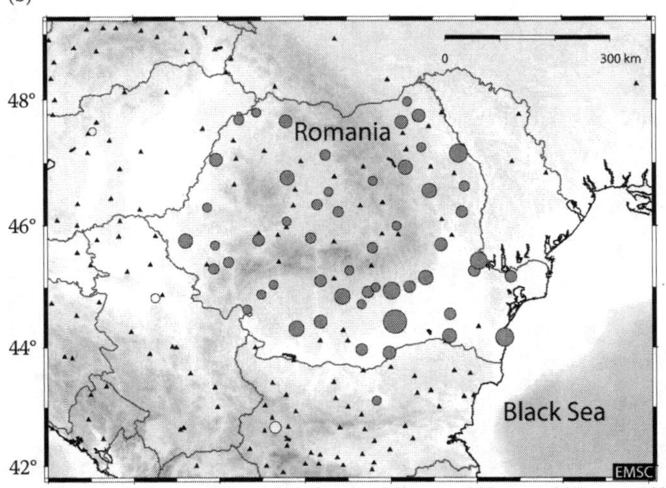

Figure 11.8 The mention of the EMSC Web site during a television program in Romania on May 7, 2009, led to the unique surge (a) not associated with an earthquake occurrence and a significant increase in the number of Romanian visitors of the Web site from all over the country (b).

on November 18, 2008 (Figure 11.6), is not significant and is unlikely to be felt. This is why low-magnitude earthquakes are not publicly reported in real time by the majority of seismological observatories. Most of them restrict their real-time programs to larger events that are more likely to be felt or cause damage. Felt mapping automatically delineates the area where an earthquake is felt within 5 to

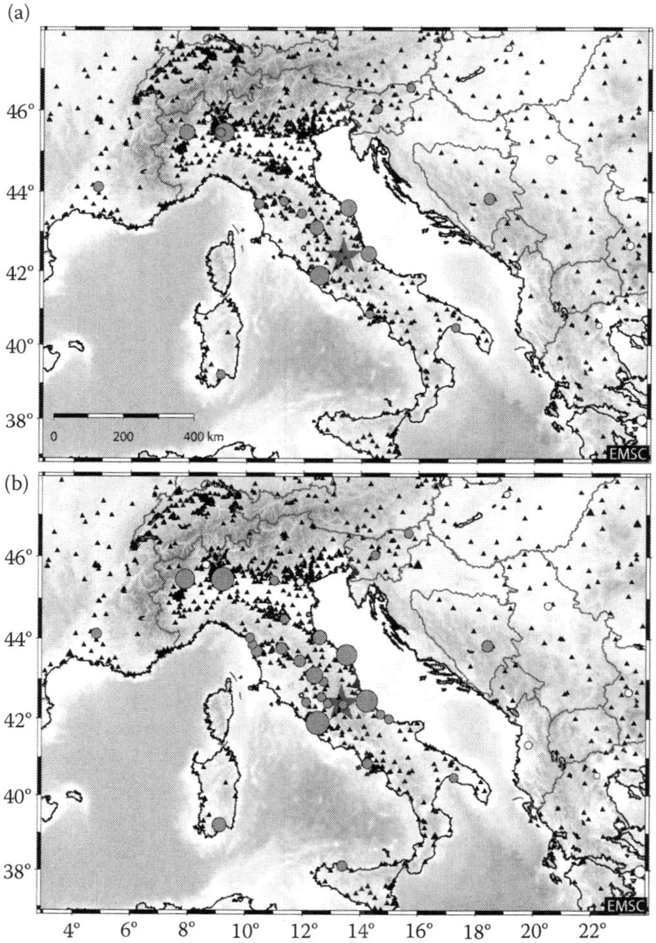

Figure 11.9 Felt maps (a) obtained for the l'Aquila, Italy, earthquake. The initial absence of visitors from the epicentral area (a) disappeared within 8 minutes (b) of the earthquake's occurrence despite the existence of significant damage. The overall felt area in our maps compares with the macroseismic map produced by the Instituto Nazionale di Geofisica e Volcanologia (INGV).

(c)

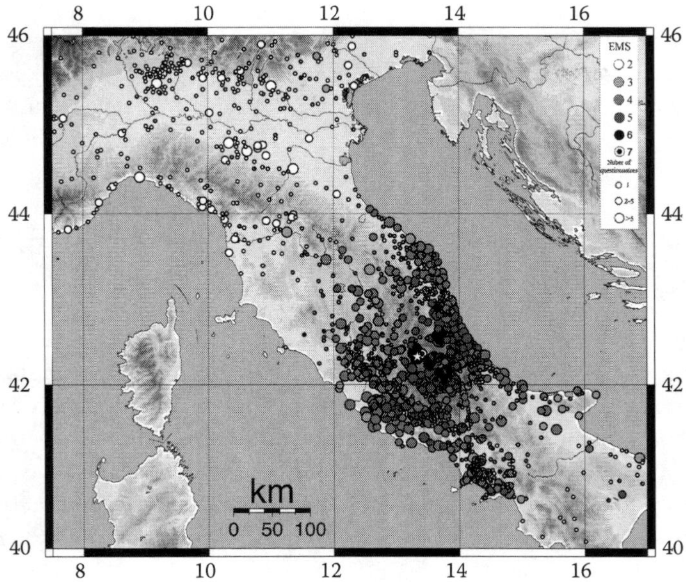

Figure 11.9 (Continued) (c) where the higher the value of the intensity ("intensita") the more violent the shaking.

10 minutes of its occurrence; it is the fastest way to collect in situ information on earthquake effects. Absence or a significant lack of visitors should map regions of widespread damage—an expected performance not confirmed during the l'Aquila earthquake in Italy on April 6, 2009 (Figure 11.9). The exact reasons are still under investigation and the data processing requires some improvements because the average number of visitors, when broken down to the city level, does not exceed a few to a few tens per hour in the middle of the night. Thus, it is not statistically significant to have no visitors from a given city in a 5- or 10-minute window.

Finally, felt maps provide some insight regarding the reaction of a population and how fast information can spread from an epicentral region. Our felt-map data confirms that citizens literally rush to the Internet within minutes of a felt earthquake, highlighting the need for information providers such as the EMSC to further improve time performances. It also demonstrates that the audience of a Web site can change overnight, especially if the Web site is identified by mass media. Some time characteristics seem to be shared for all felt earthquakes, regardless of location or magnitude: demand for information within the first 20 minutes is followed by an exponential decay for up to approximately 90 minutes. These phases are compared with the collection rate of online questionnaires (below), with possible explanations.

Lessons Learned from Online Macroseismic Questionnaires

Online macroseismic questionnaires are becoming an established tool that national seismological observatories use to collect information about earthquake effects and damage. Online questionnaire implementation at the EMSC focuses on Euro–Med earthquakes and significant global earthquakes. Its main specificity is to implement our questionnaire in 32 different languages, covering the majority of official languages in seismically active Euro–Med countries.

Three different phases exist in the collection of questionnaires (Figure 11.10), which seems to correlate with the characteristic times in the hit rate curve (Figure 11.4). During the first 20 minutes (i.e., the observed rise time), less than 5% of the total questionnaires are collected (Figure 11.10). The highest collection rate is observed between 20 to 90 minutes before exhibiting an inflexion. The decrease in the collection rate after 90 minutes probably correlates directly with the traffic rate, which is already half its peak value (Figure 11.4). The low collection rate in the first 20 minutes could possibly be linked to the delay in the publication of earthquake information on our Web site, typically 8 to 12 minutes from the occurrence of significant Euro–Med earthquakes. We surmise this initial low rate is a result of the witnesses first attempting to find the information they have been looking for before reporting their own observations. This hypothesis is supported

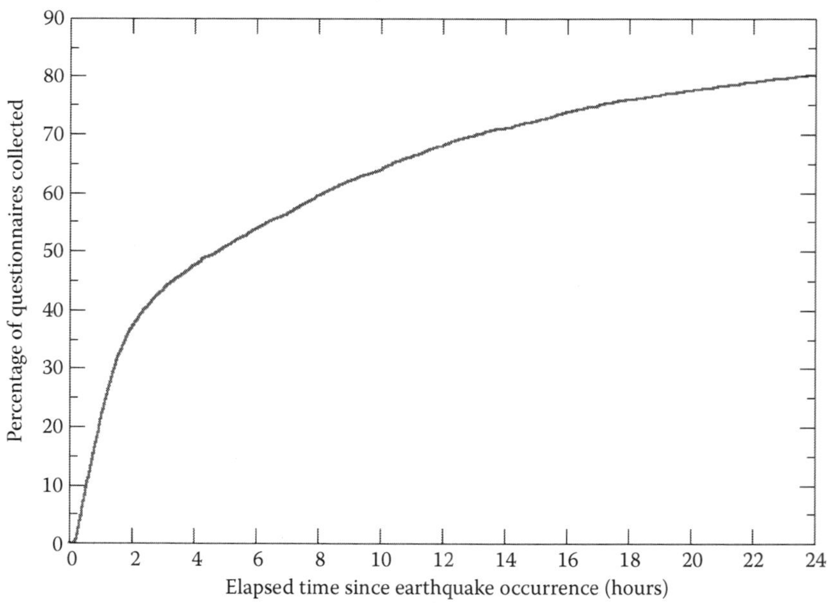

Figure 11.10 Percentage of responses to the macroseismic questionnaire with respect to time (cumulative curve for all earthquakes).

by the higher initial collection rate observed at the U.S. Geological Survey [Dave Wald 2009, personal communication], which also collects questionnaires for global earthquakes, but generally publishes preliminary earthquake information more rapidly than the EMSC. We speculate that the number of visitors builds as they wait for publication of earthquake information and only starts to decrease as they explore the given information and then complete the questionnaire. Therefore, the time characteristics observed on our hit-rate curve would be a result of publication time and average time on the Web site.

An interesting feature of our questionnaire is the use of different languages. The English questionnaire is used by 40% of the respondents, although the proportion of questionnaires collected from Euro–Med English-speaking countries, such as Ireland or the United Kingdom, are marginal. Seasonal changes can also affect responses within the same country due to tourist season, the best example being in Greece. Although 81% of the respondents used Greek and 17% used English to report their observations of the M6.2 earthquake that occurred during the low tourist season (on January 8, 2008), 36% used Greek and 38% used English for a M6.4 earthquake that occurred at the beginning of the summer on July 15, 2008. These values suggest that foreigners and tourists are more likely to use the EMSC Web site rather than the national institute in the case of an earthquake and that multilingual communication for disaster management is essential in touristic regions.

The percentage of witnesses who actually complete the macroseismic questionnaire is an important indicator of the reliability of derived information, as with any type of survey. We estimate the number of witnesses visiting our Web site by interpreting the hit-rate curve (Figure 11.1) as the number of extra unique IP addresses originating from the felt area during a given time window, starting at the earthquake occurrence. Although these estimations are crude, the results are consistent for different time windows, spanning from 2 to 24 hours, and for different earthquakes, with values ranging from less than 1% to no more than 10%. This tiny proportion of respondents highlights a strength of the felt-map approach: it characterizes the whole population of witnesses available on the site rather than sampling it.

To increase the proportion of respondents and the reliability of the macroseismic map by extension, we must understand and answer the needs of our visitors. The fact that witnesses rush to find information relating to an earthquake demonstrates their concerns and fear. They want to understand what happened, what the possible impact is, and they want to share their emotions in their own words. The need for catharsis is not fully answered by a macroseismic questionnaire, where information must be collected following stringent procedures to remain scientifically exploitable. This leads to messages blooming on discussion forums, where expression is not constrained by objective strictures. To encourage questionnaire usage, comments are now anonymously published following a simple validation. They are then sorted by location and epicentral distance so that visitors can find other's reactions in their own city or neighborhood. Too many parameters exist to evaluate the possible consequence of comment publication in terms of respondent rate, but comments are left by up to a third of questionnaire respondents, depending on the earthquake.

Felt maps and macroseismic maps are complementary tools in terms of time scale and type of information offered. Macroseismic maps describe the level of shaking or damage and are typically available on the EMSC Web site within a few hours. Felt maps delineate the felt area automatically in no more than 10 minutes after the event, in a time range where no other information on the earthquake impact is available. One of our goals going forward is to better understand the public's reaction and motivation so we can optimize the questionnaire respondent rate, which is currently low.

Lessons Learned from Collection of Photographs and Videos

Photographs and videos are principally available after less-frequent earthquakes, which produce permanent effects (e.g., damage and surface rupture). No accurate statistics or final conclusions based on photographs or videos can be drawn at this stage due to the low number of participants in this application. However, a dozen to approximately a few tens of photographs were received for each of the last three damaging Euro–Med earthquakes, which occurred since January 2008 in the United Kingdom, Greece, and Italy. Although this submission rate may change for future earthquakes, the collected photographs thus far are not decisive for characterizing an earthquake's actual impact. They were collected in a time frame ranging from 12 hours to a few days after the earthquake, and they mainly confirmed information already available in the media. Collected photographs are, however, proving valuable for documenting earthquake phenomena. This was the case for the dust cloud observed above a mountain during earthquake shaking in

Figure 11.11 Picture of the explosion of ammunition in Novaky, Slovakia, on March 2, 2007, provided to the EMSC by a witness. There were two explosions 70 seconds apart, the first visible as a plume of smoke, and the second and largest visible as a fireball.

Chile (Figure 11.3). We also received photographs documenting an explosion in an ammunition factory in Novaky, Slovakia, on March 2, 2007, that caused a major industrial accident [Kristekova et al. 2008]. The main explosion had been reported on the EMSC Web site as a magnitude 2.2 event. In the following hours, a witness sent us several photographs of the explosion itself (Figure 11.11) and of its effects [Bossu et al. 2007].

Discussion and Future Evolution

For the EMSC, our developed tools present two major advantages: they provide rapid data on earthquake impact, and they help us document earthquake phenomena at low costs. This allows us to provide a quality service to the public, the authorities in charge of earthquake response, and the scientific community. One of our challenges is to develop user-friendly tools while still collecting reliable, necessary information for scientific use, such as the causative earthquake and the location of observations. Improved Web-site ergonomics can reduce erroneous reporting, especially when a witness must choose the earthquake that caused the shaking they felt. The number of witnesses who use the tools is low, but the resulting data have been surprisingly reliable, with no example of misuse for the collection of photographs: not a single one had to be rejected for inappropriateness. Collected data cannot compete with scientific observations but can complement them in some specific areas, notably transient phenomena observed during shaking.

The level of traffic on a Web site such as the EMSC Web site is a direct measurement of the public desire for information. It is essential for the credibility of the scientific community to fully and rapidly respond to the needs of many, even when the earthquake in question has little seismological value or did not attract the interest of seismologists. Monitoring Web-site traffic can improve our earthquake service by bridging the gap between the needs of seismologists, who are usually interested in the phenomena itself, and the needs of citizens, who are primarily concerned by the earthquake they feel, irrespective of its magnitude.

The success of citizen seismology depends upon understanding the reactions and needs of witnesses when facing an earthquake. Witnesses are not amateurs who would be involved in citizen seismology as a hobby or passion. Witnesses first want answers and then only a fraction of those seeking answers will contribute to an improved earthquake response by describing their experience. Being part of the process implies a dimension of ownership for witnesses, and can help in reducing their occasional defiance toward official prevention policies. Involvement strengthens much-needed direct communication between the scientific community and the public at a time when vocations for scientific careers are plummeting.

Perhaps the most surprising result of our Web site's presence is the rapidity of reaction, which is detected within a couple of minutes of an earthquake. It is a challenging aspect in terms of hardware, which must be able to support massive increases in traffic. It also drives information providers to further improve their time performances. Finally, it is driving development at the EMSC. No short-term possibility exists to drastically reduce the required time to publish preliminary earthquake location data. Because the massive surge of traffic following a felt earthquake is the first evidence of its existence, to inform the witnesses more rapidly we are testing the automatic publication of a temporary message triggered by the surge itself, which mentions the region where the event has just been felt. Automatic translation of comments has been implemented, both for comment validation and publication. The felt-map algorithm must be further developed to better detect and map damaged areas. For example, further investigation is required to study possible geographical variations in the reaction of a population between low-hazard and higher-hazard regions, or the specific type of reaction when facing a destructive earthquake.

Concluding Remarks

The Internet offers new ways to communicate and exchange information. The scientific community cannot remain outside this fundamental societal phenomenon where citizens become active participants. The EMSC simply applies the ideas of Web 2.0 and of *wiki* processes to seismology for the benefit of both the public and the scientific community, creating citizen seismology.

Felt maps based on real-time hit rates and the location of IP addresses offer new insights on the reaction of the population when confronted with an earthquake. No significant regional variations were observed, although more investigations are required to fully understand our observations. However, the felt-map approach can be readily adapted to serve those who deal with any rapid-onset, potentially disastrous events.

Today, we are far from fully exploiting the Internet's potential to develop the culture of risk among a population. Our primary focus in this respect is on developing risk awareness in areas of low seismic hazard. In such areas, the perception of risk is low due to the rare occurrence of earthquakes. Our challenge is to develop dedicated communication tools which could be automatically offered at the right time and in the right language to citizens in these areas. This will ensure better receptivity of prevention messages because they will then relate to reliable information.

Acknowledgment

The EMSC acknowledges the funding provided by the French Ministry for Sustainable Development for its support for the citizen seismology project and the essential and long-term support it has been receiving from its host, the *Département Analyse et Surveillance de l'Environnement* of the CEA (France).

References

Bossu, R., G. Mazet-Roux, and S. Gilles. 2009. *The importance of in-situ observations for rapid loss estimates in the Euro-Med region*. EGU General Assembly 2009, Geophysical Research Abstracts Vol. 11.

Bossu, R., G. Mazet-Roux, V. Douet, S. Rives, S. Marin, and M. Aupetit. 2008. Internet users as seismic sensors for improved earthquake response. *American Geophysical Union* 89:225–226.

Bossu, R., and A. Walker. 2009. New tools for alerts and capturing immediate effects of rapid-onset events across Europe. *Chemical Hazards and Poisons Report, Health Protection Agency* 14:35–37.

Bossu, R., V. Douet, S. Godey, G. Mazet-Roux, and S. Rives. 2007. On the use of Internet to rapidly collect earthquake impact information. *EMSC Newsletter* 22:31–34.

Cochran, E. S., J. F. Lawrence, C. Christensen, and R. S. Jakka. 2009. The quake-catcher network: Citizen science expanding seismic horizons. *Seismological Research Letters* 80:26–30.

Deards, H. 2009. Twitter first off the mark with Hudson plane crash coverage. Retrieved from http://www.editorsweblog.org/multimedia/2009/01/twitter_first_off_the_mark_with_hudson_p.php (accessed January 21, 2011).

Ginsberg, J., M. H. Mohebbi, R. S. Patel, L. Brammer, M. S. Smolinski, and L. Brilliant. 2009. Detecting influenza epidemics using search engine query data. *Nature* 457: 1012–1014.

Grünthal, G., R. M. W. Musson, J. Schwarz, and M. Stucchi. 1998. European macroseismic scale EMS-98. *Cahiers du Centre Européen de Géodynamique et de Séismologie* 15:101.

International Telecommunication Union. 2008. Worldwide mobile cellular subscribers to reach 4 billions mark late 2008. Retrieved from http://www.itu.int/newsroom/press_releases/2008/29.html (accessed January 21, 2011).

Kristekova, M., P. Moczo, P. Labak, A. Cipciar, L. Fojtikova, J. Madaras, and J. Kristek. 2008. Time-frequency analysis of explosions in the ammunition factory in Novaky. *Slovakia Bulletin of the Seismological Society of America* 98:2507–2516.

Mandrescu, N. et al. 1999. Macroseismic field of the Romanian intermediate-depth earthquakes. In *Vrancea earthquakes: Tectonics, hazard and risk mitigation*, eds. F. Wenzel et al. Dordrecht: Kluwer Academic Publishers. EMSC report. Retrieved from http://www.emsc-csem.org/Doc/EMSC_DOCS/EMSC_RT_activities_2009.dpf (accessed January 21, 2011).

Mazet-Roux, G., R. Bossu, E. Carreno, and J. Guilbert. 2009. Report on 2008 real time activities. Arpajon, France: European Mediterranean Seismological Centre.

Wald, D. J., V. Quitoriano, L. Dengler, and J. W. Dewey. 1999. Utilization of the Internet for rapid community intensity maps. *Seismological Research Letters* 70:87–102.

Wald, D. J., V. Quitoriano, T. H. Heaton, and H. Kanamori. 1999. Relationship between peak ground acceleration, peak ground velocity, and modified Mercalli intensity for earthquakes in California. *Earthquake Spectra* 15:557–564.

Winerman, L. 2009. Social networking: Crisis communication. *Nature* 457:376–379.

Chapter 12

Social Capital: The Missing Link in Coping with Environmental Disaster

Sotiris Chtouris and Flora Tzelepoglou

Contents

Introduction

This chapter examines social capital as a resource in risk management in disasters—manmade and natural. Our frame of reference will be the recent Greek experience of devastating forest fires in the summer of 2007. This experience revealed that neither technical means nor cooperation among public civil protection services proved adequate in confronting massive fires all over the Greek territory. Risk prevention, mainly through information and mobilization of local residents, emerged as a key factor to confronting natural hazards in an effective way, mitigating environmental damage and community problems.

Furthermore, climatic and concomitant environmental changes render preventive actions indispensable (i.e., residents acting in an independent, decentralized context, organized in collective local associations). Because interacting social and cultural networks produce social mobilization and social capital, they are in a position to create conditions that favor long-term cooperation and strong ties, thus transcending cultural, geographical, and bureaucratic obstacles.

Risk Perception: Cultural and Social Prerequisites

It is generally accepted that modern citizens confront a host of increasingly acute risks emanating from natural environmental changes caused either by human intervention or global climatic change. Wealth production and distribution are increasingly linked to negative repercussions, disasters, and large-scale hazards. Risk recognition leads to a negatively reproduced environment due to the fact that relevant knowledge and information refer to it in a negative way. "The risk society is in this sense also the science, media and information society" [Beck 1992, p. 46].

It is vitally important to understand in which ways community members manage to confront small- or large-scale environmental risks at both the individual and collective levels, especially when there is such a wide variation in social behavior (i.e., some communities fare well, being active and socially participative, whereas others present an image of apathy and inaction). Our considerations are directly linked to such a range of civic participation in the context of natural hazards and disasters. We intend to interpret social behavior in response to risk perception and management on the basis of four different and mutually complementary frames of reference (i.e., an interpretative network that has been developed around the issues of risk and environmental disaster).

The first frame of reference is individual choice. An individual selects a range of means and practices that will cause the least possible damage and will give him or her maximum benefit when dealing with the risk. The second frame of reference involves individual integration in a social context or a competitive arena. The third frame of reference involves individual participation in collective action and the community's cultural context. The fourth frame of reference involves examining networking and social capital ties, through which we may observe the development of bridging network ties in relation to risks of various intensity and temporal continuity. Natural hazards and environmental issues are increasingly interlinked with crucial social problems with a new complex social structure—namely poverty, marginalization, and lack of social justice [Harvey 1996].*

In our attempt to make a synthesis, we suggest that risk assessment exercises a functional impact on individual and social behavior and that this is usually the result of a complex selection process, which in turn is defined by the multifunctional dynamic embeddedness of a person in multiple social contexts. The decision to confront the risk and the choice of means is largely defined by the way and the intensity with which individuals interact within the framework of multiple social ties and affiliated networks [Chtouris 1997, 2004; Read 2002]. It is within this integrating context that commitments and opportunities are established.

Setting Granovetter's theoretical reflection on weak and strong ties and embeddedness as a starting point [Granovetter 1973, 1985], we consider that a multilayered embeddedness is an approach more apt to interpret the reasons why a wide range of different decisions are taken in the context of risk management, especially in countries undergoing a social transformation process. Individuals tend to confront risks in a multitude of different ways, depending on their level of awareness or mobilization, on their social context or reference network, and on the overall context where they find themselves embedded at a certain point in time.

As far as individual and collective practices are concerned, the reference context of commitments and obligations is defined by individual multilayered embeddedness in the framework of environmental risk perception and management. Multilayered embeddedness is an object not only of an inner individual process, but also of a competitive external process developing in a specific social and cultural environment. Appreciations, judgment, and decisions are increasingly distanced from a theological type of discourse, namely the community's commands and moral imperatives, as well as the conflict dynamics in various social fields [Bourdieu 1984]. The emergence of individuality and the establishment of secular beliefs [Bauman 1990; Giddens 1990, 1991; Lyotard 1979] are well established phenomena that characterize modernity and postmodernity; however, in many cases, they are subjected to networks of the past (social, political, and local) that still remain in force and preserve their legitimacy.

More specifically, in southern Europe, we witness a social dialectic process of mutual substitution between tradition and modernity, which is evident in fields such

---

* Nature, it seems, is no longer natural. By jettisoning the artificial dualism between humans and nature (a schism manifested in the unfortunate bifurcation between physical and human geography), theorists in the discipline working within the broader domain of social theory and political economy have called for renewed attention to issues in which social and ecological concerns are intertwined, including matters of social and environmental justice [Harvey 1996].

Table 12.1 Levels of Action and Risk Perception

Reference Unit	*Risk Perception*	*Practice*
1. Individual	Individual	Individual
2. Field—arena	As competition means	Competitive interaction
3. Group—community Social capital, bonding capital, local networks	Collective—symbolic	Collective practice Dense interaction
4. Network—bridging Social capital (local/supralocal)	Open—differentiated	Distributed practices

as environmental conservation and agricultural ownership [Chtouris 1987]. Dealing with the natural environment and possible hazards often involves interpretations according to a traditional rural community mentality, with strong symbolic reference to land and land domination. While under different circumstances in a context of individualistic financial calculations, one may consider natural environment merely as an expendable natural resource that can be used for benefit and considerable economic profit; in another context, emerging ecological awareness and sensitivity ascribe to the natural environment almost mythological proportions (Table 12.1).

In a classical paradigm of individual choice in the context of risk perception and risk management, individuals have the opportunity to assess alternative options to organize their actions. Some modern theoreticians propose, as a basis of our sociological analysis, the "purposive action of individuals" [Coleman 1994, pp. 17–18] in view of an enriching exchange with other disciplines in relevant fields (i.e., economics, political sciences, and law studies). More specifically, recent studies have revealed that several decisions made in response to risks and their causes are directly influenced by the fact that individuals have the possibility of alternatively choosing among various options before taking any responsibility [Petersen 1996; Wikstrom 2002, 2004]. We suggest that when only few options are available, a more passive attitude is developed regarding environmental hazards and natural disasters. According to another interpretation "information overload makes rational responses to risk practically impossible. Combined with a growing force of individualization, in which more and more essential decisions are thrown back onto individual responsibility, it becomes difficult to develop rational strategies for dealing with risks" [Van Loon 2002, p. 188]. Generally speaking, most individualistic theories [Schwarz and Thompson 1990] tend to emphasize the importance of the individual capacity to choose more effectively when confronting menacing risks as well as the means to confront them.

It is also interesting to explore risk perception and risk management using the concept of competition developed in social and institutional contexts. The arena theories give us the possibility to comprehend the political debate on risk; in this case, individuals decide to participate or abstain according to their forecast on possible benefits or damage.

However, by integrating social context in the overall risk perception and management research, we provide a richer theoretical framework [Otway and von Winterfeldt 1982]. Indeed, in many cases natural disaster management acts as a surrogate to political antagonism because political actors renegotiate their positions in the political arena [Renn 1992]. Recent experience of catastrophic forest fires in Greece in 2007 amply proves that such an event quickly becomes a key factor in shaping political and social dialogue nationally because it indeed happened during the 2007 electoral campaign; moreover, it has a direct impact on citizens' options regarding possible solutions and mitigation of damage. A similar interpretation applies to mainstream media and how they use natural disasters. In Greece during the blazes, we were witnesses of sensationalist reporting amid hard competition for the highest possible ratings. Competition among audiovisual media develops, in a large degree, in the context of a social drama where both the media and local and national politicians address an audience [Turner 1974] already emotionally charged and highly vulnerable to solutions offered by a Deus ex machina (god from a machine, as in Greek drama). Generally speaking, mass media create their own interactive field with their viewers whenever they deal with natural disasters and ensuing damage; as a result, a special dynamic is created in relation to risk perception and management.

A cultural theory of risk provides us with the possibility to explore, in groups and communities, how everyday interaction in private and public space generates a specific pattern of practices, recurrent routines, and rituals at a microsociological level [Chtouris 2004; Wynne 1982]; thus, community members become integrated in a specific action plan in response to risks and disasters. In this case, the way real risks are perceived may be comprehended as a social construction involving moral and value judgments and political bias [Douglas 1992; Douglas and Wildavsky 1982]. A cultural theory of risk contains certain elements of determinism, which do not necessarily lead to uniform structural explanation of everyday practice. Although it accepts that strong biases and commitments exist and consequently limit individual margins of action, namely to opt for a solution or an action plan, it does not limit an individual's overall capacity to take a different course of action. The cultural theory of risk can also foster a better understanding of standard cultural representations on the subjects of nature and its hazards. As Heidegger [1977] aptly observed, "Modern technology has turned nature into a resource, standing reserve for whatever use 'we' may have in mind, particularly adept in clarifying the devastating ecological implications of so many of our modern technologies" [p. 20].

At this point it is particularly interesting to evoke the theory of social capital [Coleman 1994], social networks, and their spatial and environmental dimension [Chtouris 2009]. Making use of social capital resources in local development, particularly when there is need to strengthen local societies and communities in times of crisis and natural disasters, reveals the importance of its spatial dimension and its potential to become an indispensable part of environmental planning and civil protection policies.

Collectivities that generate social capital create the necessary preconditions to assess not only risks, but also natural hazards. Social capital is, of course, a complex concept to which a variety of content has been ascribed. According to this presentation, a series of factors and attitudes tend to produce or reproduce social capital [Onyx and Bullen 2000; Putnam 2000]. We observe particular factors that produce and preserve social capital, namely, trust, local identity, and direct social ties; furthermore, we observe structural factors, such as social networks, institutional structure, and the community's regulatory effectiveness.

Figure 12.1 distinguishes between collective (structural resources of social capital) and individual and personal resources (attitudes, perceptions, and individual purposive practices). These two pillars of social capital embrace and determine transformations of various capital forms at local and regional level (middle column), taking the form of financial, human, cultural, and social capital. The above model proposes an interactive relationship between social capital resources. We believe that it better reflects dynamic situations that occur in cases of natural disasters and environmental risks. Furthermore, in Figure 12.2, regarding social capital production, we observe the simultaneous appearance of interactive relations among different dimensions, circumstances, and actors that may have a strong impact on the growth and development of social capital. In a positive context, we may observe

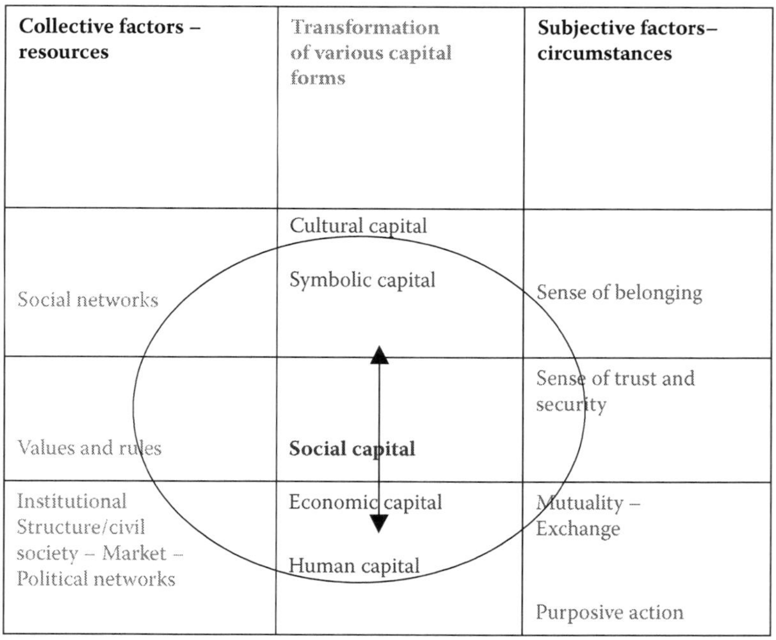

Figure 12.1 Social capital context.

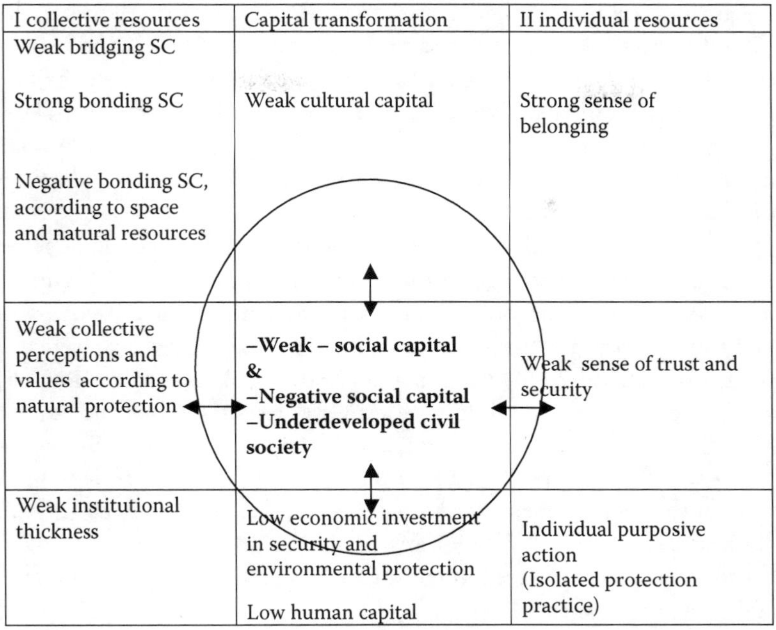

Figure 12.2 Social capital (SC) in Greece, in connection with protection performance and management in communities and areas at risk (2007).

a sense of belonging, life perspectives, and trust [Sarason 1974; Morrow 2002], whereas negative relations limit and degrade social capital, namely, oppressive ties and commitments that trap individuals in deviant behavior such as illegal land occupation, trespassing, or illegal waste disposal.

Putnam [2000, 2002] suggested a differentiation between community-oriented social capital and that which transcends local boundaries, introducing the difference between bonding social capital and bridging social capital. In this way, Putnam ascribes to social capital the ability to develop social relations beyond local boundaries; however, the boundaries or land scale that differentiate the two types of social capital are not defined precisely. More specifically, on the issue of bonding social capital, which is more locally defined, there is a view expressed by various authors that consider it a negative factor because it tends to lead to a consolidation of traditional relations, thus resisting further positive change or innovation [Woolcock 1998]; furthermore, it may lead to deviant behavior, namely illegal land appropriation or environmental polluting practices.

Our aim is to identify social capital in its limitations and possibilities and its positive or negative forms in Greek communities devastated by large-scale forest fires in 2007. We would like to explore the possibility whether this important resource can be a substitute for various interventions and economic or technological investment or whether it can reinforce their effectiveness. This can be achieved through social capital transformation into other forms of capital, such as cultural capital, financial capital, and human capital. At the same time, social capital becomes stronger or weaker under the impact of a range of different theoretical approaches as we briefly mentioned above, namely individual choice, which considers the natural environment part and parcel of market forces, as an integrating symbol of communal cohesion according to cultural theory, or as a means of political and social antagonism according to the arena theories.

In this multilayered analysis model, our theoretical references do not stress certain dimensions that concern scientific views on the notion of risk and the role played by experts in risk negotiation and management. In the current framework of environmental risk analysis, understanding the role played by science, expert knowledge, and decision analysis and the types of risk assessment is highly significant. However, Greek and Mediterranean examples show that such dimensions have not yet been fully developed as to critically shape important decisions and attitudes. This was amply proved in the framework of public debates deployed in the aftermath of the catastrophe, where experts were used by politicians and financial circles in the arena of public opinion manipulation. It is for these reasons that we consider that introducing these dimensions at this point is indispensable for all the theoretical references of this text.

2007 Devastating Forest Fires in Greece

In the summers of 2007* and 2009,† massive forest fires broke out in several areas across Greece,‡ marking a turning point in Greek public dialogue on manmade

* Detailed Satellite Photographs of the 2007 fires in Peloponnesus and the 2009 fires in Attica can be seen at the Centre for Satellite Based Crisis Information (ZKI), a service of DLR's German Remote Sensing Data Centre (DFD).
  - http://www.zki.dlr.de/applications/2007/greece/140_en.html [wild fires 2007]
  - http://www.zki.dlr.de/applications/2009/greece_fires/177_en.html [wild fires 2009]
  - http://www.zki.dlr.de/media/download/applications/2009/greece_fires/2009_08_Greece_fires_overview_low.jpg

† This paper does not dwell on the devastating fires in the North-eastern region of Attica in the summer of 2009. According to the relevant communication, the afflicted area measured 13,336 hectares, as depicted in the relevant satellite images of the European program GMES/SAFER, as well as the International Charter on Space and Major Disasters of UNOSAT.

It is worth noting here that the provision of this information, by the above-mentioned international agencies, to communities and the media, provided an overall objective and holistic overview of the major disaster areas in Attica during August 2009.

‡ An excellent photographic presentation of the disaster areas in Peloponnese is Annie Angelopoulou http://www.photo.aa-informatics.gr/Ilia.html and the public domain site GreekFires.com, which aims to "become the point of a new beginning for the Greek people, by challenging the traditional way of thinking and by providing information on the methods that other nations use successfully, in order to achieve quality of life."

environmental hazards. According to research performed by the Agricultural University of Athens and the Aristotle University of Thessaloniki,* the death toll in August alone was 67 people. In total, 84 people lost their lives in the blazes, including several firefighters. Ten forests were completely scorched, resulting in 278,212 hectares of forest trees, olive groves, and farm land destroyed. In 502 municipal departments, 35% of land was destroyed, bringing the economic damage to 3.5 million Euro (€). In all of these areas, 29% of farmers suffered a severe economic blow because they lost their livestock, farmland, houses, and businesses. The 2007 fires brought devastation to protected natural habitats, sweeping through Olympia, the site of the ancient Olympics and a World Heritage Site.

These fires were unprecedented, both in scale and intensity, and for the first time Greek public opinion, deeply affected by the horrific scenes watched on television, mobilized in its strong criticism and anger against the government as they observed the ineffectiveness and lack of coordination by all relevant state services involved in extinguishing the blazes, namely the Fire Brigade, Civil Protection Agency, prefectures, and regional authorities. Additionally, and perhaps more significantly, a new realization came as a collective awareness in relation to risk and catastrophe caused by environmental degradation and climatic change. It was a well-established fact that prolonged and severe drought acted as an aggravating factor combined with high temperatures and unusually strong winds. Fire Brigade experts† declared that no technical means whatsoever—not even firefighting aircrafts considered by fire victims as the only solution—could possibly deal with similar fires in a new environmental and climatic reality.

It became evident that communities and municipalities were greatly lacking in preventive measures and were unable to protect their villages, olive groves, farmland, and forests. At this point, it also became evident that local societies characterized by a low degree of voluntarism and self-organization as well as indifference and ignorance found themselves in a difficult position when called to confront such large-scale environmental disasters in the general framework of climatic change effects.

* Research material published in newspapers and electronic media. The research was funded by the Ioannis S. Latsis Foundation. [http://www.inekathikonmas.gr/site/content.php?sel=8].

† Christos Lambris, representative of the Fire Brigade http://www.e-cology.gr/DiscView.asp?mid=1410&forum_id=5&.

Comparison among Three Regions of Greece Confronting Catastrophic Fires

Peloponnese Research in 2007

A quantitative study by Chtouris et al. [2008] was performed across Greece by the sociology department of the Aegean University using 1,667 structured questionnaires. Of the overall sample, 10% of involved villages in the region of Laconia, in Southern Peloponnese, suffered great damage during the 2007 massive summer forest fires.*

These researchers also performed 158 brief qualitative interviews across fire-stricken areas with the intention of identifying issues that local residents consider important to them. These answers contributed to a better understanding and processing of accumulated quantitative data. In our questionnaire, we wanted to explore the relationships between trust, cooperation, and participation linked to social capital.

Our main themes involved identification with local space and physical environment, mobilization, and awareness by local residents and state agencies. Additionally, we undertook field research and media analysis. Our aim was to establish some alternative scenarios at the local level on civil protection in the event of natural and environmental hazards.

Comparisons and Findings in Relation to Problems Persisting in Fire Disaster Areas

In their responses to the questionnaire, the majority of inhabitants in fire disaster areas report that they identify their local community with their individual and family space (98; 7%). On the contrary, in areas where no fire disasters occurred, answers to the same questions report a larger distribution to other items. Indeed, among those questioned, 28% report that they consider their local community

* The research was performed by sociology students of the University of the Aegean, in the framework of the course Urban Sociology. They performed their research in their places of origin over the course of the Christmas break of 2007, three-and-a-half months after the catastrophe. Due to the fact that our students come from all over Greece, a wide geographical distribution of our researchers made a relatively random sample possible, including fire-stricken areas. Thus, 10% of those questioned come from communities in Laconia that suffered considerable damage. The small size of the total sample, namely 1,667 participants, is the result of the limited time researchers had at their disposal and the very low budget for the whole effort, deployed by the students themselves and two sociology professors. Nevertheless, it is a fact that most Greek poll companies, when conducting surveys across Greece on political or social issues, use roughly the same sample size (1,500 to 2,500) of participants [Vernardakis 2005], based on "some of the rough rules of thumb" [David and Sutton 2004, pp. 154–155].

mainly as public social space, 10.45% see it as a natural environment of ecological interest, 6.25% as business activity, 9.4% see aesthetic value, and 7.4% refer to its symbolic/sacred value. Only 37.3% report the same answer as fire disaster inhabitants, namely stating as their first option that they identify their community with individual and family space. Overall, this marks a significant differentiation, showing that there is a tendency toward individual identification with community space in fire disaster areas, as well as a development of close and strong bonding social ties at the cost of more general or environmental attitudes expressed elsewhere.

The majority of fire disaster residents (64.9%) ascribe a high value or a very high value (27.3%) to their community space. Corresponding percentages in areas without fire disasters are 36.5% and 26.4%. According to our research hypothesis, the whole issue of value ascribed to local space and community is a dependent variable determined by the experience of natural or other disasters, risk perception, volunteer participation, and social capital in general. The differentiation between fire disaster and no fire disaster seems to corroborate in part this assumption. Following a disaster, identification with one's community space leads to an enhanced appreciation of its real value.

The majority of fire disaster inhabitants (64.3%) (see Table 12.2) report that the most important individual response to natural disaster would be community mobilization, effective organization, and valid information, whereas a smaller percentage (34.4%) considers mobilization of state agencies as important. It is worth noting that the reported percentage in regard to individual participation and personal involvement in disaster management is very low. On the contrary, in areas without fire disasters, only 18.9% propose community mobilization in response to natural disasters and possible preventive measures; whereas 16.3% propose state agencies undertake the task. Furthermore, 23.1% agree with personal involvement in protecting their community, 6.1% with raising media awareness, and 6.5% with mobilization of ecological organizations. In our opinion, differentiations revealed among first, second, and third ranked selections are equally important. There is a marked shift with regard to individual intention to act, which covers a range from community and state agencies mobilization to individual participation and action-taking. Throughout ranking question responses, respondents were asked to provide three ranking selections (categories 1 to 3).* A similar percentage reported in the first selection in category 2 (community mobilization, 64.3%) is also reported in the second selection in category 3 (mobilization of state agencies, 70.4%). Furthermore, a similar percentage to the one reported in the first selection in category 2 (community mobilization, 64.3%) appears in the second selection in category 3 (mobilization of state agencies, 70.4%); whereas in the third selection, it moves to category 4 (personally involved in protecting my community and its environment, 72.0%). A low percentage is reported in relation to category 5

* "Ranking question responses are used when the researcher wishes to collect data on how respondents rank a list of items in relation to each other" [David and Sutton 2004, p. 168].

Table 12.2 Awareness to Cope with Community Environmental Problems: What Would You Be Willing to Do in Order to Be Able to Cope with Environmental Problems?

Selection	*1. Undertake no action (%)*	*2. Organize my community's mobilization (%)*	*3. Mobilize state agencies (%)*	*4. Become personally involved in protecting my local community and its environment (%)*	*5. Raise media awareness (%)*	*6. Mobilize environmental NGOs (%)*	*Total (%)*
Rank 1							
FD	1.3	64.3	34.4				100
NFD	29.0	18.9	16.3	23.1	6.1	6.5	100
Rank 2							
FD			70.4	28.8	0.8		100
NFD	1.6	17.8	16.6	30.0	17.2	16.8	100
Rank 3							
FD				72.0	26.0	2.0	100
NFD	2.7	18.2	16.0	12.9	24.0	26.2	100

Note: FD = communities who experienced recent fire disaster events; NFD = communities with no recent fire disaster events; NGO = nongovernmental organization. N = 1667 in 2007.
Responses were ranked on a scale of 1 to 3.

(media awareness, 26%). Figure 12.2 data could possibly point to an interpretation that there is a latent tendency for volunteerism and mobilization. This trend is not evident throughout respondents' initial intentions, but becomes readily accepted by a smaller number of respondents who propose their personal involvement in risk and disaster management. The number of these in the third selection who opt for their personal involvement in protecting their community and its physical environment is approximately half of those who simply wish for community mobilization.

A different picture is revealed in areas that have not recently suffered from natural disasters. In first selection responses, approximately 30% report that they have no intention to undertake any action in relation to their community's environmental problems. This percentage becomes approximately zero in subsequent selections (1.6% in second and 2.6% in third selection). The second dimension of this answer, namely mobilization and information activities, remains stable in all selections in contrast to more marked differentiations reported in fire disaster areas. Finally, personal involvement and participation increases in the second selection (30%) and decreases in the third selection answers. It is worth noting that in areas without fire disasters, there is a significant increase in second and third selection answers, namely of respondents who are interested in mobilizing environmental nongovernmental organizations as well as the media, which is not at all evident in fire disaster areas. In our opinion, this is particularly interesting and could perhaps be interpreted as a reflection of conservative political attitudes of fire disaster inhabitants and the absence of any environmental or volunteer nongovernmental organizations in these areas. It is possible that the media have played a role in this situation because they regularly report a negative picture of these nongovernmental organizations as being against farmers' interests as they propose less intensive farming and animal-breeding limitations in mountainous regions.

Respondents of fire disaster areas propose strengthening local collective bodies as their main means of confronting community problems, opting for farmers' cooperatives (83.8%) and local government authorities, i.e., municipalities (16.2%), so that these bodies undertake the necessary actions to protect and preserve their

Table 12.3 Fire Disasters Areas in 2007: What Should Inhabitants Do?

Selection	*Strengthening Cooperatives (%)*	*Mobilization of Municipalities (%)*	*Mobilization of Politicians (%)*	*Legal Means (%)*	*Total (%)*
Rank 1	83.8	16.2			100
Rank 2	82.5	15.6			100
Rank 3		40.9	38.3	18.8	100

N = 1667 in 2007.

community's environment. In contrast, they seem distrustful of politicians and official state agencies, which offer possibilities of compensation in case of damage or legal means for their protection (Table 12.3). It is only in the third selection that 38.3% of respondents opt for state agencies mobilization. An even smaller percentage in the third selection (18.8%) is in favor of legal means. It is impressive to note a marked lack of interest in relation to official civil protection authorities and official natural disaster agencies. It is also worth noting that the mobilization of local government authorities (i.e., municipalities or communities) is favored as a first selection by 16.2% of respondents, second selection by 15.6%, and third selection by 40.9%. These findings show a marked confidence in local government authorities, given the fact that municipalities and communities in fire disaster areas, as well as across Greece, suffer from numerous problems because they usually lack qualified services and personnel and necessary infrastructure.

As previously mentioned, our research team conducted a series of brief qualitative interviews in fire disaster areas; our findings confirm that one-third of fire disaster inhabitants reassert their confidence to local government authorities and wish for their operational and infrastructure improvement. Respondents also emphasized the need for the development of a volunteer movement that would cooperate with local and civil protection authorities (namely forest inspection) (Table 12.4).

A considerable percentage of respondents (23.4%) ask for political change, referring mainly to communication and cooperation channels between citizens and politicians, because they believe that a lack of those things contributes to important environmental problems in their region. At the same time, respondents hold nationwide civil protection and security systems, such as the fire brigade or the police force, in rather low-esteem. This low-esteem in these public authorities is due to their ineffectiveness in fire disaster areas during the blazes, something that both the media and the Greek public opinion amply criticized (Table 12.2).

Qualitative interview respondents mark a strong preference for volunteerism, which goes hand-in-hand with the absence of any real collective institutions that are responsible for the community's civil protection. Following qualitative data processing, lack of social capital and absence of volunteerism were also confirmed. These findings help us understand the devastating events of the 2007 fires: local

Table 12.4 Qualitative Interview Summary—Coding (158 Short Interviews): Fire Disasters Communities in 2007

Proposals to Cope with the Environmental Problems (%)					
Improve Community Infrastructure	*Improve Forest Inspection Authority*	*Participation in Voluntary Associations*	*Poltical Change*	*Better Police and Fire Brigade*	*Other*
33.8	13.6	26.0	23.4	2.6	0.6

communities were tragically unprepared and disorganized in coping effectively with such a disaster, lacking in basic preventive measures with no previous civic education and proper information.

Furthermore, respondents expressed their distrust in representatives of the central political authority. They generally perceived politicians to be distant from their everyday problems and needs and expressed their wish for local resources and civil protection technical means to be put under local control (Table 12.3). At the same time, only 3.6% of respondents in FD communities could mention any type of local initiatives with regard to risk management and disaster intervention. A recurrent theme across respondents' interviews involved proposals for the establishment of local volunteer groups that would organize forest patrols in cooperation with forest inspection officials; one should examine these proposals in conjunction with a local society's low awareness and preparedness levels in relation to natural disasters.

Moreover, as shown above, political parties, market economy, and cultural agents (individual/collective) are given low percentages as relief and protection agencies, something that goes hand-in-hand with a lot of similar remarks expressed during interviews: "Politicians should come closer to the citizen and his problems" as well as a general feeling of insecurity concerning risks threatening their communities.

Fire disaster areas report almost all their answers in items of high risk (33.8%), medium risk (37.0%), and normal risk 23.4%, whereas only 5.8% report that their community or area has normal protection (see Table 12.6). Differentiation between fire disaster areas and areas without fire disasters related to risk perception is not very significant; 77% of respondents in the areas without fire disasters respondents consider that their area or community runs some kind of risk. These rates tend to confirm our theoretical hypothesis that risk perception, with regard to natural disaster, is placed in a framework of generalized risk environment where there is a diffused sense of insecurity at play. During the next few months following the devastating fires during the summer of 2007, an electoral campaign was performed leading to the 2007 national elections. The political debate was centered on recent disaster events and political antagonism focused on risk issues.

Table 12.5 Protection/Relief Agencies: Who Is Actively Responsible for the Community's Protection and Damage Relief?

	Local Authorities (%)	*Local Initiatives (%)*	*Nobody (%)*	*Market Economy (%)*	*Political Parties (%)*	*Artistic and Cultural Agencies (%)*	*Total (%)*
NFD	58.6	12.4	17.4	4.7	3.9	3.0	100
FD	96.4	3.6					100

Note: FD = fire disaster communities; NFD = nonfire disasters communities. N = 1667 in 2007.

Table 12.6 Community Risk Scale: Position Your Area/Community in a Scale According to Your Risk Perception and Your Community's Security

	Risk Scale[a] (%)							*Total (%)*
	–3	*–2*	*–1*	*0*	*1*	*2*	*3*	
NFD	36.3	17.8	22.4	2.5	6.4	5.4	9.2	100
FD	33.8	37.0	23.4	0	5.8	0	0	100

Note: FD = fire disaster; NFD = no fire disasters. N = 1667 in 2007.

[a] Scale was measured according to the following grading system: –3 = high risk; –2 = medium risk; –1 = normal risk; 0 = neutral; 1 = normal protected; 2 = medium protected area; 3 = very well protected.

This particular political arena influenced the overall sense of risk in Greece, as it was eventually confirmed by pre- and postelectoral polling surveys. More recently, following the October, 2009 national elections, the new prime minister of Greece, Mr. G. Papandreou, brought these issues to the forefront of the political debate again. He proposed a pilot model of environmental development in a new context of sustainable governance, proving that both the 2007 and 2009 fire catastrophes had a significant impact on political and government change (the previous conservative government was succeeded by a socialist one) (Table 12.4).

The Mountains of Attica: A Competitive Arena

The Attica basin—where Athens (the Greek capital) is situated—is surrounded by significant forested and nonforested mountains, Parnitha, Penteli, Hymettus, Pikilo Oros, Parnitha, and Penteli mountains. The basin suffered extended damage from forest blazes during the summers of 2007* and 2009 (Attica map†). The heavily forested Mount Parnitha and the other five mountains have been under strong pressure by real estate developers because of the city's continuous expansion [Chtouris et al. 1993, 2007]. The 2007 blazes scorched Mount Parnitha, the largest forest bordering Athens and a European Union protected nature reserve, a forest considered the lungs of heavily polluted Athens (Table 12.5).

The sociology department at the University of the Aegean undertook a research study to document natural disaster management models and local residents' reactions in the aftermath of the disaster. A common trait that characterizes these three mountains, Parnitha, Penteli, and Pikilo Oros, in contrast to other fire-stricken

* Afroditi Paggidi (2009), *Social Capital and Cultural Issues in Parnitha Mountain*—Attica, University of the Aegean, Mytilini (unpublished thesis).

† Attica Map, Forest fires in Greece, 2009 http://www.zki.dlr.de/media/download/applications/2009/greece_fires/2009_08_Greece_fires_overview_medium.jpg.

areas, namely Peloponnese and Evia, is that their protection is directly linked with the Greek capital's quality of life. Positive social capital in these three areas has been constructed gradually during the past decades, mainly through environmental nongovernmental organization, climbing and trekking associations, forest friend associations, and camping clubs. These groups are slightly different from more modern ecological initiatives that act in the framework of volunteerism and forest protection in their effort to avert risk threatening not only forests, but also Athenian quality of life. Social mobilization, periodically triggered by the city's acute environmental problems, such as air pollution, car pollution, and heavy traffic, involves volunteer activities and various conservation initiatives taken for the protection of periurban environment and mitigation of metropolitan problems (Table 12.6).

However, according to field research, overall social capital is particularly low and, more important, completely inadequate for everyday problems. Furthermore, environmental policy is immersed in a maelstrom of conflict and rivalry in a complex framework of local politics, residents' interests, real estate developers, housing cooperatives,* building cooperatives, and the much weaker conservation and volunteer organizations. To site a characteristic example of this confrontation, in the aftermath of the 2007 fires—right after the scorching of Mount Parnitha's forest—several nongovernmental organizations and local government authorities clashed with the powerful ELPA (Automobile and Touring Club of Greece), backed by politicians and the average Greek consumer,† to stop the annual organization of Rally Acropolis on the mountain. The university's research documented more than 200 different agencies, organizations, and local initiatives that have been active periodically in the framework of protection and conservation activities, reforestation projects, and information and awareness campaigns in these forests and mountains around Athens. In some cases, particularly in reforestation projects, there is an impressive synergy developed among nongovernmental organizations, the local government, environmentally friendly businesses, and some media representatives.

The Negative Social Capital of the Area

Mount Parnitha and the other forests and mountains around Athens present a high concentration of negative social capital in relation to its natural environment in the guise of urban building cooperatives. Maximum pressure has been exercised on Mount Penteli, where there are currently 28 building cooperatives vying for 1,100 hectares of the mountain. A large part of this forest was scorched during the 1995 and 1998 fires. It is revealing that one of the building cooperatives named New Constantinople claims 550 hectares from the area burned during the 1995 fires, whereas the Saint Spyridon cooperative claims the declassification of 1,200 hectares

* A housing cooperative aims to integrate land in the city plan. The land is often illegal occupation of public property.

† http://indy.gr/analysis/rally-akropolis-stin-parnitha.

from the same area. More importantly, in the whole area surrounding Mount Penteli, these extremely active building cooperatives are in collusion with powerful professional corporations and political networks. The following are a representative sample of these cooperatives: Bank of Greece Employees Association, Social Security Foundation (IKA) Employees and Pensioners Association, Journalists Association, Association of Piraeus Lawyers, Policemen Association, Association of Forest Service Employees, and Renaissance: Association of Military Officers, to name but a few.*

In total, these building cooperatives, in conjunction with individual interests, lay claim to 5,600 hectares. The Penteli Monastery, which usually sells part of its land to prospective residents at will, has undertaken a legal fight over 4,500 hectares that it claims as its own. As far as land cost is concerned, according to real estate evaluators, the amount can go as high as 1 million euros per 0.1 hectare. If we do the math, we observe that, in total, claims of land value in this relatively small part of the capital's periurban space are more than 60 billion euros. This intense pressure for more house construction explains the recurrent fires caused by arsonists in the same areas that have been recently declared as reforestation land (Kiousis 2007).† It is characteristic that a recent political scandal in Greece involved the Vatopedi Monastery, a Monastery on Mount Athos, and the Minister of the Merchant Fleet, who was obliged to resign from his post after serious charges of corruption for illegal transfer of state land to private investors, creating the general impression in Greek public opinion that there is a closely knit network of collusion among monasteries—usually owners of prime land, corrupt politicians, and building companies that systematically convert considerable parts of the natural environment to urban dwellings.‡

The 2007 fires brought to the forefront of political debate the issue of all the surrounding mountains near Athens (i.e., Parnitha, Hymettus, Pikilo Oros, and Penteli). These areas have acquired an increasingly high environmental value because their conservation as green mountains has become a vital priority for the 4 million people inhabiting Athens, which is approximately half of the total Greek population; their quality of life has to be safeguarded. Pikilo Oros, another important mountain at the western edge of the city, has also suffered in a similar way, drawing the interest of building companies and prospective buyers and giving rise to continuous political discussions and, at times, conflict among residents, municipality officials, and state authorities. These state agencies, along with municipal interventions, constantly strive for more land on which to build more houses, and this elicits strong reactions from the local residents.§ Local residents reacted strongly against multiple interventions and pressure by municipal authorities and

* KATHIMERINI,19-08-07, http:// eyploia.aigaio-net.gr/modules.php?name=News&file=article&sid=1163.

† G. Kiousis (2007), ELEFTHEROTYPIA, 17/08/2007.

‡ http://www.enet.gr/?i=news.el.article&id=32986&ref=search.

§ http://www.scribd.com/doc/10394355/XARTA-POIKILON-OROS.

state agencies because the latter deployed all of their resources in efforts to acquire more public land for house construction. This brought the perennial problem of public land usurpation by private interests—namely, illegal constructors, illegal owners (with fake property titles from the past),* and housing cooperatives—to the surface (Chtouris et al. 1993, 2007).

Furthermore, these periurban areas suffer from heavy environmental degradation, with their natural environment and forests slowly perishing due to aggravating factors such as pollution caused by car traffic, illegal waste disposal, and illegal landfills that hinder any reforestation or renewal efforts.

In Athens and its neighboring municipalities, the situation today is characterized by competition and rivalry between positive and negative social capital, with residents and prospective buyers—who traditionally want to take as much land as possible for their housing needs—on one hand and conservation associations and local environmental initiatives on the other. It is worth noting here that these new residents, vying for prime land in these areas, are upper-middle class people who tend to build luxury houses, thus legitimizing other individual or collective interests that follow similar practices, with catastrophic results on the environment. These interests are extremely powerful, both economically and politically, and it is difficult for ecological groups and conservation associations to confront them without adequate financial or legal means. There are, however, exceptions; the Firefighting Model Programme, coordinated by 22 municipalities around the Mount Penteli area in 2007, helped extinguish 64 out of 67 blazes in record time,† saving valuable forest land. This is a characteristic example that technical means become available and environmental resources are saved when there is concerted action by citizens and public authorities. However, we observe an increasing pressure from politicians, making it necessary to appeal to judicial and other independent monitoring authorities. In 2008, a relatively small independent ecological group appealed to the State General Supervisor, denouncing the misspending of 200 million euros that had been earmarked for the endowment of the Special Forest Fund, according to the state budget. These funds, mainly from legitimate economic forest enter-

* Among other characteristic examples, we can mention the Supreme Court's decision (April 2007) recognizing as legal property titles of 2,000 hectares to the Vatopedi Monastery in Mount Athos, according to a donation made by the Byzantine Emperor Ioannis Paleologos in 1351. [http://www.enet.gr/?i=news.el.article&id=32596&ref=search. However, we know from historical sources that this emperor never really reigned.

† A program that was designed 8 years ago by the Association for the Protection and Rehabilitation of Penteli with the philosophy that prevention is the best course of action. Indeed, its effectiveness was evident because this association, representing 22 communities and municipalities, helped extinguish 64 of 67 blazes occurring between May and July of 2007. Chrisitna Papastathakopoulou, *ELEFTHEROTYPIA*.

prises, would be used exclusively for reforestation, prevention measures, and forest protection. The state supervisor is still searching for this money.*

Social and Cultural Capital: Safeguarding Ikaria's Natural Environment†

Ikaria is a relatively isolated island in the North Aegean region. During the past few decades, it has suffered through catastrophic fires, with devastating effects on the island's forests. More specifically, in 1993 a blaze scorched more than 2,000 hectares of forest land and 12 people perished despite efforts to save their land and property because they had few means of firefighting equipment at their disposal. Three young men were burned while helping old people abandon their burning houses. This event traumatized Ikaria's people and they decided to take things into their hands.

Currently, Ikaria presents a significant difference compared to the other Aegean islands. Collective action and cultural activities go hand-in-hand to protect the island's natural and manmade environment. A significant event, called a panegyri (meaning feast and holiday), is the main focus of local mobilization, as well as a beacon for tourist attraction. In Ikaria, these panegyria (plural of panegyri) are organized by the island's local cultural associations, although the church was responsible for them in the past.‡ This transition from ecclesiastic authority to secular cultural associations is directly linked to the pressing need of local people to undertake their own initiatives to provide basic infrastructure for their communities, as well as effective environmental protection.

As a result of this initiative, all of the funds collected during the summer season—when the panegyria occur all over the island's villages and attract thousands of visitors and tourists—are earmarked for infrastructure needs such as road construction and repair, forest clearing, and fire protection zones. These funds are managed by the cultural association's chairman and the management board. Local cultural associations and their members are extremely proud of their collective action and are conscious that they have achieved something that makes Ikaria a markedly different island, an important pole of attraction for tourists and visitors.

---

* The state financial supervisor is responsible to control the expenditures of state institutions. *ELEFTHEROTYPIA*, 18/4/2009: "200 million Euros for Forests: Where Have They Gone?"

† The University of the Aegean researched cultural practices and musical culture in Ikaria during the summer of 2008; the Social, Cultural and Digital Documentation Lab was responsible for the research. Professor Sotiris Chtouris was the scientific manager and Christos Bakalis, Malama Rentari, and Dimitris Papageorgiou were the researchers. The research involved 20 qualitative open interviews with musicians, members of cultural associations, and inhabitants of Ikaria.

‡ Data collected through 20 qualitative open interviews with the Ikaria inhabitants in 2008.

The panegyri is a year-round event; its organization runs throughout the year, with specific tasks and responsibilities being assigned to the cultural association's members and management board on a volunteer basis, in good spirit, and with no official obligation. The panegyri is the community's most important annual event: "All year through we live for the panegyri, for the coming together, the moment that we will all celebrate together!"* It is difficult to distinguish the most important motive from comments: the coming together, the dance, the music, the celebration, or the revival of their local identity. This dynamic cultural symbolism—the panegyri—fills the gaps in the island's social and institutional capital caused mainly by mass migration after the 1950s and aging. Following World War II, another significant aggravating factor was the oppressive policy of the leftist population, which lead to a mass exodus of young people in search of a better future. Indeed, responses to our qualitative interviews reveal a general agreement that only the panegyri expresses solidarity and collective spirit among local inhabitants. The panegyri, as it stands, links primary communal sources with collective meaning and symbolism, making real contact and interaction redundant. Thus, members of local society produce a social context highly visible to all others—the Ikariote Diaspora—and, yearly, thousands of visitors are attracted to an alternative way of life, a nonutilitarian, solidarity-oriented, and communal everyday life.

More interesting, the attraction exercised by Ikaria's panegyri functions as an aesthetic and political symbol, appealing to urban young people outside the narrow limits of the island of Ikaria. Indeed, young people from Athens and other big cities come every summer to Ikaria not only for holidays, but also for a brief, unconventional, alternative break from everyday routine. Around this symbolic focus of celebratory collective function, environmental issues become all the more pertinent, being the object of lively debates and practical action. Young people, being more sensitive to modern environmental concerns about possible risks,† meet with older generations in the common ground of their love for the island. Local communities try to mobilize state services because they tend to be more indifferent and bureaucratic regarding environmental risks. During our research in the summer of 2008 (the year after the devastating 2007 fires in Pelpoponnese and Evia), we noticed volunteers patrolling daily in forest areas, seasonal firemen in full alert, and water cisterns full and ready for use. In contrast to other summer cosmopolitan islands of Greece, Ikaria stands alone as a paradigm of alternative development within an ecological context that works.

Despite this, Ikaria's communities are still vulnerable to new risks because the island is extremely mountainous, with almost inaccessible villages and a relatively small population that is responsible for planning and executing preventive and rehabilitating actions in the event of natural disasters.

---

* Ο.π. (2008).

† Data collected through qualitative interviews with the island's inhabitants.

In Ikaria, local communities undertake the task of managing environmental risks that threaten them with the relatively low funding that they annually collect, suggesting a moral obligation to both their members and foreign visitors not to depend on outside help (e.g., fire extinguishing planes) as most communities did during the summer of 2007, but rather to rally around community action and solidarity.

National and Global Responses to 2007 Natural Catastrophes

Following the devastating fires in 2007, there were a series of initiatives at the national and global levels to support fire-stricken communities. These initiatives went beyond usual civil protection measures and could be classified into the following categories:

Financial, Technical, and Material Support to Fire-Stricken Populations

The Greek government gave direct financial aid of 3000 € to those who suffered any kind of damage caused by fire. An extra 1000 € was allocated to families with more than three children and to people with disabilities. This direct financial aid was given relatively quickly after the blazes without any bureaucratic delay or previous control and with no consideration to the importance of the damage suffered. This was unusual by Greek and European standards and it offered considerable relief to devastated people, who were grief-stricken and disoriented. It also compensated for the state's inability to offer specific material aid to each community according to its loss. On the other hand, the lack of substantial controls in relation to rightful beneficiaries led to some unorthodox practices, such as money being given to people with no need for financial support. Subsequently, there were serious problems with poor and vulnerable population groups that were denounced as having illegally taken the relative financial aid.*

A direct financial aid of 10,000 € was allocated to those who suffered damage to their property, main or secondary residence, furniture, or household items, again following relatively simple procedures. This aid covered first expenses to repair or replace damaged household items. At the same time, monthly housing benefits of 250 € to 400 € were allocated for those who suffered the total loss of their house for a period of up to 2 years. Those who wanted to restore their damaged houses

* The reimbursement process, in relation to illegally received benefits, was very brief and at times ineffective, especially for population groups who had no access to information such as Roma people. Retrieved from http://www.1555.gr/UserFiles/File/pdf/220540_diadikasia_epistrofis_boithimatos.pdf.

received 500 € per square meter, two-thirds of which was given as a loan with no interest.*

At that time, the Greek Ministry of National Economy established the Solidarity Fund, a special fund for emergencies (http://www.etaea.gr/anakoinwseis.html)† that undertook, in a coordinated way, the task of collecting donations from all over Greece and abroad to alleviate hardship for fire-stricken populations. In February of 2009, the Fund's chairman announced that the total sum of 205,297,228.89 € was collected from 92,000 donors from Greece and abroad. It was also reported that 60,006,018.04 € of the above-mentioned sum was spent during the first year, namely 45 million € for house reconstruction and restoration and another 15 million € for other uses. Currently, 788 houses that were badly burnt or damaged are being reconstructed or restored, using partial funding from the Fund. The Greek Ministry of the Environment and Public Works monitors these works and, according to its announcement, 1,425 claims for reconstruction benefits have already been submitted by the interested, out of which 1,139 have been approved, as they were duly justified.

At a parallel level, private initiatives also played a substantive role in the relief effort. Businesses, charity organizations, individuals, artists, and sponsors offered all kinds of aid and support. Across Greece, a strong wave of emotional support dominated everyday life for a long time after the disaster; the media played a crucial role in keeping the issue at the forefront. In relation to information and public awareness, national media, especially television news reporters and the press, adopted an ecological discourse, appealing to their public's ecological sensitivity and charity, creating a new communication environment for political parties and big corporations.

We should also note the important role played by the European Union solidarity policy, both in providing firefighting planes and specialized personnel as well as relief in the aftermath of the catastrophe. The symbolic focus of their mobilization was, of course, the endangered site of Ancient Olympia, given that the Beijing Olympic Games were to occur the following year (Figure 12.4).

The European Union has established a Community Civil Protection Mechanism, with the goal of facilitating and supporting the provision of European civil protection assistance in the event of a major disaster, be it natural or manmade. In the Community Civil Protection Mechanism, the European Commission and 30 participating states (European Union European Economic Area States) actively participate, supported by the Monitoring and Information Centre (MIC). Figure 12.3 represents the MIC's activation process that was used during the 2007 emergency.

According to relevant European Union statistical data, from June 27, 2007 to July 2, 2007, the MIC mobilized seven firefighting planes in regions of central

* http://www.1555.gr/default.asp?siteID=1&pageID=34&langID=1.

† http://www.1555.gr/default.asp?siteID=1&pageID=31&langID=1.

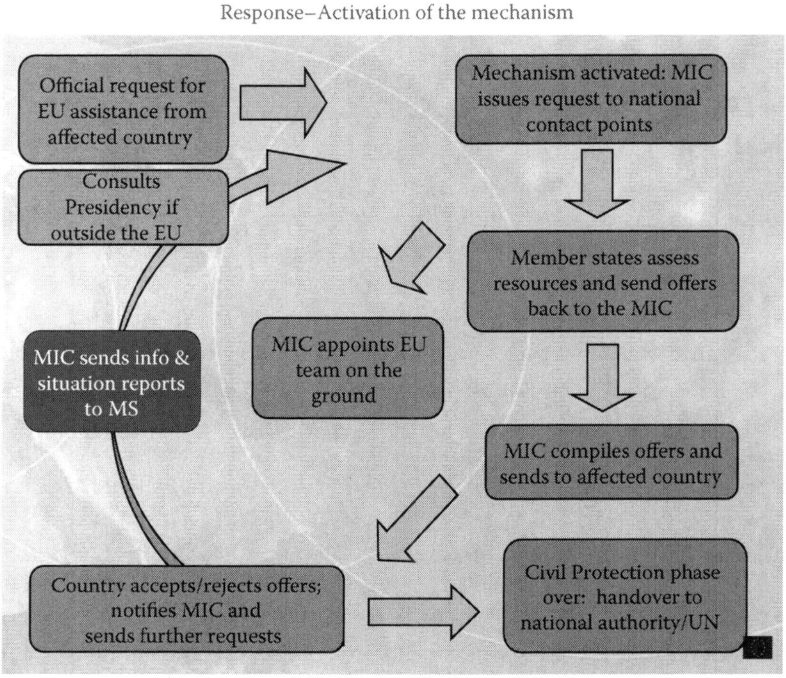

Figure 12.3 Response and activation of the European mechanism: Monitoring and information Centre (MIC) Source: http://ec.europa.eu/environment/civil. Copyright European Communities, 1995–2009. With permission.

Greece, originating from Italy, France, and Poland. From July 5, 2007 to September 1, 2007, four firefighting planes of the Canadair type were mobilized in Western Greece from France and Italy. Finally, from August 24, 2007 to September 5, 2007, covering the most crucial period of fire disaster, the MIC mobilized 13 planes, 22 helicopters, and 400 firefighters and specialized personnel from 13 countries. During the same year, the MIC mobilized to confront emergencies caused by forest fires in Italy, Cyprus, Bulgaria, Albania, and the former Yugoslav Republic of Macedonia, providing mainly firefighting planes and helicopters.*

It is worth noting here that in relation to information, documentation, and data interpretation, in the specific cases of the 2007 and 2009 natural disasters in Greece, remote sensing via satellite was provided by European and United Nations specialized services, making a valuable contribution to the overall effort.† The leading space agencies have made their satellites available for humanitarian purposes by

* http://ec.europa.eu/environment/civil.

† http://www.unitar.org/unosat_international_charter.

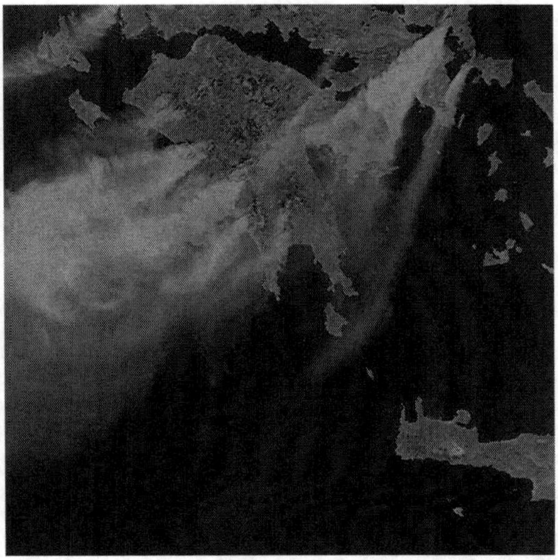

Figure 12.4 Aerial view of the region of ancient Olympia and western and southern Peloponnese fires in August 2007. (From MODIS Rapid Response Project, NASA/GSFC.)

signing the International Charter on Space and Major Disasters.* More specifically, national services offering gratis operational service on wild and forest fire detection on a daily basis using the NASA-owned MODIS sensors on board the Terra-1 and Aqua-1 space crafts, such as the Modis Fire Service of the German Remote Sensing Data Center in the German Aerospace Center, constitute a valuable help for communities, regions, and even small volunteer groups that want to monitor directly disasters and prepare a timely reaction in case of imminent danger. This objective geoinformation on environmental events seems to constitute an important factor in developing social capital and awareness at a local level, as well as swift and timely reaction to danger.

When considering both national and European responses to massive forest fires in 2007, one can observe that there was a significant mobilization at both levels. Moreover, financial and material aid provided to affected populations was considerable not only in size, but also in the wide range of relief measures that were undertaken in the immediate aftermath of the disaster. Nevertheless, the media have repeatedly reported disappointment, frustration, and anger expressed by people in the stricken areas who strongly criticized the Greek state's overall response. This criticism denounced state aid as inadequate, fragmented, and often unjust.† At this stage, it is still too early to objectively evaluate the real effectiveness of the relief measures or whether the provided aid and support helped those in need. In almost

* The member Space Agencies are ESA, CNES, CSA, ISRO, NOAA, CONAE, USGS, JAXA, and DMCCI.

† http://www.enet.gr/?i=news.el.article&id=3220&ref=search.

all cases, financial aid and support was provided via the banking system and the state bureaucracy; this did not encourage local mobilization of the afflicted population nor did it provide the necessary transparency for the whole process.

Comparative Analysis of Research Data in Relation to Environmental Risk Perception and Natural Disaster Management

Research data confirm that environmental risk perceptions and risk management are crucial social issues not only in Greece, but also in southern Europe because they involve low levels of social participation and social capital [Putnam 1993]. On the other hand, highly developed social networks based on family (nuclear and extended) ties preserve their cohesion [Chtouris 2004], but fail to function as cooperation networks when emergencies break out and risk management is needed.

An equally important problem involves lack of cooperation among local, regional, and national prevention and protection systems when natural disaster strikes. At the same time, we notice that at the local level—be it a community or a municipality—there is a certain apathy and indifference in relation to environmental risks. An aggravating factor is radical population decline in rural areas that has led to a general weakening of social cohesion and of those institutions that generate social capital.

Furthermore, we observed that social capital stock is low and weak in all research areas in relation to protection and rehabilitation measures. Despite the emergence of increasingly acute environmental problems and risks, there has not been a concomitant increase of awareness that would enhance both social participation and social capital.

This negative development is further worsened by the state's inability or unwillingness to provide local communities with the necessary financial and technical resources so that local residents themselves may undertake the responsibility of protecting their community collectively, in a spirit of cooperation and transparency. As we mentioned before, there is a marked lack of concerted action by local administrative and financial institutions that fail to coordinate their operational capability with environmental protection mechanisms and risk management agencies. As a result, existing social capital is not duly mobilized in emergency situations, as was amply proven during the 2007 crisis.

People who experience this weakening of social participation in response to risk management also hold all relevant state agencies—either administrative or technical mechanisms that pretend to support them in cases of emergency—including prefectures, regional authorities, compensation mechanisms, and civil protection systems in very low esteem.

In reference to land ownership, there are competitive aspects of positive and negative social capital at play that lead to more acute environmental problems and risks. Consequently, many more illegal practices—individual and collective—dominate land appropriation, degrading even more an area's environmental quality. Furthermore, powerful financial and corporate interests have a strong grip on public land appropriation, mainly through illegal means, so that state control mechanisms find it increasingly difficult to exercise effective controls. It seems that negative social capital is easily transformed to financial capital, thus consolidating illegal practices.

The field of environmental risk management has increasingly become an arena of political and financial conflict, thus obscuring both real problems and viable solutions. Communities that want to protect their human, natural, and material capital should aim at a risk management model that would concretely distribute competencies among local agencies and state authorities of environmental policy, civil protection, and civil society networks, so that maximum efficiency and cooperation can be assured in emergency situations. It seems that the absence of social capital at a local level leads to a host of negative reactions that aggravates the situation further, such as inefficiency and unwillingness to readily respond to relief and compensation measures, misinformation and bias, as well as authoritarian behavior on behalf of local authorities.

Following our research findings and referring to Figure 12.2, we observe that positive and negative relationships in the circle of social capital are directly affected by special circumstances and the experience of recent disasters. It is important to keep an active social and cultural memory when confronting natural disasters or social risks, not only in Greece but in neighboring countries as well [Glade et al. 1993]. Natural disasters have often led to the establishment of best practices at a local or national level. Documentation of collective memory and relevant successful practices constitute significant cultural capital that may contribute to the development of valuable new initiatives and useful measures.

References

Baron, S., Field, J., and Schuller, T. 2000. *Social capital: Critical perspectives*. Oxford: Oxford University Press.

Bauman, Z. 1990. Modernity and ambivalence. In *Global culture: Nationalism, globalization and modernity*, ed. M. Featherstone. London: Sage Publications.

Beck, U. 1992. *Risk society: Towards a new modernity*. London: Sage Publications.

Boeck, T., and Fleming, J. 2005. Social policy: A help or a hindrance to social capital? *Social Policy and Society* 4:259–270.

Bourdieu, P. 1984. *Quelques proprietes des Champs, Question de Sociologie*. Paris: Minuit.

Chtouris, S. 1987. *Tradition und moderne in der europaeischen Peripherie*. Kassel: GhK Kassel.

Chtouris, S. 1997. *Metaviomichaniki Kinonia ke i Kinonia tis Pliroforias* (Postindustrial Society and the Informational Society). Athens: Ellinika Grammata.

Chtouris, S. 2004. *Rational symbolic networks*. Athens: Nesos Editions.

Chtouris, S. 2009. I choriki diastashi tou kinonikou kefaleou [A spatial approach to the concept of social capital]. *Synhrona Themata (Modern Issues Magazine)* 104:72–82.
Chtouris, S., Heidenreich, E., and Ipsen, D. 1993. *Von der Wildnis zum Urbanem Raum: Zur Logik der peripheren Vestaedterung.* New York: Campus.
Chtouris, S., Malkogiourgou, E., Mastrogianopolous, G., and Tzelepoglou, F. 2008. *Research on urban areas facing increased environmental risk in Greece, 2007–2008*, unpublished manuscript. Mytilini: Sociology Department, University of the Aegean, Mytilini.
Coleman, J. S. 1994. *The foundations of social theory.* Cambridge, MA: Belknap Press.
David, M., and Sutton, C. D. 2004. *Social research: The basics.* London: Sage Publications.
Douglas, M. 1992. *Risk and blame: Essays in cultural theory.* London: Routledge.
Douglas, M., and Wildavsky, A. 1982. *Risk and culture: An essay in the selection and interpretation of technological and environmental dangers.* Berkeley, CA: University of California Press.
Giddens, A. 1990. *The consequences of modernity.* Cambridge: Polity.
Giddens, A. 1991. *Modernity and self-identity.* Cambridge: Polity.
Giddens, A. 1998. Risk society: The context of British politics. In *The politics of risk society*, ed. J. Franklin. Oxford: Polity Press.
Granovetter, M. 1973. The strength of weak ties. *American Journal of Sociology* 78: 1360–1380.
Granovetter, M. 1985. Economic action and social structure: The problem of embeddedness. *American Journal of Sociology* 91:481.
Harvey, D. 1996. *Justice, nature and the geography of difference.* Oxford: Blackwell.
Heidegger, M. 1977. *The question concerning technology and other essays.* New York: Harper & Row.
Heidenreich, E., Chtouris, S., and Ipsen, D. 2007. *Athens: The creation of a Mediterranean metropol.* Athens: Kritiki.
Lyons, M. 2002. *Non-profit organisations, social capital and social policy in Australia.* Melbourne, Australian Institute of Family Studies.
Lyotard, J.-F. 1979. *La Condition postmoderne: Rapport sur le savoir.* Paris: Editions de Minuit.
Onyx, J., and Bullen, P. 2000. Measuring social capital in five communities. *Journal of Applied Behavioral Science* 361:23–42.
Otway, H., and von Winterfeldt, D. 1982. Beyond acceptable risk: On the social acceptability of technologies. *Policy Sciences* 14:247–256.
Petersen, A. 1996. Risk and the regulated self: The discourse of health promotion as politics of uncertainty. *Australian and New Zealand Journal of Sociology* 321:44–57.
Putnam, R. 1993. *Making democracy work: Civic traditions in modern Italy.* Princeton, NJ: Princeton University Press.
Putnam, R. 2000. *Bowling alone: The collapse and revival of American community.* New York: Simon & Schuster.
Putnam, R. 2002. *Democracies influx: The evolution of social capital in contemporary society.* Oxford: Oxford University Press.
Read, D. W. 2002. A multitrajectory, competition model of emergent complexity in human social organizations. *Proceedings of the National Academy of Science USA* 99:7251–7256.
Renn, O. 1992. The social arena concept of risk. In *Social theories of risk*, eds. S. Krimsky and D. Golding. Westport, CT: Praeger.

Sarason, S. B. 1974. *The psychological sense of community: Prospects for a community psychology*. Oxford: Jossey-Bass.
Schwarz, M., and Thompson, M. 1990. *Divided we stand: Redefining politics, technology, and social choice*. Philadelphia: University of Pennsylvania Press.
Turner, V. 1974. *Dramas, fields, and metaphors: Symbolic action in human society*. Ithaca, NY: Cornell University Press.
Van Loon, J. 2002. *Risk and technological culture: Towards a sociology of virulence*. London: Routledge.
Vernardakis, C. 2005. *Public opinion in Greece*. Athens: Savallas.
Woolcock, M. 1998. Social capital and economic development: Toward a theoretical synthesis and policy framework. *Theory and Society* 27:151–208.
Wynne, B. 1982. *Rationality and ritual: The windscale inquiry and nuclear decisions in Britain*. Chalfont, St. Giles: British Society for the History of Science.

Chapter 13

Supporting State Agencies in Providing Early Warning Services: A Case Study from Ireland

Ioannis M. Dokas, John Feehan, Stephen C. Fortier, Franclin Foping, and Syed Imran

Contents

Introduction

Water is essential to life; however, drinking water demand and the worlds' limited available resources have made the United Nations and others identify a condition of instability known as the water crisis [Gleick 1993; United Nations 2006a]. Sound indicators of the so-called water crisis are, among others, the scarcity of usable water for a large proportion of the world's population and the increasing numbers of water contamination and pollution cases on a global scale. Quoting the United Nations (UN) [2006a], "For some, the water crisis means having to walk long distances every day to fetch enough drinking water—clean or unclean—just to get by. For others, it means suffering from avoidable malnutrition or disease caused by drought, flood or inadequate sanitation. Still others experience it as a lack of funds, institutions or knowledge to solve local problems of water use and allocation."

Raw water absorbs particles of anything it contacts: minerals, vegetation, and chemicals. It also contains microbes and must be purified before consumption. Although most of these substances are harmless, some are hazardous to the public and result in drinking water contamination with disastrous consequences. Waterborne disease outbreaks have affected communities in many parts of the world, including those in some of the world's most developed countries. For example, in Europe, 7,960 confirmed cases of *cryptosporidiosis* were reported during 2005. Ireland, with 13.75 cases per 100,000 cases, and the United Kingdom, with 9.26 cases per 100,000 cases, has the highest reported incidence rates [Semenza et al. 2007]. Ireland with 565 and United Kingdom with 5,561 confirmed cases, have the highest reported incidence rates [Semenza et al. 2007].

In North America, a significant number of waterborne disease outbreaks have been reported. There were 155 outbreaks that resulted in 431,846 cases of illness in public and individual water systems from 1991 to 2000 in the United States [CCDACC 2003]. The most dramatic case of a waterborne disease outbreak in Canada was in Walkerton in May of 2000, where more than 2,300 people became ill and 7 died. An inquiry committee was established after the outbreak. It defined the contributing factors to that tragedy: improper operational practices at the water treatment plant, lack of training of personnel, and shortcomings in the approvals and inspections programs of the Ministry of the Environment [O'Connor 2002]. It exposed the sociotechnical dimension of manmade disasters related to the domain of water treatment and supply. One of the many recommendations made by the commission was the need for a quality management standard for all agencies involved in the provision and regulation of drinking water. Such a system, if implemented fully, should ensure that issues are addressed in a proactive way before a major problem occurs. Eleven months after Walkerton, another outbreak occurred in North Battleford, Saskatchewan, Canada. It was caused by *cryptosporidium* and was estimated to have affected between 5,800 and 7,100 people. The drinking water treatment plant was vulnerable to contamination because it was 3.5 km downstream from the community's sewage effluent outfall. The North Battleford outbreak infected more people than Walkerton, but the health consequences were less severe than Walkerton and no deaths occurred.

Another example that highlights the sociotechnical dimension of disasters related to waterborne disease outbreaks is the cholera outbreak in Zimbabwe in August of 2008. The outbreak was attributed to the collapse of the urban water supply, sanitation, and garbage collection systems as well as the economic problems of the country and to the rainy season, during which feces containing cholera bacteria washed into water sources [IRIN 2008]. The adverse effects of these outbreaks have heightened public awareness that the threats to water quality and quantity can have a profound impact on our health, the environment, and the economy [CCME 2004]. Since August, at least 3,623 people have died and 76,127 people have been infected by cholera [CNN 2009].

Water Treatment and Supply

To address the risk of waterborne disease outbreaks, governments and state agencies have adopted strategies and standards aimed at supplying potable water of good quality. From source to tap is a well-known strategy employed to deliver safe drinking water. The main idea of this strategy is that the components of the water supply system must be understood and managed as a whole. The main elements of a water supply system are the source, the treatment plants, and the distribution system.

Shallow wells, rivers, natural lakes, springs, artificial lakes, and reservoirs are common water sources. The water distribution system is usually composed of three major components: pumping stations, distribution storage, and distribution piping [Mays 2004]. Water treatment plants are engineering systems that purify raw water to specific safety levels that are defined by chemical and microbiological standards. The raw water passes through a series of treatment phases where it is processed and purified to existing safety and potable levels. After its purification, the potable water is distributed to the consumers through a network of pipes, pumps, and reservoirs.

The purification phases in a typical surface water treatment plant are shown in Figure 13.1. Small particles of solids and organic matter, as well as other impurities, are suspended in the raw water entering the water treatment facility. The settling time of the colloidal particles is decreased rapidly with a flocculating agent. During coagulation, a flocculating agent is thoroughly mixed with the water and this causes colloidal particles to coagulate, forming larger particles known as flocs. The water containing the flocs is passed to large settlement/sedimentation tanks. This occurs slowly to allow heavy solid particles and flocs to settle to the bottom of the tanks due to gravity. This process is known as sedimentation, where the clear water continues its route through the plant and the remaining sludge is handled and disposed of according to regulations.

After sedimentation, any small quantities of solids remaining are removed as the water passes through sand filters. A typical sand filter consists of layers of gravel, coarse sand, and fine sand. The water may contain some harmful bacteria or microorganisms. During disinfection, a sufficient quantity of chemicals, usually chlorine or chlorine compounds, are added to kill most pathogenic microorganisms in the water and to prevent recontamination before the water reaches the taps. Finally, for fluoridated water supplies, fluoride is added to the water to reduce tooth decay for its consumers.

Water treatment plants are important components of the water supply system. Their main role is to act as barriers, eliminating contamination agents from the raw water before consumption. Some recorded waterborne disease outbreaks illustrate the importance of water treatment plants in water supplies systems. In a surveillance study of waterborne disease outbreaks in the United States between 2005 and 2006, 20 cases were found to be associated with inferior drinking water quality [Yoder et al. 2008]. From analysis of the 20 cases, 23 deficiencies of water supply systems were cited, 10 of which fell under the classification of deficiencies at the source, the water treatment, and the distribution system. The most frequently cited deficiencies of that type were associated with a treatment

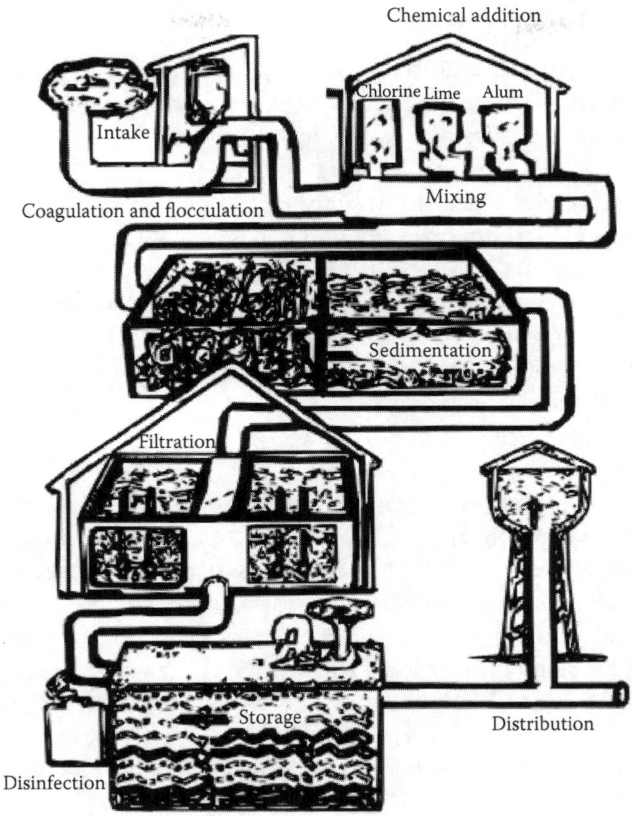

Figure 13.1 Purification phases in a surface water treatment plant. (Adapted from U.S. EPA, Water treatment process, http://water.epa.gov/learn/kids/drinkingwater/watertreatmentplant_index.cfm.)

deficiency (e.g., temporary interruption of disinfection, inadequate or no filtration, or chronically inadequate disinfection) and with the inexistence of any type of treatment of ground water sources.

A characteristic example of a waterborne disease outbreak due to a failure of a water treatment plant is the 1993 *cryptosporidium* outbreak in Milwaukee, Wisconsin caused by *cryptosporidium* oocysts that passed through the filtration system of one of the city's water-treatment plants [MacKenzie et al. 1994]. It was estimated that over the span of approximately 2 weeks, 403,000 residents in the Milwaukee area became ill with stomach cramps, fever, diarrhea, and dehydration caused by the pathogen. Unfortunately, 69 deaths were attributed to this outbreak [Corso et al. 2003]. This case indicates the magnitude of the adverse effects of failure of a water treatment plant.

Early Warning Systems

Assuring the safety and security of water treatment and distribution systems is a primary objective of drinking water management. Key elements of a risk management approach to drinking water safety include the following [Rizak et al. 2007]:

- Being preventive rather than reactive
- Understanding the entire water supply system and the hazards that can compromise drinking water quality
- Distinguishing greater risks from lesser ones and developing effective measures to manage significant risks first
- Investing resources in risk management appropriately to maximize the intended outcomes
- Taking time to learn from experience

Systems that provide early warning services in the water supply domain can be useful instruments in supporting proactive risk management approaches and strategies. Some types of sensor and computer systems can enhance the collection and interpretation of monitoring data of critical water and operational parameters. Some types of information technology systems can enhance feedback control mechanisms for maintaining good water quality and can support the effective and reliable dissemination of important information to the stakeholders in a timely manner. The challenge is to design and develop an early warning service, composed of a complete set of components or systems that, together with their proper interaction and coordination, will enhance monitoring of critical parameters, alert stakeholders, and enhance the decision-making process of domain experts and the dissemination of the appropriate tasks—or the execution of tasks—that must be performed to prevent a failure that is expected to have severe adverse effects.

A characteristic of the drinking water contamination problem is that it is often detected only after a health crisis, when people have fallen ill or have died as a result of drinking unsafe water [Hartl 2004]. This implies that more effective and reliable proactive risk management strategies need to be incorporated, which must be supported by more effective early warning services. Unquestionably, deviation from normal operational procedures and failures in water treatment utilities are considered among the causes of this problem. In light of the adverse affects of contaminated drinking water and with the increased threat of terrorist activities in some countries, the need for a proactive risk management approach to water treatment and distribution systems is more important now than ever before.

This chapter is focused on an innovative multidisciplinary approach consisting of knowledge modelling and representation, risk and accident analysis, water purification engineering, software engineering, and artificial intelligence methods and technologies supporting the Irish Environmental Protection Agency (EPA) in providing early warning services for water treatment utilities. The approach is the core effort of a 5-year research project, funded by the Irish EPA, which began developing a prototype Web-based early warning system for water treatment plant operations in January of 2008. The early warning system will be delivered to the Irish EPA to set the foundations for early warning services in this domain.

The ultimate goal is to design and develop a novel Web-based early warning service that will be used by the personnel of institutions (i.e., *institutions* is used here in the broadest sense) involved in drinking water delivery and the governance model of Ireland, such as the Irish EPA and local authorities. The aim of this service is to make visible, as early as possible, disturbances of the performance of drinking water treatment, critical problems, and accidents at the water treatment plant level and at the wider sociotechnical system to achieve the following goals:

- Protect the personnel and the infrastructure in the water treatment facilities
- Protect the environment
- Inform and guide personnel and the public accordingly, in case of imminent anticipated adverse effects
- Protect the public from drinking contaminated water

This chapter aims to satisfy the following objectives:

1. Highlight the problem of drinking water quality on a global scale and in the Republic of Ireland in particular
2. Briefly explain the role of governance in drinking water safety
3. Describe our innovative approach in developing an early warning system service and the role of state agencies to our proposed solution

Responses of State Agencies

Responses in Relation to Water Security

Developed countries are concerned about the threat of terrorist attacks to critical infrastructure, especially water utilities. State agencies have been working on this issue for several years. A review of several nations' policies on protecting water sources highlighted similarities in approaching the problem. Typically, state agencies are providing the following:

- Mutual aid and cooperative agreements between neighboring nations. For example, Canada and the United States share common bodies of water and the EPA works closely with Health Canada's Water Quality and Health Bureau.
- Water supply systems analysis.
- Review of water quality data.

- Vulnerability analysis and risk assessments.
- Preventive measures.
- Critical control points.

At the local level, protecting drinking water sources usually requires the combined efforts of many partners such as public water systems, communities, resource managers, and the public.

State agencies are also planning to respond to threats and incidents of intentional contamination of drinking water. The potential for water contamination is of great concern considering the potential consequences [Allgeier 2004], including a potential loss of life, property, and efficacy:

- Short-term and long-term health issues with the potential for death within the affected population
- Disruption of water system operation and the unavailability of safe drinking water
- Physical damage to the water system infrastructure
- Loss of trust by the general population in the local water authority
- Potential for long-term denial of water, which could lead to exorbitant costs for remediation and replacement of infrastructure

In the following sections, we will discuss how the United States, Canada, Germany, and New Zealand address the threat of terror attacks to water utilities.

United States

The U.S. Environmental Protection Agency (EPA) is working in partnership with state and local governments to protect the nation's drinking water supply from terrorist attack. Under Presidential Decision Directive (PDD) 63, issued in May of 1998, the U.S. EPA was designated as the lead agency for the water supply sector [PDD 63 1998].

In September, 1998, the EPA established a public–private partnership with water-related organizations and subsequently designated the Association of Metropolitan Water Agencies (AMWA) as the water sector liaison to the federal government on critical infrastructure. According to U.S. EPA data, in 2004, there were approximately 53,000 community water systems in the United States. U.S. EPA's data demonstrates that ownership of these systems is evenly split—approximately half are publicly owned by state and local authorities and the other half are privately owned. According to the U.S. EPA, the majority of the U.S. population gets its water from publicly owned systems.

In November, 1998, a preliminary plan, *National Infrastructure Assurance: Water Supply Sector*, was drafted. Although the November, 1998 preliminary plan showed a scheduled completion date of the end of 2003 for these activities, the schedule was accelerated in response to the terrorist acts of September 11, 2009.

In October 2001, the Water Protection Task Force was established to ensure that activities to protect and secure water supply infrastructure are comprehensive and are carried out expeditiously. The U.S. EPA disseminated to America's water utilities useful information about steps they can take to protect their sources of supply and their infrastructure, which include pumping stations, treatment facilities, and computer systems.

The water sector consists of two basic components: freshwater supply and wastewater collection and treatment. The focus was on community water systems within supply because the primary focus of critical infrastructure efforts is on these systems. The water sector is regulated by the Bioterrorism Act of 2002,* which requires community water systems serving more than 3,300 people to perform vulnerability analyses of their facilities. The act further required systems to prepare or revise an emergency response plan incorporating the results of the vulnerability assessment within 6 months after completing the assessment.

The U.S. EPA provides tools and technical assistance to aid community planners, local source water protection teams, and public water supply operators [U.S. EPA 2006]. The tools include coverage for protection, susceptibility determination, delineation, and potential contaminant source inventory. One valuable tool is the *Response Protocol Toolbox: Planning for and Responding to Drinking Water Contamination Threats and Incidents* [U.S. EPA 2004]. It was created to help the water sector effectively and appropriately plan for and respond to contamination threats and incidents.

Canada

The responsibility for making sure drinking water supplies are safe in Canada is shared between the provincial, territorial, federal, and municipal governments. The day-to-day responsibility of providing safe drinking water to the public generally rests with the provinces and territories, while municipalities usually oversee the day-to-day operations of the treatment facilities.

Health Canada's Water Quality and Health Bureau provides a leadership role in science and research. Its mandate and expertise lies in protecting the health of all Canadians by developing the *Guidelines for Canadian Drinking Water Quality* [Health Canada 1996] in partnership with the provinces and territories. These guidelines are used by every jurisdiction in Canada and are the basis for establishing drinking water quality requirements for all Canadians. Canada works closely with the U.S. EPA and the Department of Homeland Security to monitor threats to shared water sources.

* Public Law No. 107-188, 116 Stat. 594 (2002).

Germany

The Federal Environment Agency (Umweltbundesamt—UBA) is the authority in charge of drinking water [UBA 2008]. Within the area of authority accorded to the Federal Ministry of Health, the UBA develops the scientific foundation and benchmarks for a safe drinking water supply. The UBA evaluates potential health risks related to the abstraction, treatment, and distribution of drinking water and develops schemes to avoid or control risks.

The staff of the UBA department Drinking and Swimming Pool Water Hygiene is responsible for the following:

- Development of schemes for the protection of ground and surface waters used to abstract drinking water
- Assessment of the processing substances and disinfection procedures as relevant to their use in drinking water treatment
- Development of procedures for the evaluation of substances and materials used in drinking water distribution
- Assessment of the incidence of pathogen presence in drinking water and development of detection methods
- Assessment of health-relevant properties of chemical substances in drinking water
- Development of procedures to detect organic and inorganic pollutants in drinking water

New Zealand

New Zealand does not perceive the threat of a terror attack to its water utilities as much as the United States does. The Health (Drinking Water) Amendment Bill is currently before Parliament [NZ 2009]. The legislation will require each water supply to be covered by a quality assurance program contained in a public health risk management plan. There is currently no legal requirement for water suppliers to have a public health risk management plan in place, but the government strongly suggests them. At the end of 2005, 256 (11.3%) suppliers had initiated work on a plan.

The annual review of drinking water quality in New Zealand is part of an organized campaign started in 1992 to improve the quality of the country's drinking water. The review covers the fourth year that the microbiological and chemical quality of drinking water was assessed using the Drinking Water Standards for New Zealand 2000 [DWSNZ 2000]. Other activities to ensure drinking water safety that New Zealand has developed since 1993 include the following:

- Drinking Water Standards for New Zealand 1995, 2000, and 2005
- External surveillance by health protection officers

- Register of community drinking water supplies
- Public health grading of community water supplies
- Annual Report on the Microbiological and Chemical Quality of Drinking Water Supplies in New Zealand
- Guidelines for Drinking Water Quality Management in New Zealand
- Registry of Ministry-of-Health-recognized laboratories for drinking water compliance testing
- National electronic water quality information database (WINZ)
- Public health risk management plans for drinking water supplies

Improving the quality of drinking water and active involvement in vulnerability and threat analysis was a common theme with all of the countries researched for this chapter.

Responses of State Agencies in the Republic of Ireland

Cryptosporidium *Outbreak in Galway City*

In March 2007, Galway City, Ireland, experienced the adverse effects of having the drinking water contaminated with the *cryptosporidium* parasite caused by human sewage [Bradley 2007], resulting in at least 180 serious cases of cryptosporidiosis. This outbreak was not the first on the island of Ireland. There were three waterborne outbreaks in Belfast between 2000 and 2001, and outbreaks in Westmeath [2002], Ennis [2005], and Carlow [2005].

Following the outbreak of cryptosporidiosis in Galway City, an independent review of the actions of the local authority (Galway City Council), in whose functional area the outbreak occurred, was initiated. The review team interviewed a wide range of people who were directly or indirectly involved and affected by the outbreak to ascertain its impacts on the general public, on the business and tourism interests in the city, and on the wider community. They also reviewed the performance of the local authority before and during the outbreak and the background to and causes of the outbreak.

Input was also invited from elected members of the local authority, staff and management of the local authority, staff and management of the neighboring local authority, the Corporate Policy Group in Galway City Council, representatives of local business and tourism interests, the EPA, the Health Service Executive (HSE), and the Department of Environment, Heritage and Local Government (DELHG).

They concluded that there was a complex combination of factors at local, regional, and national levels that together led to the problems with water quality in Galway post-March, 2007 and led to a serious diminution of public confidence in the ability of the City Council to supply safe drinking water to the public [IPAI 2009].

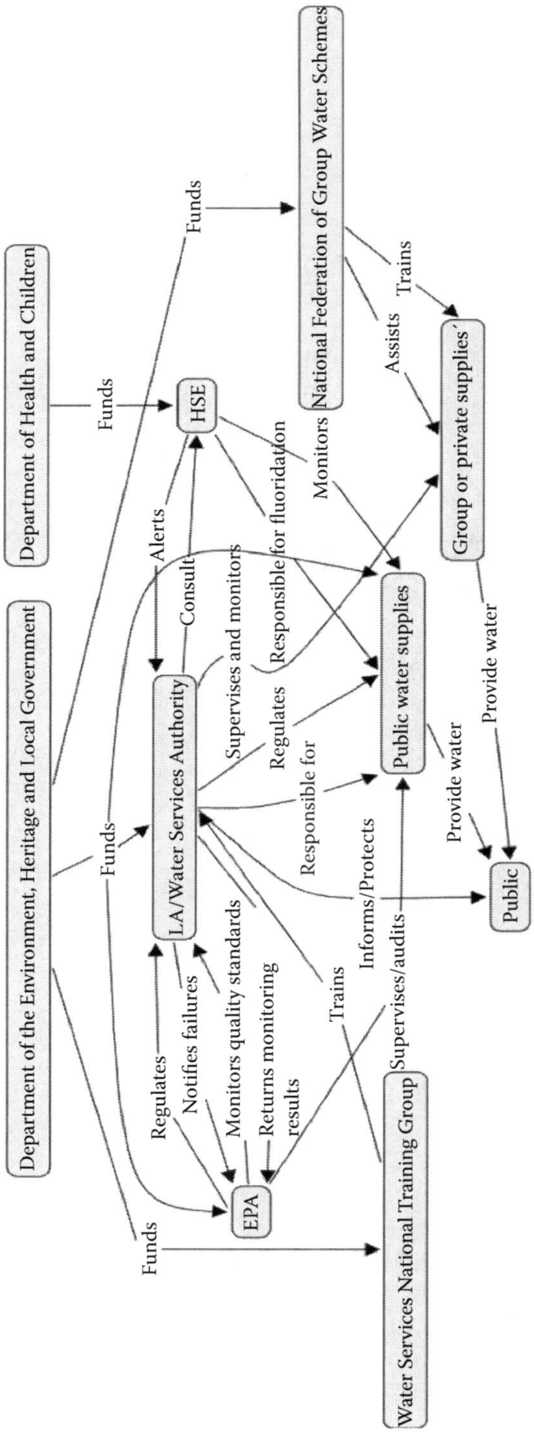

Figure 13.2 Governance model.

The Drinking Water Governance Model in the Republic of Ireland

There are a broad range of government departments and agencies involved in the provision and regulation of drinking water in the Republic of Ireland. The primary responsibility for the provision of safe drinking water rests with the local authorities, the Committees of Group Water Schemes, and those who individually provide their own drinking water. The main agencies responsible for the regulation of drinking water in the Republic of Ireland are the EPA and the HSE. There is also a range of other agencies involved in the protection of drinking water sources and the regulation of drinking water quality, which are illustrated in Figure 13.2.

Government Departments

The DEHLG develops and implements government policy in relation to drinking water. The DEHLG is responsible for legislation in relation to drinking water and, with local authorities, is responsible for the implementation of a range of water related EU Directives. The DEHLG funds the local authorities and also administers and funds the major drinking water schemes.

The Department of Health and Children develops and implements Government policy on health and personal social services. It is responsible for the overall performance of the health services in the Republic of Ireland. The Department of Health and Children funds the HSE. The local authorities must seek the advice and agreement of the HSE where there may be implications for public health.

State Agencies

Environmental Protection Agency

The EPA is an independent public body and its sponsor in government is the DEHLG. In March of 2007, new drinking water regulations were enacted by the DEHLG. Although these regulations made no changes to the monitoring required and only one minor change to the quality standards to be achieved, they significantly changed the role of the EPA in relation to drinking water. Since March of 2007, the powers assigned to the EPA include a responsibility to do the following:

- Ensure that local authorities are taking the appropriate action to ensure that public water supplies comply with the relevant quality standards
- Review the actions taken by local authorities in public water supplies where there has been a breach of a standard

- Review and approve monitoring programs to ensure that adequate monitoring is carried out by local authorities
- Audit local authority drinking water treatment plants
- Publish guidance on how local authorities are to implement the drinking water regulations

The drinking water regulations do not provide the EPA with powers to prosecute a water supplier for supplying water that is not clean and wholesome. In general, the powers available to the EPA under the 2007 regulations relate to the performance of the local authority in respect of any EPA direction. The regulations require local authorities to notify the EPA of failures to meet the quality standards, following which the EPA can direct the local authority to take corrective action. Only where the corrective action is not taken as directed can prosecution be considered for failing to comply with the terms of a direction. In other words, the EPA may prosecute a local authority only if it fails to comply with an EPA direction. The EPA is required to collect and verify monitoring results for all water supplies in Ireland. This involves the collection of results on an annual basis from local authorities and the preparation and submission of an annual report on the quality of drinking water in Ireland to the Minister for the DEHLG.

In this regard, the EPA is obliged to verify compliance of drinking water supplied by a water services authority and to supervise the performance of each water services authority of its monitoring functions under the drinking water regulations. The assessment of the monitoring results submitted includes an assessment of whether the minimum monitoring frequencies for each water supply have been complied with. The EPA is also required to keep a register of water supplies for which it is the supervisory authority. The EPA collects this information as part of the annual reporting of monitoring results. It has developed the Environmental Data Exchange Network (EDEN) to assist local authorities in the submission of their water-related environmental data. EDEN allows monitoring data entered into the local authorities' laboratory information system to be directly uploaded to EDEN. This enables local authorities to fulfill their obligations regarding reporting under the drinking water regulation.

The EPA may issue a direction subject to agreement with the HSE, where water intended for human consumption constitutes, or may constitute, a risk to human health. It is an offense for a water services authority not to comply with a direction.

The EPA also provides guidance to local authorities on the implementation of the drinking water regulations and recommends that they adopt the World Health Organization's water safety plan approach to the management of drinking water supplies.

Health Service Executive

The Health Service Executive (HSE) is the single body responsible for meeting Ireland's health and social care needs. The objective of the HSE is to provide services that improve, promote, and protect the health and welfare of the public. Where a drinking water sample fails to meet a parametric value in the drinking water regulations, the local authority must determine whether the non-compliance presents a risk or potential danger to human health in consultation with the HSE. The only exception to the requirement for prompt notification of a failure to meet the parametric values in the regulations relates to fluoride where the supply is artificially fluoridated.

Under the 2007 drinking water regulations, where a supply of water is considered to be a risk to human health, follow-up action must be undertaken by the local authority in consultation with the HSE. Responsibility to ensure that follow-up action is taken lies with the relevant local authority in the first instance, whether in relation to its own supplies or a water supply under its supervision. The drinking water regulations oblige the local authority in the first instance, subject to the agreement of the HSE, to ensure that use of the relevant supply is prohibited or restricted, and that consumers are informed promptly and given any necessary advice. Where a public water supply is involved, the authority must also inform the EPA promptly.

Fluoridation of drinking water commenced in the Republic of Ireland in 1964 after the introduction of the Health (Fluoridation of Water Supplies) Act. Its main aim was the prevention and control of dental decay. The act provides that the HSE shall arrange for the fluoridation of public piped water supplies. The act also provides that local authorities may act as the agents of the HSE in fluoridating public piped water supplies. The HSE is responsible for the fluoridation of drinking water supplied through public water supplies. When a local authority acts as an agent of the HSE in fluoridating public water supplies, it is responsible for doing so in line with any legislation or regulations that may be in place that govern the fluoridation of water supplies. Currently, approximately 73% of the population receives fluoridated water from public water supplies. The HSE is also responsible for monitoring the fluoride content of public water supplies.

The HSE monitors drinking water supplies on behalf of some local authorities to make sure that the water supply meets the required public health standards. The HSE laboratories may undertake private sampling and testing of drinking water quality for individuals and householders. The HSE informs local authorities of an exceeded level where sampling is performed by the HSE and of an increase in notifications of illness that may be linked to a water supply.

Other Bodies

Water Services Authorities or Local Authorities

Water services authorities (29 county councils, 5 city councils) are responsible for the quality of drinking water that they supply to consumers. This water must comply with the 2007 drinking water regulations. These regulations include 48 parameters in three categories: microbiological, chemical, and indicator parameters. The indicator category includes some chemical and biological parameters together with other factors (color, conductivity, odor, taste, and turbidity) and two indicators of radioactivity.

Where a supply is found to be in breach of the drinking water regulations, the local authorities must notify the EPA of the non-compliance and take whatever action is necessary to inform and protect the health of consumers if the exceeded level could result in a public health risk. In determining whether there is a risk to public health, water services authorities are obliged under the drinking water regulations to consult with the HSE.

Although the EPA has supervisory authority over the water services authorities concerning the quality of the drinking water they are providing, the water services authorities have direct supervisory authority over the providers of other water supplies, such as group schemes and private supplies. The supervisory authority may issue a direction subject to agreement with the HSE and grant a departure from the parametric values also subject to the agreement of the HSE.

In fulfilling their duties, a water services authority must ensure that the water supplies that come under the drinking water regulations are monitored at a frequency as set out in the drinking water regulations and the national drinking water monitoring program. There are two monitoring categories in the national drinking water monitoring program: check monitoring and audit monitoring. Audit monitoring requires fewer samples but is more demanding in terms of analysis. Check monitoring provides information on the organoleptic (e.g., taste and odor) and microbiological quality of water and on the effectiveness of drinking water treatment. Only some parameters require monitoring in certain circumstances. Audit monitoring determines compliance with all parameters given. All parameters are subject to audit monitoring unless it can be established by a supervisory authority, for a period of time to be determined by it, that a parameter is unlikely to be present in concentrations that would breach the values.

The minimum monitoring frequencies in the national drinking water monitoring program are related to the volume of water distributed or produced each day within a supply zone. The number of inhabitants in a supply zone may be used instead of the volume of water to determine the minimum frequency, assuming a water consumption of 200 liters per day per capita.

Drinking water is monitored directly by most local authorities using their own laboratories or, in some cases, through the HSE laboratories or private (contract) laboratories. In some instances, a combination of monitoring procedures is in place even within the same geographic area. The water services authority return check and audit monitoring results to the EPA for the annual report of drinking water quality in Ireland.

The requirement to notify the EPA applies to all non-compliance regardless of whether the sample taken is a regular compliance sample or an operational or investigative sample. For any results that exceed the parametric value, the water services authority must consult with the HSE. Water services authorities are obliged to comply with directions from the EPA and must develop action plans for ensuring a remedy to exceedances of the drinking water regulations. Water services authorities must also ensure that a major emergency management system is in place to deal with serious breaches of the drinking water regulations. The HSE must be an integral part of this response through water liaison committees or incident response teams that should be established in each area for the management of serious water incidents.

The water services authorities ensure provision of funding for new capital drinking water schemes and upgrading works at existing drinking water treatment plants through the national water services investment program in conjunction with the DEHLG. Local authorities are generally required to undertake and update their needs for drinking water services capital works in their area every 5 years. These assessments provide the basis for drinking water scheme scheduling and the compilation of successive investment program by the DEHLG.

Water Services National Training Group

The Water Services National Training Group (WSNTG) was established by the DEHLG to provide appropriate training to staff working in local authorities, particularly in the water services area. In 1996, five regional training centers were established across the Republic of Ireland. The WSNTG currently consists of a cross-section of representatives from the DEHLG and local government, local authorities, and the National Federation of Group Water Schemes. The WSNTG provides a range of training programs for water treatment works operators, water treatment supervisors, and personnel working with group water schemes at their five regional training centers.

National Federation of Group Water Schemes

The National Federation of Group Water Schemes (NFGWS) is the representative organization for private and part-private group water schemes in the Republic of Ireland. NFGWS is the representative and negotiating organization for the group water scheme sector in the Republic of Ireland.

The NFGWS also plays a part in furthering operational training programmes provided by the WSNTG at its five regional training centers throughout the Republic of Ireland. The NFGWS also keeps group water schemes informed of developments in the rural water sector. It is funded by the DEHLG. The primary role of the NFGWS is to assist the committees of group water schemes in meeting the standards set out in the drinking water regulations by providing training and developmental support. The NFGWS has also introduced a quality assurance system for group water schemes. As of March, 2007, the local authority was designated as the supervisory authority over private water and group water schemes.

Communications and Information Exchange on Drinking Water Issues

In Ireland, the broad range of agencies involved in the safe provision of drinking water poses a challenge ensuring appropriate and timely information exchange at both the national and local levels. The sharing and exchange of relevant information between the key agencies involved in the provision of safe drinking water plays a significant role in the protection of human health.

The EPA has issued guidance to all local authorities which states that each local authority should have in place documented procedures for consultation with the HSE when a non-compliance with the 2007 drinking water regulations is detected, including contact details, information to be supplied, and details of agreed actions to be taken in the event of specific failures.

Communication and exchange of information between the HSE and local authorities is facilitated by agreed protocols and regular interaction. Local authorities (water services authorities) and the HSE need to communicate early to alert each other of actual or potential drinking water quality problems, or of cases of disease occurring at abnormal levels in the community. The Water Quality Liaison Group was established to facilitate meetings between the HSE and the local authority at least twice a year. These meetings review water quality results and any exceedances, waterborne disease trends, fluoridation, and guidelines. In addition to the Water Quality Liaison Group, some regions in Ireland also have operational group meetings regarding water quality. These exist primarily where the HSE undertakes the drinking water sampling on behalf of the local authorities. Meetings of the operational groups review issues, such as the frequency of drinking water sampling, communication of sample results, and the regular interaction between the HSE and the local authorities.

The HSE informs local authorities of an exceeded level where sampling is performed by the HSE, and of any increase in notifications of illness that may be linked to a water supply. The local authority informs and consults with the HSE where there is an exceeded level or significant event (e.g., failure of water treatment or pollution incident).

It is the statutory role of the Water Services Authority (local authority) in Ireland to advise consumers if a public drinking water supply becomes a potential danger to human health. Although the Water Services Authority has a responsibility to advise all consumers, additional advice may be provided directly by the HSE. For example, the Environmental Health Departments of the HSE may give specific advice to food premises, crèches, and healthcare facilities, whereas the HSE's Public Health Departments may contact general practitioners, consultants, schools, and colleges. Other agencies with a food enforcement role may need to be informed (e.g., Department of Agriculture, Food Safety Authority of Ireland, and local authority veterinary officers).

Interagency Incident Response Team

Some exceedances of the drinking water regulations may result in an incident that requires an Interagency Incident Response Team (IIRT) be convened. The process of convening an IIRT may be initiated by either the Water Services Authority or the HSE. An IIRT may be convened in the event of a significantly exceeded level or incident that poses a risk to public health or human illness that may be related to a drinking water supply. The IIRT would deal with the overall management of water incidents in terms of finding the source of contamination, preventing the distribution of contaminated water to the public, and providing an alternative water supply.

Drinking Water Incident Response Plan

In Ireland, the WSNTG has provided guidance for local authorities on preparing and implementing a drinking water incident response plan. This guidance proposes that a response plan be developed by water services authorities for responding to drinking water contamination incidents or other emergencies, and that water safety plans be developed at the water scheme level to ensure safety of a drinking water supply from catchment to the consumer.

Early Warning Systems in Water Treatment and Supply

Term Descriptions

Advances in technology have made it possible to use early warning system services in a wide spectrum of domains like environmental, military, medical, financial, political, industrial, and engineering. This indicates that early warning systems are playing a key role in the timely reaction and response to disturbances and failures associated with potential adverse effects. However, the concept of an early warning system can be considered to be fuzzy, without a clear and sound definition. According to Glance [2004], "A universally accepted definition of an early warning system does not yet exist. Probably one never will."

Numerous attempts have been made to describe the early warning system concept. Each one emphasizes specific aspects of a real world system that is meant to provide early warning services. For example, a description that is based on the characteristic elements of an effective early warning system is given by the United Nations. The description states that in order to be effective, early warning systems have to integrate four interrelated elements: a knowledge of the risks faced; a technical monitoring and warning service; the dissemination of meaningful warnings to those at risk; and public awareness and preparedness to act [UN 2006b].

A global survey of the capacities and gaps of early warning systems [UN 2006b] that adopted the description of the United Nations has indicated that knowledge and technological tools, methods, and instruments exist to generate and communicate predictions and warnings for almost every hazard. However, the advances in each of the four necessary elements of early warning systems are mixed. "Even where the capability exists to reliably generate and issue warnings, the other three components are too often absent or weak" [UN 2006b].

Specifically in the water treatment and supply domain, a significant number of descriptions of the early warning system concept exist. One such description emphasizes how technologies providing early warning services are used in practice, stating that early warning systems are used to detect any sudden deterioration in the quality of the source drinking water supply either just before the water goes into the distribution system or some distance upstream [Brosnan 1999].

Another description notes some important functionalities of an early warning system [U.S. EPA 2009]. The definition states that an early warning system is an integrated system for monitoring, analyzing, interpreting, and communicating water system data. It can enable decision-makers to respond quickly and effectively to protect public health in the event of intentional water system contamination. This description complies to a high degree with the description by the United Nations because it is obvious that a system must be comprised of the majority of components included in the description of United Nations to perform the functionalities described above.

Last, a report viewing early warning systems from the perspective of critical infrastructure security has reviewed the state of the art technologies and techniques for integrated early warning systems for drinking water infrastructures [Hasan et al. 2005]. The report provides a list of 13 important characteristics of an ideal early warning system. It also states that for any system to be considered an effective early warning system, it must meet each of the following set of core characteristics:

1. Provide for a rapid response
2. Screen for a number of contaminants while maintaining sufficient sensitivity
3. Operate as an automated system that allows for remote monitoring

The report has concluded that a viable, integrated early warning system that meets each of these important characteristics and can be routinely used is several years away.

In the domain of drinking water treatment, the early warning system concept is more or less synonymous to real-time monitoring of water quality parameters. Real-time monitoring of water quality parameters is achieved with the use of sensors enhanced with supervisory control and data acquisition (SCADA) systems, together forming a type of early warning service. However, it can be said that, in practice, the typical early warning process model for water treatment and distribution facilities conforms to a two-phased process approach. During the first phase, a continuous real-time monitoring of water quality parameters based on SCADA technologies detects deterioration in drinking water quality and issues a warning. Then, the warning activates a second monitoring phase during which more sensitive contaminant identification technologies, available in certified laboratories, are used. However, due to technological limitations, the laboratories' tests cannot be delivered to the decision-makers immediately.

A representative example of early warning system service in the water treatment domain and supply would be the WaterSentinel system [U.S. EPA 2005a, 2005b, 2005c]. The system is the result of a program developed by the U.S. EPA in close partnership with drinking water utilities and other key stakeholders in response to a U.S. Presidential Homeland Security Directive involving designing, deploying, and evaluating a model contamination warning system for timely detection and appropriate response to drinking water contamination threats and incidents that would have broad application to drinking water utilities.

However, the phrase used to describe the WaterSentinel system was "contamination warning system." That is because it was recognizing that a reliable system providing early warning of a contaminant prior to human exposure with public health impacts using today's technologies may not be possible. A contamination warning system involves the active deployment and use of monitoring technologies and strategies and enhanced surveillance activities to collect, integrate, analyze, and communicate information to provide a timely warning of potential water contamination incidents and initiate response actions to minimize public health and economic impacts [Hasan et al. 2005]. Some components of the system include the following [U.S. EPA 2005c]:

- Online water quality monitoring first uses data from sensors and SCADA systems to define the expected normal levels of monitored parameters. Any changes from these expected normal levels of the parameters being monitored are interpreted as warning signals of potential contamination.
- Sampling and analysis of water at a predetermined frequency are used to detect contaminants that cannot be detected by the online monitoring system. Also, water samples are collected in response to triggers from water quality monitors or other information streams to identify the potential unknown contaminants in the sample.
- Enhanced security monitoring is used to monitor, collect, and document security breaches, witness accounts, and notifications by perpetrators, news media, or law enforcement.
- Consumer complaints regarding unusual taste, odor, or appearance of the water are often reported to water utilities, which document the reports and conventionally use them to identify and address water quality problems.

- Public health surveillance can make use of data from existing syndromic surveillance conducted by the public health sector, from emergency medical service logs, emergency call centers, and poison control hotlines.
- A consequence management plan details the treatment plant's response actions, along with those of the related local, state, and federal agencies that should respond to a drinking water contamination incident.

Taking into consideration the above, one can argue that, at present, the early warning system concept in the domain of drinking water quality is used mainly in relation to the following:

- Multiparameter online sensors for monitoring water quality parameters (pH, conductivity, and chlorine residual)
- Laboratory tests
- Portable monitoring units and devices
- Customer complaints and public health surveillance data

Useful Services but Not Complete

Undoubtedly, drinking water quality monitoring systems and early warning monitoring systems are providing useful services. However, we argue that the rationale behind these systems and eventually the systems themselves, although on a path in the right direction to support a proactive risk management strategy, are not complete and thus will never be entirely effective or sufficient on their own. To prove this statement, it is necessary first to briefly explain some accident models taken from safety science to understand how organizational disturbances, failures, and accidents can occur.

A model providing an explanation of the conditions that can lead to accidents was introduced by Professor Reason [1997]. It is widely known as the Swiss cheese model. Reason hypothesizes that accidents and disturbances are occurring due to a breach or bypass of a set of defenses intended to stop hazards from reaching vulnerable environments, people, or assets. The defenses are designed to serve one or more functions; for example, to contain and eliminate the hazards, interpose safety barriers between the hazards and the potential losses, and provide alarms and warnings when danger is imminent (e.g., the monitoring and alert systems that are currently available). Based on the Swiss cheese model depicted in Figure 13.3, the defenses form successive layers of protection, each guarding against the possible breakdown of the one in front.

However, based on Reason [1997], there are three implicated factors that can breach the defenses: human, technical, and organizational. These factors become holes in the defenses that, in reality, are in constant flux. Based on Reason's [1997] model, the necessary condition for an organizational accident is the rare conjunction

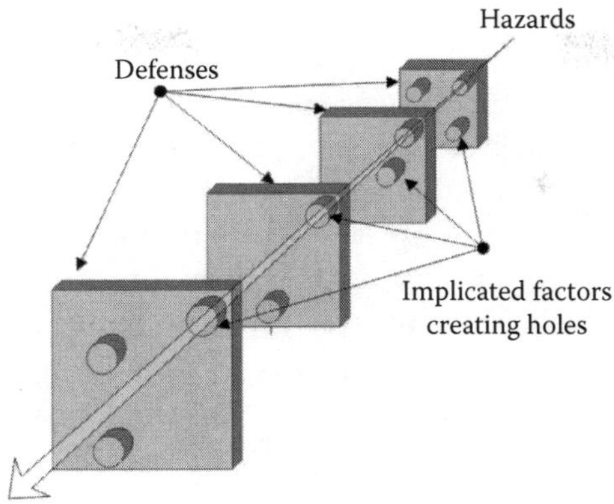

Figure 13.3 The Swiss cheese model.

of a set of holes in successive defenses, allowing hazards to come into damaging contact with people or assets.

Rasmussen and Svedung [2000] argue that risk management is basically a control problem and a proactive risk management strategy should be based the following:

1. Identification of the boundaries of safe performance
2. Efforts to make those boundaries visible to decision-makers
3. Efforts to counteract pressures that drive decision-makers toward the boundaries

They stress that accidents are, in fact, the effects of a systematic migration of organizational behavior toward the accident and were not caused by a coincidence of independent failures and human errors. Thus, the control that is needed to maintain safety must be based on a continuous monitoring of the actual level of safety; that is, the margin between the present system conditions and preconditions for safe operations (i.e., the safety constraints) defined during the design phase. Among other things, Rasmussen and Svedung [2000] highlight the importance of effective interactions and information flow among the decision-makers at the different levels of the social organization ranging from government and regulatory to management and operational levels. They make clear that many nested levels of decision-making are involved in risk management and regulatory rule making to control hazardous processes and that these social organizations depend on proper coordination of decision-making at all levels.

Leveson [2002], on the other hand, highlights the importance of taking into consideration issues like decision-making, software errors, system accidents (versus component failure accidents) and organizational risk factors during hazard analysis and safety-related risk management. She points out that accidents result when safety constraints are violated and thus the basic concept behind the understanding of accidents are not the events but rather the concept of safety constraint and of how safety constraints can be violated. In short, Leveson [2002] agrees with Rasmussen and Svedung [2000] that safety is a dynamic control problem. However, she disagrees with their approach because it is heavily focused on events that can lead to system failures and accidents and, as she points out, their approach suffers from emphasis on designing for safety and controlling or eliminating hazards in the basic system design [Leveson 2002]. Her approach, system theoretic accident model and processes (STAMP), aims at identifying and enforcing safety constraints on system development and operations that ultimately results in safe behavior as early as possible, even during the design phase.

In essence, the current approaches of early warning systems in the domain of drinking water, which are entirely based on monitoring of microbiological and chemical parameters (i.e., the hazards based on Reason's model) are not complete, mainly because their reasoning is based on a knowledge model about risks that is not complete. Figure 13.4 depicts the role of current early warning systems in the domain of drinking water, using the Swiss cheese model as a reference model of accident causation. Sensors are monitoring drinking water parameters in between defenses. When a sensor detects a breach of the value of the drinking water quality parameter, the early warning system can activate an alarm and in some cases can activate an emergency response procedure (i.e., shut down part of or the entire water treatment plant).

We argue that to be complete, an early warning service ideally should be able to monitor not just the presence of hazards in between the successive layers of defenses, but it should also be able to monitor, as effectively and reliably as possible,

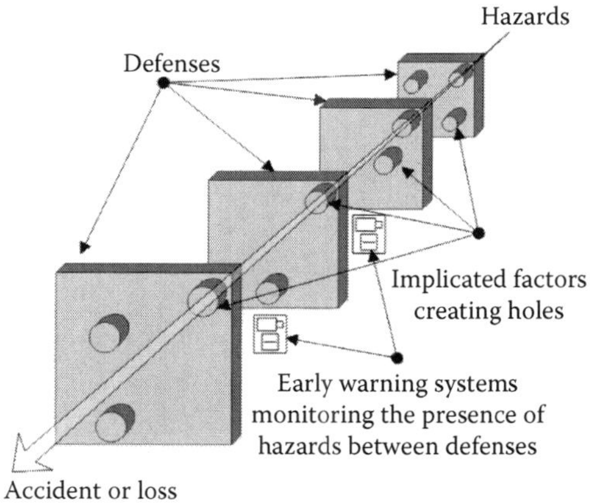

Figure 13.4 The role of current early warning systems in the domain of drinking water.

the performance of the entire organization of water utilities and of the environment around it, including the actors and decision-makers and their interactions together with the status of system safety constraints. In short, to support a proactive risk management strategy, a complete early warning service is needed that will monitor the performance of components of the entire sociotechnical system together with the surrounding environment, aiming at assessing the occurrence of incidents and failures and at making the disturbances of components performance visible to decision-makers using weak and sound warning signals.

As a result, the number of users and the sources of input data of a "complete" early warning system should not be restricted only to operators and to sensors based only at the water utilities but rather should include users and input data from agencies and bodies that belong in the wider sociotechnical system of water utilities, such as local authorities, state agencies, and regulators. This simply means that existing early warning monitoring systems installed in a water treatment plant will provide only a subset of the necessary and sufficient input data required for the complete early warning service.

Thus, using again the Swiss cheese model as a reference model of accident causation, we argue that in order to be complete, an early warning service must be able to assess, and ideally to foresee, under which circumstances the holes of the successive layers of defenses may be aligned. This generates the need for monitoring the current conditions of the implicated factors mentioned by Reason (i.e., human, technical, and organizational), including the status of safety constraints on all levels of the sociotechnical system of water utilities and the status of the surrounding environment. Ultimately, the objective of a complete early warning service should be to make the disturbances of component performance and the violations of safety constraints visible to decision-makers located at different levels of the governance model of water treatment by providing weak and sound warnings and to support their decision-making process. Its aim should be to support a proactive risk management strategy.

A Novel Early Warning Service

Aim and Goals

The novel characteristic of our early warning service, in comparison to existing systems, is that we adapt a systemic approach in analyzing failures, monitoring for disturbances, and generating warnings for potential risks in water utilities. This approach takes into consideration the need to monitor not only critical operational parameters in the technical system (i.e., the water treatment plant) but also critical parameters of the components of their wider social system of water treatment plants. By systemic, we mean that we are taking into account and treating as a whole the entire sociotechnical system of water utilities and their environment. The objective of the proposed novel early warning service is to make the boundaries of safety performance of water utilities visible to the stakeholders and decision-makers located at different governance levels and to support them in implementing a proactive strategy to reduce the forces that drive water utilities beyond their safety boundaries.

The main goal of our service is to provide weak and sound warnings to the stakeholders of the sociotechnical water supply system in a timely manner, together with an estimated likelihood of the occurrence of failure and accidents that will cause severe adverse effects. The proposed early warning service is currently in the design phase and is intended to be developed and used in the Republic of Ireland by many actors at all levels of the wider sociotechnical system of water utilities. Intended users are, among others, the managers of water utilities, the Irish EPA the HSE and local authorities.

Stakeholders' Needs and Functional Requirements

Users that may be scattered throughout Ireland must be able to change the status of the components of the sociotechnical system. In addition, real-time sensor data of critical water quality parameters must be collected to assess the possibility or likelihood of an occurrence of operational problems or failures in a timely manner. Given some evidence on the status of water treatment plant components (e.g., on or off, available or not) and about the values of critical water quality parameters, the early warning system will assess whether safety constraints have been breached and will estimate the likelihood of a drinking water contamination incident. When safety constraints are breached, warning messages (i.e., weak warnings) will be sent to appropriate stakeholders. If the likelihood of a drinking water contamination incident is above a defined threshold, then warning messages (i.e., sound warnings) will be instantly sent to the relevant stakeholders of different components in the sociotechnical system. The information embedded in the warning message should be dependent on the stakeholder's requirements. For example, the context of a warning message going to a senior engineer at a local authority will be different from the warning message going to EPA personnel. Emergency response procedures must be sent to appropriate stakeholders, such as water treatment plant managers and local authority personnel.

Stakeholders' needs determine their requirements for an early warning service. They dictate what type of services the early warning system must provide. These needs were identified during the analysis phase through face to face meetings and are shown in Table 13.1. In Figure 13.5, a unified modeling language [UML] use case diagram depicts the functional requirements as desired features of the early warning service. This diagram provides a model of the interactions between the early warning service and the entities that will interact with it; it is a part of the documentation phase of the requirements analysis. A SCADA system that is in operation in a water treatment plant will provide online data. In addition, the caretaker of the treatment plant will provide offline data (i.e., data from measurements by instruments that are not connected to the SCADA system). The caretaker will be able to change the operational status of the components of the plant. For example, a pump or a sedimentation tank can be in operation mode, in maintenance mode, or in out-of-order mode. Given the inputs, the system will check if safety constraints are breached and calculate the likelihood of having a component of the sociotechnical system—in the example, the component is a water treatment plant—in a warning state, meaning that its performance has crossed its safety boundaries. This warning can be sent via text messages, e-mail, and Web feeds to the appropriate stakeholders. If, for example, a contamination incident was confirmed by the HSE, authorized HSE personnel could raise the alert with the specific local authority to which the water treatment plant belongs and the system can automatically send alert messages to the persons assigned to implement the emergency response strategy.

Knowledge Areas Involved

To realize the early warning service from a software engineering point of view, we plan to use a software architectural pattern known as model view controller (MVC). In MVC, the model is the layer that makes use of the logic rules and algorithms

Table 13.1 Stakeholders' Needs

No.	*Need*
1	To be accessible to all stakeholders
2	To provide important and relevant information to each stakeholder
3	To allow the profiling of each and every water utility
4	To allow the modelling of knowledge about causes—failures—effects
5	To allow updating of knowledge related to causes—failures—effects
6	To allow estimation of the potential of failure of water utilities, given some evidence and to reason with uncertain data
7	To provide the appropriate warning signals (i.e., weak warnings or sound warnings) when the estimation of a potential failure exceeds the corresponding threshold values
8	To allow as inputs real-time sensor data and user updates on the status of the components of the sociotechnical system
9	To disseminate warnings to appropriate stakeholders
10	To provide advice and corrective actions to the stakeholders

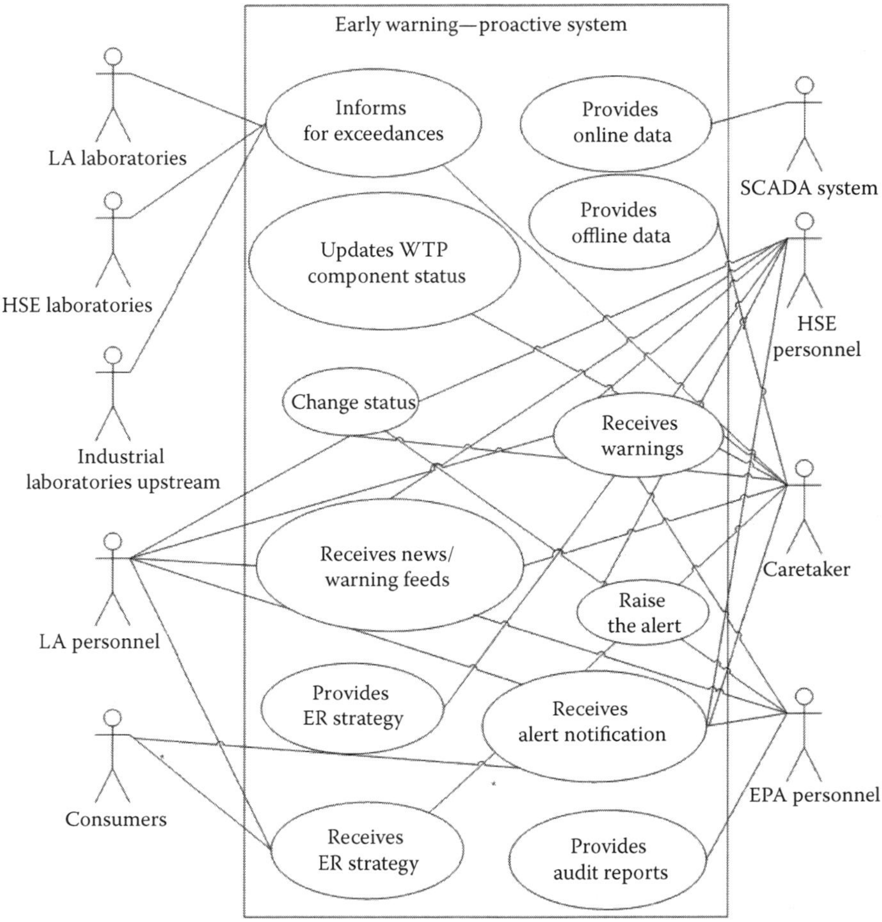

Figure 13.5 UML use case diagram depicting functional requirements as desired features of the early warning system.

that handle information exchange between a database and the user interfaces, process commands, perform calculations, etc. The view layer is where the Web pages define how data from the model layer is arranged on the Web page. The controller coordinates the early warning service. It manages the communication of data and the logic rules used to manipulate the data to and from the model. Figure 13.6 depicts the MVC architecture of the early warning service, where the model layer consists of the business logic layer and the data store layer. The business logic layer contains the logic rules of the early warning service and defines the necessaries calculations that must be activated for each task. The data store layer contains the data from the early warning service.

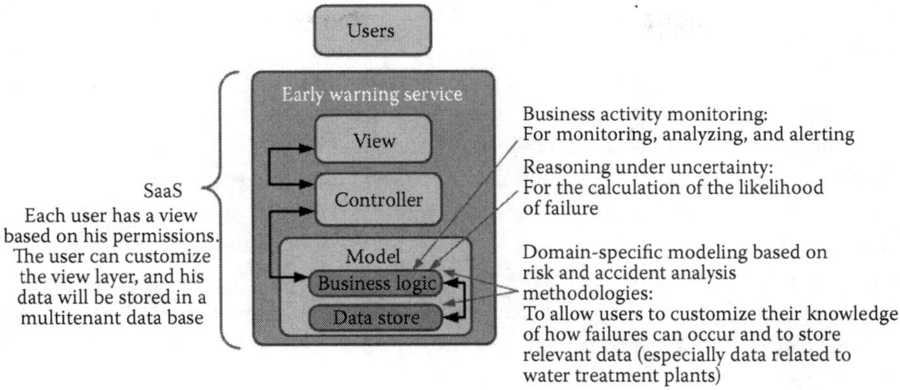

Figure 13.6 MVC architecture and multidisciplinary technologies.

The technologies, which will be used to develop the early warning services, are depicted in Figure 13.6. Their names are related to elements of the early warning service with arrows and with the side brace. These are briefly described in the following paragraphs. The arrow relations convey that some characteristics and functionalities of the early warning service were inspired from these technologies and the necessary computer code and routines that are required to represent these technologies at the system level are located at the model layer of the early warning service. The side brace denotes that the entire early warning service will be deployed to its users using a specific cloud computing technology.

Software as a Service and Business Activity Monitoring

It is necessary that the inputs to the early warning service be given by a wide range of users, as well as by IT systems in different geographical points, and that they allow or enhance monitoring and information flow between the actors in the wider sociotechnical system of water utilities. One way of addressing this need is by the use of state-of-the-art Web technologies. For example, a type of cloud computing architecture, Software as a Service (SaaS), is the most appropriate alternative for providing the required early warning services to the appropriate stakeholders.

In the traditional software deployment model, businesses or end-users purchase up-front license fees and run the applications on their infrastructure. Customers host the system, as well as their data, on their own computers. The SaaS model allows software to be deployed as a hosted service and accessed over the Internet. This means that a single instance of a SaaS application—in this case the early warning service—can accommodate several different users sharing a single database of their data. This feature is known as multitenancy. On the other hand, SaaS applications are now fully customizable so that users have the ability to tailor the user interface, settings, and functionalities of the SaaS application to meet their needs and wants.

Monitoring of critical operation parameters in water utilities and in their wider sociotechnical system is a need that must be addressed by an early warning system. In relation to this, a software technology that can be used is known as business activity monitoring (BAM). The ultimate objective of this software technology is to provide real-time insight of the status and outcomes of different key business operations, processes, and transactions. Such an analytical insight allows organizations to quickly spot internal issues and forecast potential hitches that can reduce their outcomes. BAM's importance can be explained by the increased need faced by businesses to have better business performance and an efficient measurement and analysis tool to assess their managerial tasks [Kang et al. 2008]. Businesses realize the benefits of identifying and predicting problems in order to ward off mitigating circumstances or disasters, which can be very costly.

BAM aims at monitoring, analyzing, and alerting the output of operations within a given business. However, as business size and complexity become greater, so do IT requirements. Decision-makers previously stored all important company information and data in warehouses. They eventually realized that there were several drawbacks to that approach: expensive maintenance costs; lengthy implementation cycles; and the inability to handle unstructured data [Hellinger et al. 2002].

In our case, BAM can be used to gain insight into activities performed by users of the early warning service, and thus provide a monitoring service of important variables. Information about any unusual events occurring during a given step of water treatment will be logged and made available to decision-makers and stakeholders so that they may take appropriate and timely measures to avert failures and disasters.

Domain-Specific Modelling Based on Risk and Accident Analysis Methodologies

The effectiveness of an early warning service in preventing facility or system failures depends on the modelling and adequate knowledge of involved risks. Therefore, to prevent accidental failure effectively, an early warning system must embody knowledge of hazards and vulnerabilities of the system under study, as well as knowledge about the conditions that can lead to a failure. Thus, the need for knowledge modelling—the structured use of symbols to represent real objects and the relations between them—and its representation in computer code is very much of concern.

One could argue that the problem of knowledge modelling of failures and accidents can be dealt with widely known risk analysis methods such as fault tree analysis, event tree analysis, and cause consequence analysis. This could be true in the case where the problem domain is narrow or small. If the problem domain is large and complex, then these methods do not look promising and a methodology such as the STAMP [Leveson 2002] could be of use.

When developing an early warning system, there is a need for the representation of knowledge models expressing failure and accident causation at the system level, meaning that there is a need to generate computer code and data directly from the models. This is especially true when the knowledge models of the reference system are large and complex. One solution to this problem is to use domain-specific modelling.

Domain-specific modelling provides a higher level of abstraction than typical model-driven development approaches. This prevents errors in earlier stages of development by not allowing the illegal designs or designs that do not follow the underlying architectural rules and thus producing the code free from careless mistakes and syntax and logic errors [Kelly et al. 2008] without any manual code editing. Domain-specific models can also be used to generate other models and to configure other systems, perhaps in combination with other enabling technologies by providing the domain-specific application process interfaces. In so doing, they can enhance productivity and generate solutions automatically.

Thus it can be said that knowledge of how systems or facilities can fail can be elicited and modelled appropriately using some accident analysis techniques. Further, domain-specific modelling gives us the ability to generate computer code directly from these models and allows users of the early warning service to produce significant amounts of data that are stored at the data storage layer of the early warning service.

In short, it can be said that with the utilization of domain-specific modelling, water utilities management and stakeholders can model the failures of water utilities following the principles and the notations of risk and accident analysis techniques. Then, computer code can be generated based on these models that can be stored in the database of the early warning service. This type of code generation is feasible with domain-specific modelling and can happen automatically without any intervention from the users. This can be translated into increased productivity and a reduction in the number of errors during the phases of modelling and coding.

Reasoning with Uncertain Data

Assuming that the data and the rules explaining accident causation are stored in the database of the early warning service, there will be a need to use both to estimate the likelihood of a failure given some evidence about the condition of the components in the sociotechnical system. There are several alternative approaches to achieve this task; probabilistic risk analysis, Bayesian belief networks, fuzzy logic, and possibility theory are a few of them. The selection of the most appropriate alternative in this case will depend on—among other factors—the types of data the layer will require, the amount of data available, and the quality of data.

Handling uncertainty handling and reasoning while under uncertainty are important issues. Assumptions affecting uncertainty analysis involve the means of parameters and their statistical distributions and should be clearly stated. Signal noise can lead to an increase in uncertainty and to false warnings. Model testing and expert validation of the selected method will be required before implementing the final system. A hybrid solution such as, for example, using a combination of probabilistic and fuzzy reasoning can be considered a solution.

Lessons Learned and Future Work

The aim of the early warning service that was briefly described here is to support a proactive risk management strategy of manmade disasters related to drinking water incidents. Its objective is to make the boundaries of safety performance of water utilities visible to the stakeholders and decision-makers who are located at different governance levels. Although the early warning service is currently at the design phase, there are some lessons learned that should be disseminated.

First, the early warning system concept can be interrelated with a wide range of systems performing different functions. By understanding safety science models—which explain the conditions that can lead to incidents and accidents—and looking at the water supply and treatment literature for descriptions of early warning systems, it became clear that the current approaches are not complete. Early warning systems currently used in the domain of drinking water treatment and supply are used mainly to monitor the presence of hazard in between the layers of defenses. These systems do not monitor the critical parameters and latent conditions that create the forces responsible for the creation of holes in the defenses. To address this problem, new early warning systems are needed that will monitor the condition of the interactions of the components of the entire sociotechnical system of water utilities with their surrounding environment in addition to the presence of hazards. In short, there is a need for systemic sociotechnical early warning systems.

State agencies, local authorities, private companies, and other organizations related to the provision of clean and potable water should be considered stakeholders of the sociotechnical early warning system. Unfortunately, state agencies and other bodies in the governance model of drinking water treatment and supply are not always informed the actions and decisions of other bodies in this domain. Some could even agree these bodies are encountering problems in coordinating their actions when applying a proactive risk management strategy. This is where a sociotechnical early warning system can improve the safety performance of water distribution and supply systems by providing a common application through which all actors in the sociotechnical system may be aware of what others have decided or have done, enhancing the transparency of information flow and the control between all levels of the system.

The analysis of the sociotechnical dimension of each domain in which early warning systems are applied can enhance attempts to define a solution to the problem. However, a sociotechnical analysis may reveal deficiencies in the system and some corrective actions may require the related agencies and institutions to make changes in the way that they are operating and interacting. Implementing these changes is not an easy task and, thus, these issues must seriously be considered during the design phase of any sociotechnical early warning service. A response to this is to invite all stakeholders to participate in and contribute to the analysis and design phase of the sociotechnical early warning system, as well as during the phases of testing and evaluation.

State agencies are ideal bodies for providing institutional support for socio-technical early warning services. They usually have personnel with the knowledge and expertise to validate and evaluate the results of a warning service, interpret the messages effectively, and certify the quality of such service. In addition, state agencies may facilitate the communication between the stakeholders and in an attempt to document their needs and requirements in relation to the early warning service.

An important issue to be considered during the early phases of early warning system development is the existence of information systems that are already in use by organizations. Data and information stored in these systems can be reused. In addition, some systems already in use can provide useful services for the early warning system. For example, the EDEN system that is already in use in the Republic of Ireland may be used as the main data store component for the early warning service of water utilities in the Republic of Ireland.

The development of a systemic sociotechnical early warning service requires the effective combination of technologies, methods, and techniques from many scientific fields. It is a multidisciplinary problem that depends on, among other things, the type of incident, the characteristics of the agencies and bodies forming the governance model of the sociotechnical system, the nature and characteristics of the threats and hazards, as well as the level of complexity between the components of the pure technical systems. For example, a collection of methods and technologies ranging from safety and social science, software and Web engineering, civil engineering, artificial intelligence, and computer science will be used during the life cycle of the early warning service discussed herein.

At the top of the list of things to do in the near future for the early warning service is to develop the first prototype and to organize a demonstration and feedback workshop with people from agencies and bodies that form the drinking water supply governance model in the Republic of Ireland. The prototype early warning system will be developed based on the software as a service deployment model using the model view controller architecture. The functional algorithms and the rules of information exchange of the early warning system will be defined using a combination of technologies like domain-specific modelling based on risk and accident analysis models enhanced with technologies such as Bayesian belief networks that support reasoning under uncertainty as well as business activity monitoring.

The aim of the workshop that we plan to organize will be twofold. The first goal is to demonstrate the prototype early warning service to the representatives of the group of people who will use it, to consult them, and to ask them for constrictive reviews. The second goal is to organize a round table discussion aiming at evaluating the potential for designing and implementing a proactive risk management strategy for drinking water incidents in the Republic of Ireland, which will be in accordance with the early warning service.

Acknowledgments

This work is supported by the research project SCEWA (Grant 2007-DRP-2-S5), funded by the Irish Environmental Protection Agency under the DERP grand scheme.

References

Allgeier, S. C., and Magnuson, M. L. 2004. Responding to threats and incidents of intentional drinking water contamination. *Journal of Contemporary Water Research and Education* 129:13–17.

Bradley, R. 2007. Cryptosporidium: Who is responsible? Retrieved from http://www.mlaw.ie/news/cryptosporidium-who-is-responsible (accessed January 20, 2009).

Brosnan, T. M. 1999. *Early warning monitoring to detect hazardous events in water supplies.* Washington, DC: International Life Sciences Institute.

Canadian Council of Ministers of the Environment (CCME). 2004. *From source to tap: Guidance on the multi-barrier approach to safe drinking water.* Winnipeg, MB: Federal Provincial Territorial Committee on Drinking Water and the CCME Water Quality Task Group.

Chlorine Chemistry Division of the American Chemistry Council (CCDACC). 2003. Drinking water chlorination: A review of disinfection practices and issues. Retrieved from http://www.c3.org/chlorine_issues/disinfection/c3white2003.html (accessed April 9, 2009).

CNN. 2009. Zimbabwe cholera epidemic worsening. Retrieved from http://www.cnn.com/2009/WORLD/africa/02/17/Zimbabwe.cholera.crisis/index.html (accessed August 4, 2009).

Corso, P. S., Kramer, M. H., Blair, K. A., Addiss, D. G., Davis, J. P., and Haddix, A. C. 2003. Cost of illness in the 1993 waterborne cryptosporidium outbreak, Milwaukee, Wisconsin. *Emerging Infectious Diseases* 9:426–431.

Glantz, H. M. 2004. Usable science 8: Early warning systems: Do and Don'ts. Retrieved from http://www.ccb.ucar.edu/warning/report.pdf (accessed January 20, 2009).

Gleick, P. H. 1993. *Water in crisis: A guide to the world's fresh water resources.* New York: Oxford University Press.

Hartl, G. 2004. WHO issues revised drinking water guidelines to help prevent water-related outbreaks and disease. Retrieved from http://www.who.int/mediacentre/news/releases/2004/pr67/en/index.html (accessed January 20, 2009).

Hasan, J., Goldbloom-Helzner, D., Ichida, A., Rouse, T., and Gibson, M. 2005. *Technologies and techniques for early warning systems to monitor and evaluate drinking water quality: A state-of-the-art review.* Washington, DC: Environmental Protection Agency Office of Water.

Health Canada. 1996. *Guidelines for Canadian drinking water quality*, 6th ed. Ottawa, Ontario, Canada: Health Canada.

Hellinger, M., and Fingerhut, S. 2002. Business activity monitoring: EAI meets data warehousing. *EAI Journal* July, 19–21. Retrieved from http://www.businessactivitymonitoring.com/docs/bam.pdf (accessed January 20, 2009).

Institute of Public Administration Ireland. 2009. *Review of the management of the outbreak of cryptosporidiosis and E. coli in Galway City during 2007.* Dublin, Ireland: Institute of Public Administration.

Integrated Regional Information Networks (IRIN). 2008. Zimbabwe: Cholera feeds off a perfect storm. Retrieved from http://www.irinnews.org/Report.aspx?ReportId=81699 (accessed August 4, 2009).

Kang, J. G., and Han, K. H. 2008. A business activity monitoring system supporting real-time business performance management. *Third International Conference on Convergence and Hybrid Information Technology* 1:473–478.

Kelly, S., and Tolvanen, J.-P. 2008. *Domain-specific modeling, enabling full code generation.* Chichester, NY: John Wiley & Sons, Inc.

Leveson, N. G. 2002. *System safety engineering: Back to the future.* Retrieved, from http://sunnyday.mit.edu/book2.pdf (accessed January 20, 2009).

MacKenzie, W. R., Hoxie, W. R., Proctor N. J., Gradus, M. E., Blair M. S., Peterson, D. E., Kazmierczak, J. J., Addiss, D. G., Fox, K. R., Rose, J. B., Davis, J. P. 1994. A massive outbreak in Milwaukee of cryptosporidium infection transmitted through the public water supply. *New England Journal of Medicine* 331:161–167.

Mays, L. W. 2004. *Water supply systems security.* New York: McGraw-Hill Professional Engineering.

New Zealand Ministry of Health. 2009. Steady improvement in the quality of New Zealand's drinking water. Retrieved from http://www.moh.govt.nz/moh.nsf/pagesmh/5784?Open (accessed January 20, 2009).

O'Connor, D. R. 2002. *Report of the Walkerton inquiry: Part 1. The events of May 2000 and related issues.* Toronto: The Walkerton Inquiry.

Presidential Decision Directive 63. 1998. Protecting America's critical infrastructures. Retrieved from http://fas.org/irp/offdocs/pdd-63.htm (accessed April 4, 2009).

Rasmussen, J., and Svedung, I. 2000. *Proactive risk management in a dynamic society.* Karlstad, Sweden: Swedish Rescue Services Agency.

Reason, J. T. 1997. *Managing the risks of organizational accidents.* Aldershot, Hampshire, England: Ashgate.

Rizak, S., and Hrudey, S. 2007. Strategic water quality monitoring for drinking water safety, Cooperative Research Centre (CRC) for Water Quality and Treatment Research Report 37. Salisbury, Australia: CRC for Water Quality and Treatment.

Semenza, J. C., and Nichols, G. 2007. Cryptosporidiosis surveillance and water-borne outbreaks in Europe. Retrieved from http://www.eurosurveillance.org/ViewArticle.aspx?ArticleId=711 (accessed January 20, 2009).

U.S. Environmental Protection Agency (EPA). 2004. Response protocol toolbox. Retrieved from http://www.epa.gov/safewater/watersecurity/pubs/rptb_response_guidelines.pdf (accessed April 4, 2009).

U.S. EPA. 2005a. Overview of event detection systems for watersentinel, Technical report, U.S. EPA, Water Security Division, 2005. Retrieved from http://www.epa.gov/safewater/watersecurity/pubs/watersentinel_event_detection.pdf (accessed January 20, 2009).

U.S. EPA. 2005b. Watersentinel consequence management strategy, Technical report, U.S. EPA, Water Security Division. Retrieved from http://www.epa.gov/safewater/watersecurity/pubs/watersentinel_mangt_strategy.pdf (accessed January 20, 2009).

U.S. EPA. 2005c. Watersentinel system architecture draft version 1.0, Technical report, U.S. EPA, Water Security Division. Retrieved from http://epa.gov/watersecurity/pubs/watersentinel_system_architecture.pdf (accessed January 20, 2009).

U.S. EPA. 2006. Tools and technical assistance. Retrieved from http://cfpub.epa.gov/safewater/sourcewater/sourcewater.cfm?action=Tools (accessed March 5, 2009).

U.S. EPA. 2009. Early Warning Systems—Homeland Security Research. Retrieved http://www.epa.gov/NHSRC/news/news083005a.html (accessed March 5, 2009).

Umweltbundesamt (UBA). 2008. *Water, drinking water, and water protection.* Retrieved from http://www.umweltbundesamt.de/wasser-e/themen/drinking-water/index.htm (accessed April 4, 2009).

United Nations. 2006a. *Water: A shared responsibility.* New York: United Nations Educational, Scientific and Cultural Organization.

United Nations. 2006b. Global survey of early warning systems, ISDR platform for the promotion of early warning PPEW. Bonn, Germany: United Nations.

Yoder, J., Roberts, V., Craun, G. F., Hill, V., Hicks, L. A., Alexander, N. T., Radke, V., Calderon, R. L., Hlavsa, M. C., Beach, M. J., Roy, S. L. 2008. Surveillance for waterborne disease and outbreaks associated with drinking water and water not intended for drinking: United States, 2005–2006. *Morbidity and Mortality Weekly Report Surveillance Summaries* 57:39–62.

ASIA AND THE PACIFIC

IV

Chapter 14

NGO and Community-Oriented Disaster Response: Lessons from the Humanitarian Response to the Asian Tsunami in India and the Jogyakarta Earthquake in Indonesia

Patrick Kilby and Kim Williamson

Contents

Introduction

In this era, where disasters are recognized as human phenomena and are linked to preexisting vulnerabilities and risk, disaster risk reduction is seen as an integral part of disaster management, in part to get away from the reactive nature of many (if not most) disaster responses. In reality, however, every disaster is different and it is hard to manage for every risk, so an effective disaster response is an essential and ongoing component of disaster risk management. Having a strong disaster response strategy can drastically reduce the number of lives lost and repair a damaged infrastructure more quickly, speeding a community's return to everyday life.

In practice, many disaster risk reduction theories and the systems in place quickly stop working, often within minutes of a disaster happening. In their place unfolds a cacophony of community, government, humanitarian, civil society, and industry actors from the spectrum of the international to local, representing a range of values, objectives, and mandates.

Within this environment, it is easy to lose sight of the overarching humanitarian purpose of disaster response: to return communities to a state of normalcy and to make them more resilient than before. In other words, the overall goal should be oriented around community goals to support communities on their path of recovery as defined by them, rather than being directed primarily by the values and priorities of those who wish to help. This chapter is an exploration of what community-oriented disaster response means in practice. Two case studies are used as examples (one of local nongovernmental organizations [NGOs] in Southern India in response to the Asian Tsunami of 2004, and the other the international NGO's response to the Jogjakarta earthquake of 2006 in Indonesia) and draws some lessons about the importance of involving local processes and institutions.

NGO Involvement in Disaster Response

There has been some controversy surrounding the styles of disaster responses following several discordant and poorly planned interventions over the past 30 years, with the most notable being the Rwanda crisis of 1994 and those in Biafra in 1969, Bangladesh in 1972, and Ethiopia in 1985. As a result, a set of humanitarian principles were developed by the Sphere Project, comprising a humanitarian charter and minimum standards. These guidelines acknowledge first that affected communities are the initial and main respondents after a disaster and, second, that the state has the primary responsibility for assisting in recovery [Sphere Project 2004]. These guidelines recognize that there will sometimes be a gap in fulfilling this obligation due to a lack of capacity or deliberate neglect, and that humanitarian agencies often play a significant and complementary role in relation to these primary responsibilities. NGOs are one set of humanitarian agencies that come into play during times of disasters. Of course, the NGO sector itself is not homogenous, comprising organizations that vary in scale from global, regional, national, and local organizations to community-based groups, such as religious or youth groups. Each of these different NGO actors brings different skills and capacities to disaster response. In this chapter, we make a simple distinction between local NGOs and international NGOs and compare the involvement of each through the two case studies we have identified.

Humanitarian Principles and Community-Oriented Disaster Response

A disaster response is characterized by a broad group of actors motivated by a myriad of objectives ranging from humanitarian and political to economic [Dover 2004]. Within this "seething cauldron" [Katoch 2006, p. 158] of disaster actors, one hopes that the overarching objective of an intervention is grounded in basic humanitarian principles. According to the Sphere minimum standards, disaster response "should reflect the needs, concerns, and values of disaster-affected people" [Sphere Project 2004, p. 28; Handmer and Dovers 2008]. We argue that the strongest reason for this shift is that community-oriented intervention lies at the heart of the sector's oft-repeated motto of "build back better"; that is, build more resilient communities [Kennedy et al. 2008].

Resilience represents a shift away from an approach of focusing on the vulnerability and needs of disaster-affected communities toward emphasizing their capacity to adapt and recover [International Federation of Red Cross and Red Crescent Societies 2004]. Like the term *sustainability*, *resilience* has become somewhat ubiquitous and has been used in a variety of ways to mean a community being better prepared for disaster so that they are less affected after such an event by implementing disaster risk reduction and preparedness strategies [United Nations 2005] and encouraging sustainable livelihoods with the idea of "layers of resilience," including natural, financial, human, social, and physical capital [International Federation of Red Cross and Red Crescent Societies 2004, p. 18]. However, for the purposes of this chapter, we refer to the specific usage, meaning the ability to absorb and accommodate disturbance and still retain basic structures and function, in other words, to maintain the same identity [e.g., United Nations 2009; Walker and Salt 2006].

The key point here is that resilience refers to the maintenance of an existing system: the ability to receive external shocks and not be fundamentally changed by them. Therefore, resilience is about stability, not change. Thus, in contrast to much development aid, humanitarian aid is about returning communities to their predisaster state. Although this may seem obvious, in practice, it is not; this problem is not confined to the often intrusive international NGOs, but also to all levels of government, industry, and local organizations. There is a tendency to think that it is a good idea to take advantage of the destruction and chaos to install a new order ranging from the relatively benign, such as a different style of housing, to the exploitative, where the poor can lose access to land in favor of larger interests who have the ear of the government through changing land zoning on the grounds of safety. An example of this was the attempt to move poor fishing communities back from the (valuable) shoreline following the tsunami in India and Sri Lanka.

Thus, if the goal of disaster response is to build resiliency, humanitarian assistance should support the community's own recovery to a state of normality as defined by that community:

> In a disaster our job isn't to change the way people think. But lots of people come in thinking it is. It is our job to promote sustainable, durable, and culturally appropriate solutions in harmony with the desires of the local population [Dave Hodgkin, international disaster worker, interview, July 2008].

It follows that community-oriented disaster response should be the means to build resilience and maintain humanitarian integrity, and part of this is for communities to define the terms of humanitarian solutions as the drivers of disaster recovery and with that the role of the humanitarian sector as facilitators of community recovery.

Of course, there are several obstacles in practice as compared to top-down emergency management. For example, community participation may be more costly in terms of time and resources, and there is a tension between the need to orient solutions around community needs and values and the simple logistical challenge of rolling out assistance on a mass scale following a major disaster. Another issue is that the needs of disaster-affected communities are highly time-specific, and that different degrees of community engagement are appropriate at different times. For example, in the urgency of the emergency phase of a disaster, there is little time or need for extended community consultation when all of those involved with the response are under pressure to save human lives. As the humanitarian response moves into recovery and rehabilitation, there is more time and a stronger imperative to shift into community-oriented programs. For example, community engagement may be much more important with regard to infrastructure, such as permanent housing, compared to the temporary arrangements of tents and tarpaulins.

The term *community-oriented* also raises the obvious question as to which community or which part of the community? A community can be defined in multiple ways, and there may be numerous or overlapping communities in a single place [Handmer and Dovers 2008]. Furthermore, not all parts or members of a community have common interests, hence their interests may be at odds with one another [see Cooke and Kothari 2001].

There is also the tension between the need to orient solutions around community needs and values and the simple logistical challenge of rolling out assistance on a mass scale following a major disaster. The following two case studies explore different ways of negotiating this tension through the involvement of the NGO sector. The first case study discusses a network of 12 local NGOs who undertook disaster relief and rehabilitation programs on the east coast of India after the 2004 tsunami, playing a significant role in community recovery and rehabilitation. The second case study examines the part played by international NGOs after the Jogyakarta earthquake in 2006, where there was a large presence of international NGOs on the ground compared to the case study of tsunami-affected regions of India, in which international NGOs were largely discouraged. It is important to keep in mind that all disasters are different and how the principles explored in these case studies might be applied in practice will vary. The case studies are presented as explorations of examples of community-oriented disaster response and are not solutions to the complex and messy problems of disasters.

The Strength of Networks: The Local NGO Response to the Tsunami in India

On December 26, 2004, the Asian tsunami hit the east coast of India and left a trail of destruction in its wake: 2,000 km of the Indian coastline were submerged for a distance of up to 2 km inland; 12,400 people were killed; 46,600 houses were lost or damaged in 559 villages; and 650,000 people were homeless. In all, 3.2 million people were affected [Walls 2005], of whom 2.7 million lost their jobs [Sugiyarto and Hagiwara 2005]. The most seriously affected (comprising one-third of the total) were from the fishing sector, losing 88,000 boats and nearly a year's worth of production [Stobutzki and Hall 2005]. However, within this sector it was women, children, and *dalits* (once known as the untouchables) who were the most vulnerable and who arguably suffered the worst impacts and were the most neglected in official responses [Choo 2005; Krishnakumar 2005; Walls 2005]. In addition to human losses, fishing grounds were damaged and large swathes of land were rendered useless due to salt-water inundation. The response to the tsunami was swift and comprehensive, and local people, groups, and networks were at the heart of it. At the local level, the response encompassed a combination of endeavors by governmental bodies, NGOs, and local community initiatives, with a relief program providing food and shelter mainly for the first 3 months, followed by a major rehabilitation and reconstruction drive.

A network of 12 local NGOs—the East Coast Development Forum (ECDF)—reached approximately 20% of the affected population along the east coast of India, responding to the tsunami using a three-pronged approach: providing immediate relief, restoring livelihoods, and advocating for improved and more responsive government services. The basis of the ECDF was an existing network, the East Coast Fisher's Forum, an advocacy network of 10 local NGOs that had been working for fisher's rights for more than a decade. The network had extensive community links and members among the fishing community along the east coast of India, as well as strong links with government [PREM 2003]. Because the tsunami hit the fishing community hardest, the East Coast Fishers Forum was ideally placed to take a lead in coordinating a local level response. However, in addition to fisher communities, many dalit and tribal communities living along the east coast were also seriously affected by the tsunami and were either being overlooked or actively discriminated against in much of the official response [Choo 2005; Walls 2005]. Consequently, two NGO networks that focus on dalit and tribal communities were also included in the ECDF.

The People's Rural Education Movement (PREM), an NGO from Orissa, was the common link within the diverse membership of the ECDF; and although the tsunami did not directly affect its own constituency in Orissa, it had connections through its association with local NGOs. PREM brought practical disaster management skills gained from its work in the super cyclone of 1999, which devastated parts of Orissa [Das 2005], to the group. It was the strategic support offered by PREM, the strength of the local NGO networks, their strong links with local fisher communities, and a long record of trust and credibility that enabled these NGOs to successfully manage the delivery of emergency relief and later rehabilitation programs to affected communities.

Each of the 12 member NGOs was responsible for their existing constituency, made up of small self-help groups that they had supported for some years via small savings and microfinance programs, as well as advocating for fisher rights. The relief program used a common set of procedures introduced by PREM, which involved distributing a standard package of supplies to the members of these groups. Likewise, the rehabilitation program used a simple model to provide cash support to existing self-help groups under long established systems. These approaches reduced the administrative load, avoided duplication, and enabled the adoption of a decentralized approach. It was the informal systems based on trust, confidence, and credibility that enabled the ECDF to access the affected communities to determine what was required and how to proceed. These informal systems were complemented by more formal accountability and distribution systems for the work itself. The main change made within the ECDF was the adoption of common systems and procedures while maintaining the existing decentralized flexible response framework.

All ECDF members had been active in development and had worked together in advocacy campaigns for at least 10 years, with approximately one-half active in the area for more than 20 years. East Coast Fishers Forum members had worked together on issues that were common among themselves and their constituencies, and therefore had developed the necessary trust and leadership for the network to function well. Overall, approximately 80% of the support went to preexisting self-help groups, and hence the risk of poor or nonperformance within the relief and rehabilitation program was low. For the first 2 to 3 days after the tsunami, local communities were entirely responsible for the response—helping each other out, rescuing those in need, and offering temporary shelter, food, and water to the victims. Next came support from governmental and NGO agencies with programs in place with a variety of objectives, including rescuing people and establishing camps for displaced persons. However, it is the critical role of local communities in the first few hours of a disaster that is often overlooked in many formal (or agency-produced) accounts of disaster responses, and it is local communities' links with governmental agencies and NGOs that seem to be a strong indicator of successful recovery, as this case will demonstrate.

The NGOs tended to fill gaps not covered by government programs. They met short- to medium-term demands for food and shelter and later became involved in long-term rehabilitation, particularly of livelihoods, as well as for assuming a watch-dog role to ensure respect for people's rights and equal treatment for all groups in governmental and nongovernmental programs. Finally, they proposed rehabilitation approaches geared toward sustainability and reducing the risk of catastrophic consequences of natural disasters in the future. It can be argued that the role of the networks themselves, the high level of trust, the leadership style, and the developed culture of mutuality that is inherent in them were keys to the effective coordination, beneficiary selection, and overall management [Hudson 2001; Liebler and Ferri 2004].

The relief work went ahead without a hitch largely due to the presence of village organizations throughout the process, the local knowledge of network members, and the experience and systems offered by PREM [Prakarama 2005]. In April of 2005, the rehabilitation work began, focusing on livelihoods, particularly those of women. Here, relatively small loans of 10,000 rupees (approximately 250 USD) were provided to affected families through established women's self-help groups rather than engaging in large-scale infrastructure rehabilitation, such as of fishing boats and housing, which were being covered by other agencies. This was a flexible approach that gave the family a large say in the use of funds and allowed the community to maintain a high level of control. In official programs, there had been serious gender disparity [Choo 2005; Krishnakumar 2005] in which women were left out of official livelihood programs, which mainly concentrated on restoring the infrastructure of the fishing industry itself rather than recognizing the diverse roles played by all family members. In addition, dalit and tribal women suffered discrimination not only by virtue of their gender, but also because of their caste [Choo 2005]. This straightforward approach to rehabilitation suited network-oriented relief management. Working with existing self-help groups did not require technical expertise and, more importantly, it was sufficiently simple to streamline administration so that payments were made directly to the self-help groups from the ECDF secretariat rather than through the NGOs. Instead, the NGOs provided nonfinancial institutional and other support to the self-help groups to help them make it work.

Advocacy was regarded as an integral part of the response because the government either conducted the majority of relief work or was responsible for the policy framework within which relief efforts occurred. Advocacy concentrated initially on issues directly affecting the relief and rehabilitation drive but then broadened to the related matters of fisher livelihoods, dalit and tribal rights, and coastal protection to minimize the impact of future disasters. Underlying the approach of the ECDF's advocacy work was the principle that relief and rehabilitation programs should be seen as preserving and respecting rights rather than as privileges. The ECDF and community representatives met with state and federal ministers and others to report the findings of their survey of deaths and other losses and identified people who were not included on government lists, such as dalit and tribal communities, and sought to ensure that they received proper compensation [Krishnakumar 2005]. The ECDF also made information available to the community on the government and NGO packages that were available so that all of those in need had access to relief, thereby limiting the extent to which vested interests could direct the relief packages to their own constituencies. The ECDF was also able to mediate between the government and fisher, dalit, and tribal communities on discrimination issues because it represented and enjoyed credibility among all three communities. For example, when some huts belonging to dalits were washed away, the government said that it did not have to pay compensation or replace them because the huts were unauthorized and were located farther than 200 m from the shore. The ECDF argued that the location of a hut should not be used to deny dalits their right to shelter and in the end they were allowed to rebuild at least 50 m from the shore.

The ECDF's response to the tsunami along the east coast of India was regarded as successful due to its experience with other disasters in that part of India and the ability of local NGOs to come together to address a local crisis. This example offers important lessons for NGO cooperation and networking in disaster situations, both in India and in the region more broadly. The trust and cooperation built up though joint advocacy initiatives may provide the foundation for better cooperative work by NGOs in future disasters. At the root of the ECDF's success in the rehabilitation program is the fact that its institutional form emerged out of three existing networks, and its 12 members could draw on a strong tradition of collaboration to promote issues affecting fishing and other communities along the east coast. The key lesson that emerges is that the characteristics of NGO advocacy networks can be applied to disaster situations. The ECDF drew on elements of three successful advocacy networks: a shared leadership approach encouraging people to be innovative and flexible;

a decentralized approach to the work; and the incorporation of common systems into this decentralized approach. In other disaster areas, it may be possible to identify and build on existing locally based advocacy networks so that any response is more effective in reaching local communities and in having their trust and experience to be able to work together in these high-pressure environments.

Shelter Response to the Jogyakarta Earthquake

The second case study was characterized by significant international NGO presence on the ground. International NGOs bring a different set of skills to the table, including expertise and capacity specific to disaster response, technical inputs, coordination of international actors, policy discussion in the international context, such as the Hyogo Framework for Action, capacity building of local staff, and access to international funds [Willitts-King 2009]. They also bring a perennial set of problems; a key one being that international actors have a tendency to put themselves at the center of the aid process, including their relationships with beneficiaries, their attitudes toward local NGOs, their involvement with government, and their willingness to collaborate as part of a larger response:

> The provision of aid is a top-down, externally driven, and relatively rigid process that allows little space for local participation beyond formalistic consultation. Much of what happens escapes local scrutiny and control. The system is viewed as inflexible, arrogant, and culturally insensitive. This is sometimes exacerbated by inappropriate personal behaviour [sic], conspicuous consumption, and other manifestations of the "white car syndrome." Never far from the surface are perceptions that the aid system does not deliver on expectations, is expat-heavy and "corrupted" by the long chain of intermediaries between distant capitals and would-be beneficiaries. In other words, seen from below, the enterprise reflects the expectation that humanitarian theatres should adapt to it, rather than the reverse [Feinstein International Centre 2008, p. 11].

The inward-looking tendencies of the international agencies, the donor-centered provision of aid, and the lack of accountability to beneficiaries are fundamental problems of a humanitarian system that is driven by the goals of people rather than the goals of communities. However, this does not have to be the case. The following case study, like the previous one, explores humanitarian action that worked with communities to develop locally appropriate solutions that built on existing community skills.

The post earthquake community construction of a house in South Jogyakarta. (Photo credit, Rikki Zulkernaen.)

Jogyakarta Earthquake 2006

In the early hours of May 27, 2006, an earthquake struck just to the south of the city of Jogyakarta in Central Java, Indonesia. Although Java is located in the Ring of Fire earthquake zone, there had not been a major earthquake in living memory and, over time, the quality of local construction had markedly decreased, resulting in massive and widespread damage from this relatively mild earthquake (approximately a 6 on the Richter scale). The earthquake damaged or destroyed nearly half a million houses, leaving 1.6 million people without shelter (The Consultative Group on Indonesia 2006). Although the final death toll of approximately 6,000 people was far lower than the Indian Ocean Tsunami 17 months earlier, it set the scene for a major shelter crisis.

The earthquake response was characterized by extensive NGO involvement. The international community arrived within the week; some agencies were already on alert for a potential eruption of the Mt. Merapi volcano to the north of the city and were therefore already in Jogyakarta and prepositioned with supplies. Jogyakarta also has a vibrant civil society, with local NGOs numbering in the hundreds. In total, more than 500 organizations provided some assistance as part of the earthquake response, including over 100 international NGOs. The majority of the NGO sector worked together under the coordinating structure of the Shelter Cluster, a new approach at the time to coordination and collaboration that supported strong working relationships with multiple levels of government and other disaster actors. The earthquake response was unusually collaborative, with the participation of international and local NGOs, United Nations agencies, private companies, 14 military departments, and civil societies (including 24 universities and thousands of volunteers). In September of 2006, the humanitarian response had received 57 million USD, less than half the funds needed according to the interagency earthquake response plan, although total NGO assistance a year later eventually stood at 175 million USD (United Nations Coordination Centre for Yogyakarta and Central Java 2007).

Humanitarian recovery efforts were largely focused on the shelter crisis, responding to the pressing need to replace the large numbers of houses that were destroyed, particularly of low-income communities. The Indonesian government pledged support for providing permanent housing, but after many problems with building transitional shelter camps in the tsunami-affected areas in Aceh, it now promoted a one-step housing plan—straight from emergency tents and tarpaulins to permanent housing. Although the government provided assistance for affected families much faster than was expected, with inputs into its housing reconstruction program of more than 600 million USD, the one-step plan was initially a major concern because emergency shelters (tents and tarpaulins) had an expected lifespan of only several months, whereas the construction of hundreds of thousands of permanent houses could have taken many years. Further trepidation about the upcoming monsoon season led to widespread concerns about the emerging gap in the proposed supply of permanent housing and the potential humanitarian shelter crisis as emergency shelter started to deteriorate.

This led to the Shelter Cluster advocating and winning government support for the construction of transitional shelter. In contrast to the top-down approach to much construction after the 2004 tsunami in Aceh, the Cluster's approach was strongly community-focused and process-oriented. It began with a technical working group analyzing what structures communities were already starting to rebuild themselves, and which materials and techniques that they were using. In partnership with local universities, a no-logo design for a transitional shelter, or t-shelter, made of bamboo was developed based on best practice local construction techniques. The NGO sector eventually produced 85,000 transitional bamboo shelters at a cost of approximately 200 USD per shelter, built through community-based projects, and communities themselves built two to three times that number of similar structures. This alignment with community objectives was an important factor in the success of the shelter response, with approximately 100% of transitional structures being used as the first step in a permanent structure or reused for other household purposes after more permanent structures were erected.

The bamboo shelters were adaptations of a common form of construction; a simple, familiar structure based on local vernacular architecture that used local building materials, tools, and techniques that was cheap to build and used existing community knowledge. The design further benefited from small modifications from construction experts to strengthen and improve the traditional structure. According to the technical advisor to the Cluster, those solutions involved "local technology, reusable materials, low environmental impact, expandable, renovatable, reusable and, finally, disposable" [Dave Hodgkin, international disaster worker, interview, July 2008]. The process of building the structures was also community-oriented. The Cluster's approach was centered around the design guidelines for the t-shelters, not around prescriptions for how construction should proceed or to direct funding to external businesses to build the shelters. The guidelines encompassed basic principles for strong and safe housing but were flexible and were used in multiple ways. For example, the Indonesian branch of the International Federation of the Red Cross and Red Crescent Societies gave small cash grants to neighbor groups to buy tools and materials, supporting them with training and teams of volunteers who went to live with the communities. Other agencies, such as the International Organisation of Migration, supplied communities with materials as well as volunters and technical assistance. The common thread running through the construction of the shelters was to support communities with enough additional resources to build back these essential structures as quickly as feasible in the postdisaster environment, but for the process to be performed by the communities. A traditional spirit of volunteerism in Javanese culture, "gotong royong," was one element that made this process possible.

The unusual but appropriate choice of bamboo as the primary building material was another factor in the success of the intervention. Although bamboo has

International Red Cross and Red Crescent providing tools. (Photo credit, Eko Bhirwo.)

not historically been a common choice of construction material for international humanitarian interventions, it can be an excellent choice of material in regions to which bamboo is endemic. Bamboo can be cheap and fast to build with; it is grown in communities and harvested at the community level. Because it is essentially a community resource, the use of bamboo across the earthquake recovery efforts strengthened the local economy. There are also environmental benefits—if harvested correctly, bamboo is the ideal sustainable material, with a significantly lower environmental impact than the usual humanitarian solution of timber and corrugated iron.

In other respects, the environmental impacts were serious. The provision of bamboo for 85,000 traditional bamboo shelters has a huge impact on bamboo ecology and other parts of the local economy. Although it is possible to harvest bamboo in a sustainable manner, the scale of the demand was so high that many local suppliers "clear-felled" it (i.e., took the whole bamboo clump out, including much of the root material), with the medium-term effect being that the regrowth time for bamboo clumps in some areas stretched to a decade or more. The other issue is the effect on the bamboo economy and those who need a supply of bamboo for their livelihoods. It is hard to judge this effect, but certainly there would be serious shortages of bamboo stock for at least a year, and an ongoing shortage for several years while stocks recovered. This does not mean that bamboo or local materials shouldn't be used, but rather that the need to recognize the broader local economic and ecological effects should be considered. It would seem that one of the important issues in using local resources is the scale of its use and the long-term impact that large-scale exploitation of resources can have on both the ecology and the communities that are dependent on that resource. Overall, despite the medium-term ecological impact, the shelter program was seen as a community success because the designers of the transitional bamboo shelter guidelines had internalized the connections between the choice of shelter and the local context. It seemed like the most reasonable and beneficial option was to find a culturally appropriate

Preparing a bamboo thatch wall. (Photo credit, Eko Bhirwo.)

solution, which was to some extent environmentally friendly and community-based, despite the inherent problems of using local resources on a large scale.

Community Responses to Disasters: Some Lessons

In disaster response, the critical test is whether a community can recover its essential structures and functions and not just the extent to which it suffered damage. The fact that houses have fallen down or that farming or fishing practices have been disrupted is not important in itself; it is the ability of communities to eventually recommence those activities that is more important. However, whether a community can return to something approaching its prior patterns and organization is not clear immediately after a disaster. If, for example, an international NGO believed that it had the right solution and backed up that belief with enough resources, community transformation could occur in a direction different from community recovery, which may merely achieve change, not resilience. For these two case studies, it becomes clear that if a humanitarian actor is sincere about supporting community resilience, they need to be committed to recreating the community, not replacing it with the actor's priorities and preferences. For an institution characterized by globalized, context-independent knowledge, the way that this can be achieved is by ensuring that community knowledge drives the process (in other words, by allowing communities to define normality and their personal trajectory toward recovery).

Resilience also applies in its more prosaic sense: improving a community's ability to be more resistant to disaster. In terms of the international shelter sector, resilience may refer to developing local capacity through improving construction skills, training, etc. and the idea of 'build back better' to counter the fact that communities generally tend to have fewer resources and so 'build back worse' after a disaster. The humanitarian sector can help communities become more resilient by offering the resources and technical support to build back better. At the onset of disaster, the community is in a state of chaos and "the future is up for grabs" [Walker and Salt 2006, p. 78]. The ability to influence a community's trajectory is greatest at this critical point when the community is damaged and traumatized, and it will be easier to cross community thresholds for changes for improvement. The international humanitarian community is mandated to assist only when the community's capacity to recover and the government's capacity to assist recovery have been exceeded. In other words, humanitarian aid is required when communities in the national context lack the resources to return to their predisaster state. Therefore, the humanitarian sector can be considered a transient but integral component of a local resilient system. It means that listening to communities is crucial.

The second lesson is related: Being close to the community is important. International NGOs cannot do this and must have different approaches as outlined in the Indonesia case study. But the question is: Is that enough? It is the local

NGOs—embedded in the community and with local links—that can be most relevant in their response in terms of assessing needs and establishing local capacity. In the Indian case study, the local NGOs had the history with local communities and therefore had access, their trust, and the ear of government to ensure that the government programs were equitably implemented. An issue that often emerges is that international NGOs fail to recognize the local knowledge, capacity, and systems in place. This could be because they do not have time to learn those systems themselves or they might not speak the local language, both literally and in terms of processes and analysis. Finally, they believe that processes and approaches can be easily transported from one disaster or emergency to another and be effective and that the local context is less important in the face of an emergency. Of course, this is generally not the case and the lesson from both of these case studies is that the local linkages are the key to success.

Conclusion

The two case studies show how adaptive local systems can be made to work for broader community improvements. In the case of the Indian tsunami, local organizations and their credibility enabled a successful rehabilitation of fisher communities and the fruitful engagement with government to ensure that resources were able to reach the marginalized groups in society. In the case of Indonesia, the notion of bringing international NGOs into clusters with local NGOs and then not only successfully lobby the government, but also adapt to local solutions led to a successful shelter program. In both cases, scale issues did emerge. In the case of India, when long-term development work was being undertaken later and larger funding was being channelled through relatively small NGOs, problems of capacity to deliver on the scale expected emerged. In the case of Indonesia, using local materials for building had its own ecological and environmental impacts. It is not only adapting local systems that is important for successful disaster management, but it is also the framework that we enter these situations. It is here that the notion of resilience is important because it recognizes that seeking to ensure that people are resilient and agents in their own destiny and that the role of the outsider is to facilitate some of these processes, accelerate them if possible, and add to the local systems so that the communities are more resilient in the future.

References

Choo, P. S. 2005. Women in the December 26 tsunami: How they have coped; how can we help. *NAGA World fish Center Newsletter* 28:13–16.

Consultative Group on Indonesia, The. 2006. The 15th Consultative Group for Indonesia Meeting, report, ADB Manilla. Retrieved from http://www.adb.org/Documents/Conference/ino-cons-mtg-IN180-06.pdf (accessed May 7, 2009).

Cooke, B., and Kothari, U. eds. 2001. *Participation: The new tyranny*. London: Zed Books.
Das, B. P. 2005. *Socio-economic devastation of Orissa Coast, India: Caused by unprecedented sea level rise during October 1999 Super Cyclone*. Anchorage: World Water and Environmental Resources Congress.
Dover, S. 2004. Sustainability and disaster management. *The Australian Journal of Emergency Management*, 19:21–25.
Feinstein International Centre. 2008. *Strengthening the humanity and dignity of people in crisis through knowledge and practice*. Bedford, MA: Author.
Handmer, J., and Dovers, S. 2008. *The handbook of disaster & emergency policies and institutions*. London: Earthscan.
Hudson, A. 2001. NGOs and transnational advocacy networks: From legitimacy to political responsibility. *Global Networks* 1:331–352.
International Federation of Red Cross and Red Crescent Societies. 2004. *World Disasters Report: Focus on community resilience*. Geneva: Red Cross.
Katoch, A. 2006. The responders' cauldron: The uniqueness of international disaster response. *Journal of International Affairs* 59:153–172.
Kennedy, J., Ashmore, J., Babister, E., and Kelman, I. 2008. The meaning of build back better: Evidence from post-tsunami Aceh and Sri Lanka. *Journal of Contingencies and Crisis Management* 16:24–36.
Krishnakumar, A. 2005. In *Chennai, women's vulnerability exposed by tsunami*. Unpublished report. Washington, DC: Population Reference Bureau.
Liebler, C., and Ferri, M. 2004. *NGO networks: Building capacity in a changing world*. Washington, DC: Population Reference Bureau.
People's Rural Education Movement. 2003. National consultation on issues affecting east coast fisher people, 6–7 May. Report. Visakhapatnam, India: People's Rural Education Movement.
Prakarama, A. 2005. *South Asia Tsunami response: 3-month assessment*. Colombo, Sri Lanka: Oxfam.
Sphere Project. 2004. *Humanitarian charter and minimum standards in disaster response*. Geneva: Sphere Project.
Stobutzki, I. C., and Hall, S. J. 2005. Rebuilding coastal fisheries livelihoods after the tsunami: Key lessons from past experience. *NAGA World Fish Center Newsletter* 28:6–12.
Sugiyarto, G., and Hagiwara, A. T. 2005. Poverty impact of the tsunami: An initial assessment and scenario analysis. Paper presented at the 4th AEP Research Network General Meeting, July 2005, Colombo, Sri Lanka.
United Nations Coordination Centre for Yogyakarta and Central Java. 2007. *Java earthquake response: Cluster Atlas May 2006–2007*. Jakarta, Indonesia: United Nations.
United Nations. 2005. *The Hyogo Framework for Action 2005–2015: Building the resilience of nations and communities to disasters*. Geneva: United States.
United Nations. 2009. *International strategy for disaster reduction*. Geneva: UN-ISDR
Walker, B., and Salt, D. 2006. *Resilience thinking*. Washington, DC: Island Press.
Walls, M. 2005. Report: Review of the tsunami recovery. London: Tsunami Recovery Network, Development Planning Unit, University College London.
Willitts-King, B. 2009. The role of the affected state in humanitarian action: A case study on Indonesia. HPG Working Paper. London: Overseas Development Institute.

Chapter 15

Analysis of the Emergency Response by Business, Government, and the Public during and after a Sour Gas Well Blowout

Chuansheng Jiang and Yunfeng Deng

Contents

Introduction

At 10:00 PM on December 23, 2003, a blowout* occurred at the Luojia 16H well of the China National Petroleum Corporation (CNPC) Sichuan Northeast Gas Division in Gaoqiao Town, Kai County, Chongqing, China. From the time of the accident to the return of the evacuated residents one week later, four towns were affected. In all, 241 residents and 2 workers of the Luojia 16H well lost their lives, 26,555 individuals were outpatients, 2,142 individuals were hospitalized, and more than 65,000 others were evacuated; the direct economic loss was estimated at approximately 92 million yuan RMB [CASST 2007]. After the accident, officials from the State Council of China, the Chongqing Municipal Government, the

* A blowout is a sudden release of oil and gas from a well.

State Administration of Work Safety (SAWS), the State Environment Protection Administration (SEPA), and the CNPC actively participated in the response activities of the disaster. After the direct rescuing activities ended, the State Council organized an accident investigation team to investigate the causes of the accident and the response and losses to make an assessment of responsibility for the disaster.

After the accident investigation concluded and the individuals found guilty were sentenced, the China Academy of Safety Science and Technology (CASST) performed a technical field survey for the accident with the support of SAWS and the Chongqing Municipal Government in October of 2004. The field investigation was accompanied with a comprehensive quantitative analysis with a wind tunnel analog test and computer simulation. The study analyzed the emergence and evolution of the blowout disaster as well as enterprise, government, and public emergency response activities from the viewpoints of both technology and management and put forward accident prevention and control recommendations [CASST 2006]. Finally, with the support of national scientific and technological projects, CASST developed a series of professional standards to help prevent future blowout incidents and blowout risk prevention and control [CASST 2008]. This was the first time any safety science research institution in mainland China performed a significant and systematic technical field investigation for a technological disaster. In addition, Chongqing Centers for Disease Control (CCDC) and other health research institutions studied the victims' psychological health and the influences of the disaster on the environment according to the experiences of participation in the accident response activities and postevent field surveys [Chen et al. 2007; Zhang et al. 2007a; Zhang et al. 2007b].

On the basis of the above-mentioned preliminary investigation and study results, this chapter will comprehensively present the following: (1) background of the enterprise and the location of the accident; (2) evolution of the disaster and the response and recovery activities by businesses, the government, and the public; (3) lessons learned about emergency response management; and (4) the further discussion on several potential social and cultural issues.

Background and Location

The free-flow rate of Luojia 16H natural gas well is 4 to 10 million m^3/d, in which the H_2S content is up to 9.02%. This well is located in Xiaoyang Village, Gaoqiao Town, Kai County, Chongqing, and belongs to CNPC Southwest Oil and Gas Branch, which implements a unified vertical management style. Sichuan Northeast Gas Division Xuanhan Working Area, a subsidiary of CNPC Southwest Oil and Gas Branch, is in charge of the day-to-day operations of the Luojia 16H well [CASST 2006, 2007].

The well was drilled by the No. 12 brigade of the No. 2 Branch of Chuandong Drilling and Exploring Company of CNPC Sichuan Petroleum Administration.

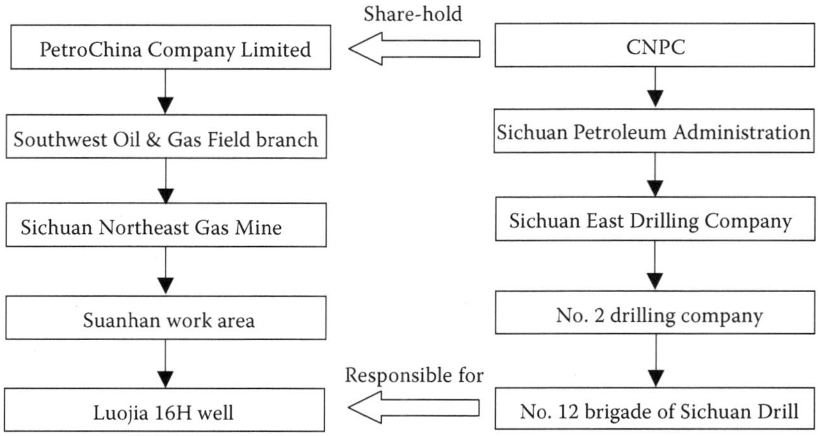

Figure 15.1 Luojia 16H well subordination relationship diagram.

Chuandong Drilling and Exploring Company is a class-two enterprise of Sichuan Petroleum Administration specifically engaged in drilling construction. CNPC is a shareholding company of PetroChina Company Limited, meaning that this enterprise is directly under the control of the central government and under the control of the State-Owned Assets Supervision and Administration Commission of the State Council. Its administrative class is equivalent to a province or a municipality directly under the control of the central government (see Figure 15.1 for the company's internal management relationships).

Luojia 16H well is located in a low-lying area in Xiaoyang Village, approximately 1.2 km away from the north of Gaoqiao Town of Kai County. On its northwest side, there is a mountain with a ridge elevation of approximately 900 m. On its southwest side, there are many small hills and valleys, and the mountains are relatively flat and gentle on its southeast side. The well site is adjacent to the main road from Gaoqiao Town to Maliu Town. The annual average air temperature at the well site is 16.7°C with an average air temperature in winter of 5.4°C.

The two villages adjacent to Luojia 16H well are divided by a river flowing from north to south. The village of Gaowang is to the east and the village of Xiaoyang Village is to the west (as shown in Figure 15.2). There are a total of 2,419 villagers, most of whom live within a 1-km range around the well site, including 552 villagers who live in houses less than 500 m away from the site. The closest house is less than 50 m away from the well derrick.* Gaoqiao Town is located in the north mountain area of Kai County with bad traffic conditions and backward economic development. In 2003, the per capita annual income was approximately 1600 yuan RMB;

* The well derrick is the tower-like framework over a well.

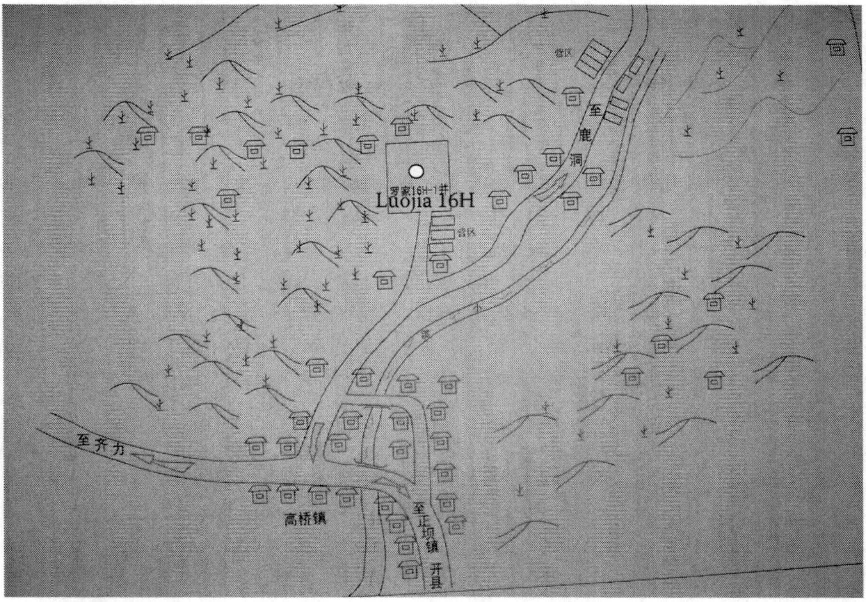

Figure 15.2 Topographic map around Luojia 16H well.

most peasant households had no land-line telephones and mobile telephone users were almost nonexistent.

Disaster Evolution and Response of Relevant Parties

The entire disaster lasted approximately 85 hours; the incident occurred from 9:55 PM on December 23, 2003, to 11:00 AM on December 27. On December 23, at 9:55 PM the mud logger reported that she found that a well-kick* happened; the driller sent out blowout warnings and prepared to take controlling measures. At 9:57 PM, the controlling measures failed and a blowout occurred. At 10:04 PM, the blowout was completely out of control and a heavy smell of hydrogen sulfide had spread over the entire well site. At 10:08 PM, the drilling crew took further controlling measures but still failed to control the blowout [CASST 2006].

At 10:20 PM, the drilling crew reported the blowout information to Sichuan East Gas Division, who forwarded the report to Chuandong Drilling and Exploring Company. The Sichuan East Gas Division and the Chuandong Drilling

* Well-kick: An entry of water, gas, oil, or other formation fluid into the wellbore during drilling. It occurs because the pressure exerted by the column of drilling fluid is not great enough to overcome the pressure exerted by the fluids in the drilled formation.

and Exploring Company began to perform response activities collaboratively. At approximately 10:30 PM, the drilling crew began to evacuate; meanwhile, nearby residents were instructed to evacuate. Between 10:25 and 10:40 PM, the drilling team reported to the government of Gaoqiao Town the accident conditions and requested that officials help evacuate the public within 1 km of the well site. Later, the drilling crew requested that the government of Gaoqiao Town enlarge the evacuation range to 3 to 5 km from the well site. Between 10:50 and 11:00 PM, the duty staff of the drilling crew called the three local emergency responder units: 110, 120, and 119* to report the accident. At approximately 11:00 PM, the Chongqing Administration of Work Safety received the report about the blowout accident from Sichuan East Gas Division, and the Chongqing Administration of Work Safety immediately forwarded the report to the Chongqing Municipal Government [CASST 2007].

At 11:26 PM, the Chongqing Municipal Government sent a fax of the report of the drilling crew to the Kai County Government and asked them to immediately organize necessary rescue teams to go to the region around the well site. The Kai County Government subsequently requested that Gaoqiao Town try its best to evacuate the residents around the well site. The officials of Gaoqiao County called the heads of the relevant villages to instruct all the villagers to evacuate. Meanwhile, the rescue teams and the officials from Kai County Government and Gaoqiao Town Government rushed into the field.

At approximately 11:30 PM, the drilling crew organized several workers to return to the well site and shut off the master gate valve of the elevated oil tank, the slurry pump, the diesel engine, and the power generator. After shutting off the master gate valve, all drilling crew staff evacuated and a security line was set up around the well site. Simultaneously, the Southwest Branch Company of Oil Gas Field and Sichuan Petroleum Administration collaboratively organized their first emergency response meeting. At approximately 11:50 PM, the officials from Kai County Government arrived at the site with the rescue teams. At approximately 12:00 AM on December 24, the first batch of rescue teams from Chuandong Drilling and Exploring Company left for the well site with one fire engine, one environment monitoring vehicle, and one ambulance. At approximately 12:30 AM, the Chuandong Drilling and Exploring Company and the Sichuan Northeast Gas Division jointly made the urgent request for help in a fax to Chongqing Municipal Government and requested that they immediately notify the Kai County Government to help with the response activities. By 12:40 AM, the first group of rescue workers from the Sichuan Petroleum Administration and the Southwest Branch Company of Oil Gas Field left for the well site. At approximately 3:00 AM, the Kai County Government established the search and rescue commanding group with the support of the drilling crew, organized the residents around the well site to evacuate, and dispatched

* In Mainland China, 110 is the emergency number for the police, 120 is the emergency number for medical help, and 119 is the emergency number for the fire brigade.

the rescue teams to search for and rescue poisoned individuals. It was not until 5:25 AM that the Sichuan Petroleum Administration reported the blowout situation at Luojia 16H well to CNPC headquarters [State Council Investigation Team 2004; Tang 2004].

At 8:20 AM, the Sichuan Petroleum Administration and the Southwest Branch Company of Oil Gas Field held a second emergency response conference to discuss the response arrangement. By 9:30 AM, the rescue team from Chuandong Drilling and Exploring Company of Sichuan Petroleum Administration arrived at Zhengba Town, a town near Gaoqiao Town, and began setting up a rescuing command center and temporary monitoring stations at Zhengba. They were also engaged in helping local residents evacuate and monitoring the H_2S concentration in the air. At 11:20 AM, they received the instructions to ignite the blowout well and then contacted the armed police from Kai County about the ignition arrangement.

At 11:30 AM, a temporary field rescue team, composed of officials from the Chuandong Drilling and Exploring Company of Sichuan Petroleum Administration, the Kai County Government, and the drilling crew, was established at Gaoqiao Town to prepare for the wellhead ignition. At 2:00 PM, a member of the rescuing staff reported that the sound from the well site became weaker. The team director immediately dispatched several rescue staff to enter the well site to observe. The rescue staff confirmed that the blowout at the wellhead had stopped, while the blowout at the outlet of No. 3 blooie pipe continued. At about the same time, a formal field rescue team was established at Gaoqiao Town, about 1 km away from Luojia 16H well site that was composed of officials and technicians from the Sichuan Petroleum Administration and the Southwest Branch Company of Oil Gas Field. To prepare for ignition, the rescue team dispatched two engineers and six rescue staff members to enter the well site to investigate the situations surrounding the wellhead and the blooie pipe so that a safe ignition location could be selected at 2:20 PM. At 3:00 PM, the above-mentioned eight individuals entered the well site again to confirm the safe ignition place. Meanwhile, to perform the ignition as soon as possible before the armed police arrived, the field rescue staff opened a shop with a pry bar and found some magic bullets (a special kind of firework). By 3:20 PM, two rescue staffs were dispatched to perform the ignition operation about 50 m away from the blowout outlet of the No. 3 blooie pipe. Due to the strength of the blowout, the ignition attempt failed. At 3:55 PM, the field workers found more fireworks in the same shop. With those fireworks, the second batch of rescue staff successfully performed the ignition about 70 m away from the blowout outlet.

At 9:00 PM, Wu Jianong, the deputy mayor of Chongqing Municipal Government, arrived at the site and organized the rescue headquarters. By 10:30 PM, officials from the CNPC headquarters arrived and surveyed the site. A meeting was held with the participation of relevant officials and technical experts from surrounding local governments, the CNPC headquarters, the Sichuan Petroleum Administration, the

Southwest Branch Company of Oil Gas Field, and the Chuandong Drilling and Exploring Company. The rescue headquarters were reorganized at the meeting. A new chief commander and deputy chief commander were appointed, and six professional rescue branches were established that were placed in charge of preparing materials and equipment, formulating technical rescue measures and emergency response schemes, collecting information, providing security and safeguards, providing medical care, and dealing with postaccident issues. On December 25, the rescue headquarters dispatched 20 rescue teams composed of 460 policemen and soldiers, all equipped with portable detectors for hydrogen sulfide, to enter the well site to perform the comprehensive rescue. These rescue teams found many dead bodies. The distribution of the dead bodies is shown in Figure 15.3. In Figure 15.4, the distribution of the villages with casualties is further compared with the distribution of H_2S peak concentration at a height of 1 m from the ground surface obtained through numerical simulation.

At 9:00 PM on December 25, the SAWS officials arrived at the accident site. At 1:00 AM on December 26, the officials from State Council arrived at Kai County with the State Council work team and decided to establish a new rescue headquarters for the blowout accident. All the rescue tasks were being managed by the Chongqing Municipal Government and supported by CNPC headquarters and relevant divisions. The chief commander of the rescue headquarters was assumed

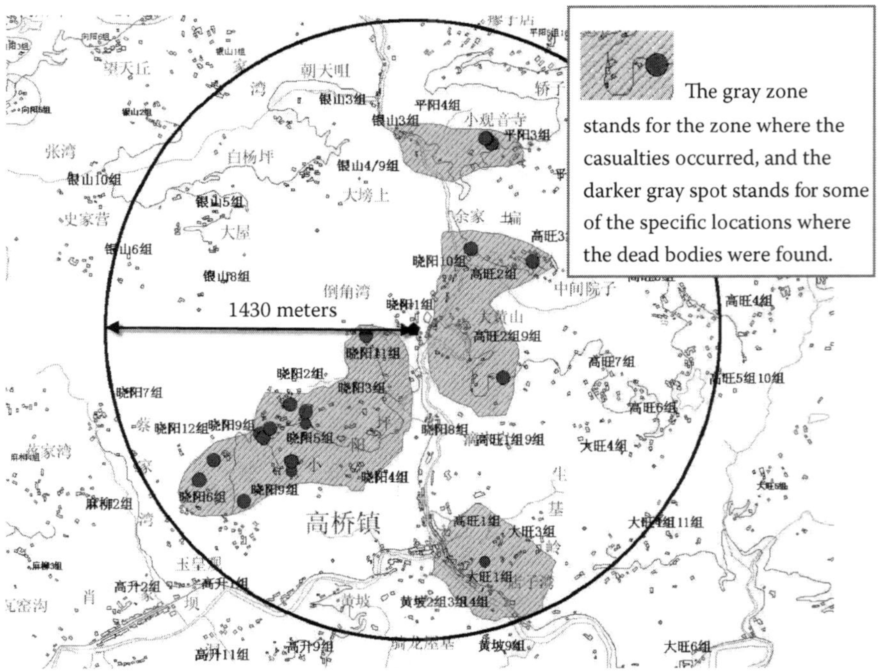

Figure 15.3 Distribution of villages with casualties.

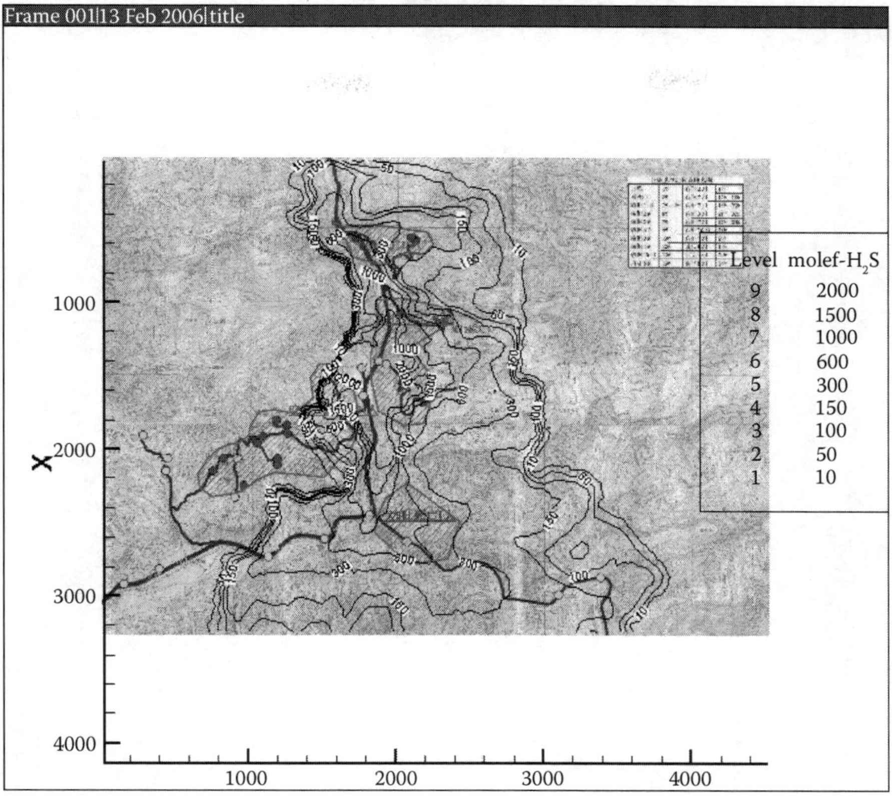

Figure 15.4 Comparison of the distribution of H_2S peak concentrations (obtained through numerical simulation at 1 m from the ground surface) with the distribution of villages with casualties. molef-H_2S stands for the volumetric concentration of H_2S in part per million (ppm).

by Wu Jianong, the deputy mayor of Chongqing, with three branches under the command headquarters:

1. The rescue team's (organizations from Chongqing City, SAWS, Ministry of Health, SEPA, and CNPC headquarters) main responsibilities included victim rescuing, treating the wounded, plugging the wellhead, monitoring H_2S levels, etc.
2. The accident investigation team's (included the Ministry of Supervision, the Ministry of Public Security, the State-Owned Assets Supervision and Administration Commission of the State Council, SEPA, the Ministry of Labor and Social Security and CNPC headquarters) responsibilities included investigation into the causes of the accident and the responsibilities of the people who were at fault.

Table 15.1 Timing of Events Measured against Time from Loss of Control of Blowout

No.	*Events During the Emergency Response*	*Time of Event*	*Time Elapsed Since Loss of Control of Blowout*
1	Blowout occurs.	9:57 PM, Dec. 23, 2003	—
2	Blowout becomes out of control.	10:04 PM, Dec. 23	—
3	The drilling crew takes further controlling measures, but fails to control the blowout.	10:08 PM, Dec. 23	4 minutes
4	The drilling crew reports the blowout information to Sichuan East Gas Division, who forwards the report to Chuandong Drilling and Exploring Company and begins to carry out response activities with them.	12:20 PM, Dec. 23	16 minutes
5	The drilling crew begins to evacuate and informs nearby residents to evacuate.	12:30 PM, Dec. 23	26 minutes
6	The drilling team reports to the government of Gaoqiao Town about the accident conditions and requests the government of Gaoqiao Town evacuate the public within 1 km around the well site.	10:40 PM, Dec. 23	36 minutes
7	The duty staff of the drilling crew calls 110, 120, and 119.	10:50 PM, Dec. 23	46 minutes
8	Sichuan Northeast Gas Division reports to Chongqing Administration of Work Safety.	11:00 PM, Dec. 23	56 minutes
9	Chongqing Administration of Work Safety receives the information about the blowout accident from Sichuan Northeast Gas Division and forwards the information to the Chongqing Municipal Government.	11:00 PM, Dec. 23	56 minutes

10	The drilling crew organizes several workers to return to the well site and shut off equipment. All drilling crew staff evacuates from the well site and a security line is set up around the well site.	11:20 PM, Dec. 23	1 hour and 16 minutes
11	Chongqing Municipal Government faxes the report of the drilling crew to the Kai County Government, and the Kai County Government subsequently notifies the Gaoqiao Town Government.	11:26 PM, Dec. 23	1 hour and 22 minutes
12	Southwest Branch Company of Oil Gas Field and Sichuan Petroleum Administration collaboratively organizes their first emergency response meeting.	11:30 PM, Dec. 23	1 hour and 26 minutes
12	Kai County government officials arrive at the site with responsible persons from relevant divisions and fire, police, and medical personnel.	11:50 PM, Dec. 23	1 hour and 46 minutes
13	The first batch of rescue workers from Chuandong Drilling and Exploring Company leaves for the well site.	0:00 AM, Dec. 24	1 hour and 56 minutes
14	Chuandong Drilling and Exploring Company and Sichuan Northeast Gas Division jointly makes an urgent request by fax for help from the Chongqing Municipal Government.	0:30 AM, Dec. 24	2 hours and 26 minutes
15	The first batch of rescue workers from Sichuan Petroleum Administration Southwest Branch Company of Oil Gas Field leaves for the well site.	0:40 AM, Dec. 24	2 hours and 36 minutes
16	Kai County Government establishes the search and rescue command group with the support of the drilling crew to organize the public evacuation and rescue operations.	3:00 AM, Dec. 24	4 hours and 56 minutes

(continued)

Table 15.1 Timing of Events Measured against Time from Loss of Control of Blowout (Continued)

No.	*Events During the Emergency Response*	*Time of Event*	*Time Elapsed Since Loss of Control of Blowout*
17	Sichuan Petroleum Administration reports to CNPC headquarters about the blowout situation.	5:25 AM, Dec. 24	7 hours and 21 minutes
18	Sichuan Petroleum Administration and Southwest Branch Company of Oil Gas Field jointly holds the second emergency response conference.	8:20 AM, Dec. 24	10 hours and 16 minutes
19	The rescue team from Chuandong Drilling and Exploring Company of Sichuan Petroleum Administration arrives at Zhengba Town.	9:30 AM, Dec. 24	11 hours and 26 minutes
20	The rescue team from Chuandong Drilling and Exploring Company of Sichuan Petroleum Administration arrives at Gaoqiao Town.	10:30 AM, Dec 24	12 hours and 26 minutes
21	The rescue team from Chuandong Drilling and Exploring Company of Sichuan Petroleum Administration receives instructions to ignite the blowout well.	11:20 AM, Dec. 24	13 hours and 16 minutes
22	A temporary field rescue team composed of officials from Chuandong Drilling and Exploring Company of Sichuan Petroleum Administration, Kai County Government, and the drilling crew is established at Gaoqiao Town to prepare for the wellhead ignition.	11:30 AM, Dec. 24	13 hours and 26 minutes
23	The temporary field rescue team confirms that the blowout at the wellhead has stopped, while the blowout at the outlet of No. 3 blooie pipe continues.	2:00 PM, Dec. 24	15 hours and 56 minutes

24	Sichuan Petroleum Administration and Southwest Branch Company of Oil Gas Field establishes a formal field rescue team at Gaoqiao Town.	2:00 PM, Dec. 24	15 hours and 56 minutes
25	The leading team director dispatches rescue staff to enter the well site to investigate the situations of the wellhead and the blooie pipe for the first time and to select a safe ignition location.	2:20 PM, Dec. 24	16 hours and 16 minutes
26	The team director dispatches the rescue staff to check the wellhead and the blooie pipes for the second time and, finally, specifies the ignition location.	3:00 PM, Dec. 24	16 hours and 56 minutes
27	The field rescue staff urgently opens a nearby shop with a pry bar and finds some magic bullets.	3:00 PM, Dec. 24	16 hours and 56 minutes
28	The leading team director dispatches two groups of rescue staff to carry out the ignition at a place about 50 m away from the blowout outlet, but the attempt fails.	3:20 PM, Dec. 24	17 hours and 16 minutes
29	The field rescue staff makes the second attempt at ignition with the fireworks found in the shop and finally succeeds.	3:55 PM, Dec. 24	17 hours and 51 minutes
30	Wu Jianong, the deputy mayor of Chongqing Municipal Government arrives at the field and organizes the rescue headquarters.	9:00 PM, Dec. 24	22 hours and 56 minutes
31	Officials from CNPC headquarters arrives at the site and the rescue headquarters is readjusted with members from the local government, CNPC headquarters, Sichuan Petroleum Administration and Southwest Branch Company of Oil Gas Field.	10:30 PM, Dec. 24	24 hours and 16 minutes

(*continued*)

Table 15.1 Timing of Events Measured against Time from Loss of Control of Blowout (Continued)

No.	*Events During the Emergency Response*	*Time of Event*	*Time Elapsed Since Loss of Control of Blowout*
32	The rescue headquarters organizes a group of 460 rescue staff to form 20 rescue teams who carry out the large area search and rescue and find many dead bodies.	1:00 AM, Dec. 25	26 hours and 56 minutes
	Officials from SAWS arrive at the site.	9:00 PM, Dec. 25	46 hours and 56 minutes
33	Officials from the State Council arrive at Kai County with the State Council work team and a new rescue headquarters is established.	1:00 AM, Dec. 26	50 hours and 56 minutes
34	The rescue command headquarters decides to fully clear the site, establishes the security line 5 km away from the wellhead, and sends 82 rescue teams to search for and rescue households within the region one by one.	1:00 AM, Dec. 26	50 hours and 56 minutes
35	The well-killing formally starts.	9:36 AM, Dec. 27	83 hours and 32 minutes
36	The well-killing is successfully completed.	11:00 AM, Dec. 27	84 hours and 56 minutes

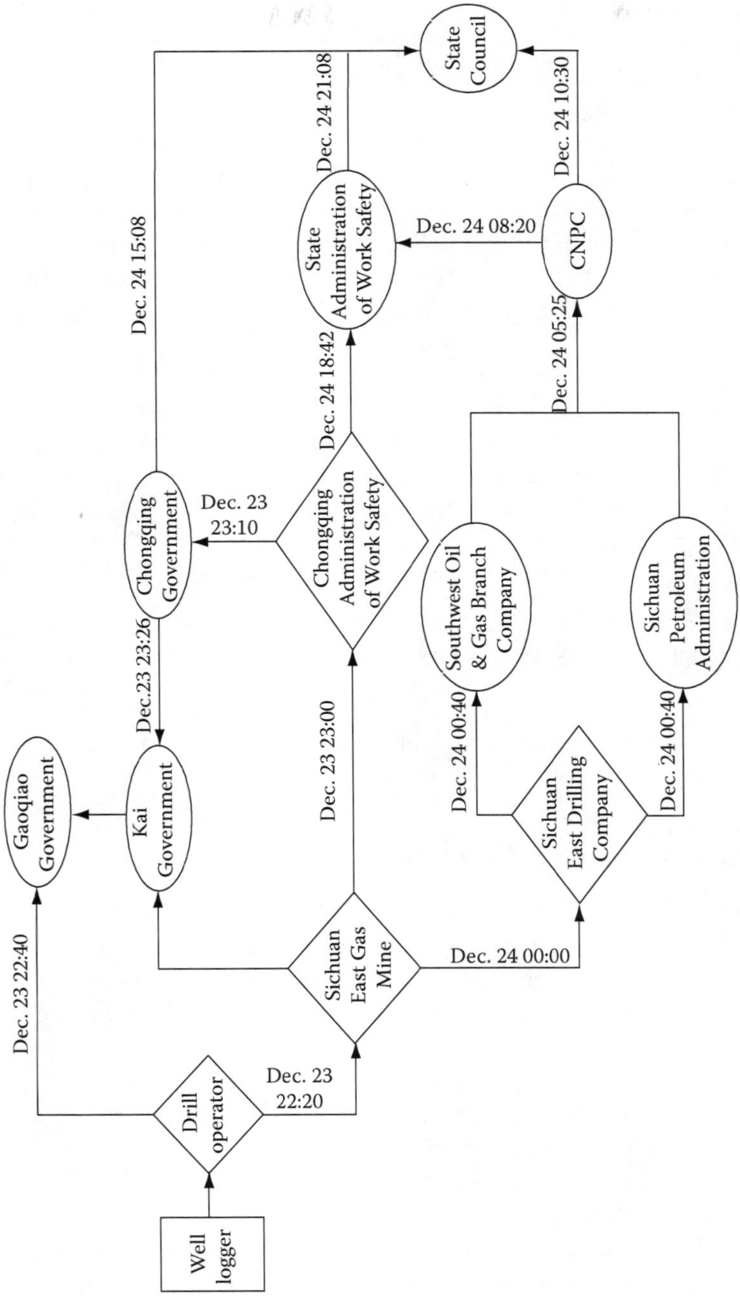

Figure 15.5 Information flow chart of 12/23 blowout accident.

3. Settlements and postaccident circumstances team's (led by the Chongqing Municipal Government and CNPC headquarters, the Ministry of Civil Affairs, the Ministry of Labor and Social Security, the State-Owned Assets Supervision and the Administration Commission of the State Council) responsibilities included evacuation organization, settlement of victims, and dealing with postaccident casualties.

On December 26, to support the well-killing work, the rescue headquarters decided to comprehensively clean up the region around the well site. A security line was set up 5 km away from the wellhead. Headquarters sent out 82 rescue teams to go door-to-door within the region to see whether there was anyone who needed aid. Finally, approximately 900 individuals originally detained in the dangerous area were evacuated to safety. During this period, the detailed well-killing scheme and the corresponding emergency response plans had been studied and formulated, and the well-killing equipment and materials had been transported to the well site. At 9:36 AM on December 27, the well-killing formally began, and at approximately 11:00 AM, the well-killing was successfully finished. This series of events is briefly listed in Table 15.1, in addition to the sequence of events that occurred, which is depicted in Figure 15.5.

Postaccident Recovery and Reconstruction

Since the accident, largely owing to the diligent work of the governments at all levels, the recovery and reconstruction of the affected area has proceeded quickly and smoothly.

Return of the Evacuated Victims

After the well-killing was successfully completed, the focus of the overall emergency response transferred from the organizing of a large scale evacuation and settlement of the victims to returning victims to their homes. The rescue headquarters took the following measures to ensure the environment was safe for people to return [CASST 2006, 2007]:

1. Organized professionals to clean up dead animals in the affected area, which were later deeply buried or incinerated.
2. Organized sanitary and epidemic prevention teams to disinfect the area and to prevent the development of epidemic diseases.
3. Assigned the environment protection department to establish six mobile monitoring stations to monitor the affected area's environment around the clock.
4. Assigned health divisions to perform random tests of grain, vegetables, meat, and other foods.

5. Printed and issued the Notification to the Victims Returned, which clarified the detailed precautions for returning victims to follow.

With all the above-mentioned measures fully implemented, officials finally acknowledged that air, surface water, and food standards had been met to allow victims to return in two stages. Stage I involved organizing those people that had originally lived more than 5 km away from the wellhead. On December 28, the rescue headquarters dispatched 125 buses and 600 workers to help 26,000 victims return to their homes (including 14,000 individuals from the government-designated refugee stations). In Stage II, the remaining victims that originally lived within 5 km of the wellhead were allowed to return to their homes; this occurred on December 29, as 486 buses and 760 workers helped 39,000 victims return home. According to the statistics of the Chongqing Municipal Government, 65,632 individuals returned to their homes: 38,923 individuals returned with help from the government and by December 30, 26,709 individuals returned by themselves. During the returning process, no one was injured, no one was lost, and no one died.

Postaccident Restoration

After the victims returned to their hometowns, the local government took the following measures to stabilize the victims' morale and reorganize the residents to restore production and deal with mortality compensation [Chen et al. 2007; Zhang et al. 2007a; Zhang et al. 2007b]:

1. Accelerated the after-accident recovery work. By January 5, 2004, all remains of the 243 dead were cremated and buried.
2. Paid close attention to medical and health care. In total, 1,076 medical individuals in 30 patrol medical teams from 19 hospitals actively participated in rescuing activities, and the medications used in these activities cost 3.10 million yuan RMB.
3. Made good use of rescuing funds and materials. The local government established a strict rescuing fund and materials management and usage regulations. They placed the rescuing fund into a special financial account, and the rescuing materials were managed by the civil department.
4. Strengthened the security and safeguard work. The local government dispatched more than 2,000 policemen, formed 8 patrol security teams, and set up 54 alarm stations.

Loss Estimation and Compensation

To ensure the appropriate loss estimation and compensation, the Kai County government performed the following work according to the actual situations.

Compensation for the Victims

According to the relevant national regulations and the features of this accident, the local government formulated the compensation plans respectively for death, injuries, and victims. The compensation expense was 35,854,997 yuan RMB, in which, the compensation for the death was 34,353,130 yuan RMB, and the child care expense was 811,363.6 yuan RMB.

Compensation of Properties

Authorities organized 273 cadres to form 24 work teams to contact 24 villages, 316 groups, and 8,407 families in the stricken area and comprehensively investigated and carefully rechecked the losses of properties. The compensation of properties was performed with all the relevant cadres, bank staff, and policemen in witness. In total, 4,298 peasant households were compensated.

Postaccident Reconstruction

Local Economic Construction

"Huge rescue funds accelerate the development of Gaoqiao Town by at least 10 years. In the past, the fund appropriated to Gaoqiao Town from the upper governments each year was only several hundred thousand yuan RMB," an official from Gaoqiao Town government said [Shen 2005, p. 23]. By October of 2004, Kai County had received 188.7451 million yuan RMB from the special rescuing fund and social donation that was partially used for losses of life and property in addition to post-disaster reconstruction projects.

The road from Gaoqiao Town to Zhengba and Kai County urban areas was formerly a 15-km uneven mountainous road. Due to years of rolling by overloaded coal trucks, the road surface had been seriously damaged. On the eve of the accident, thousands of local residents fled along this road, which has since been thoroughly reconstructed.

With fast development of the local economy, many families owned landline telephones, mobile telephones, motorcycles, cars, and even new buildings. All of the households now use tap water.* Owing to the high economic expectations accompanying the gas well development, the villagers tacitly accepted the existence of the baneful wells. When the CNPC drilling crews left, the local government organized some villagers to see them off. After 2 months, CNPC

* In many rural areas of China, there is currently no tap water. The villagers take water from a public water well for daily use.

gas recovery teams came. Before their cookhouse squad reached, all members had their meals with the villagers' families. The shadow of the disaster seems to disappear completely.

Safety Standards and Codes

After this accident, some scientific research institutions including CASST, with support from SAWS and relevant state-owned large-scale enterprises, performed the studying and compiling of professional safety standards for the development of gas wells with high sulfur content. Until now, two professional safety standards have been issued: Public Risk Classification Method of Natural Gas Well Containing H_2S [SAWS 2008a] and Public Safety Protection Distance Requirements of Natural Gas Well Containing H_2S [SAWS 2008b]. Four rules will be issued soon: Ignition Time Requirements for the Wellhead of Natural Gas Well Containing H_2S with Blowout Out of Control [SAWS 2008c], Emergency Planning Zone Calculation Method of Natural Gas Well Containing H_2S, Technical Guide of Quantitative Risk Analysis of Natural Gas Well Containing H_2S, and Public Safety Assessment Rules of Natural Gas Well Containing H_2S [CASST 2008]. All of the standards are aimed to control such blowout risks from the very beginning.

Main Lessons from This Accident

In 2001, as a government measure to strengthen work safety management, the State Council issued Regulation of State Council about Administrative Responsibility Investigation in Especially Serious Safety Accident [Central People's Government of the People's Republic of China (CPGPRC) 2001]. In 2002, the Law of The People's Republic of China on Work Safety [National People's Congress 2002] was issued, which formally established the responsibility investigation system for work safety accidents. When this accident happened, the SARS outbreak had occurred and China's governments had begun to pay more attention to emergency management. At that time, a complete set of emergency rescuing systems were not yet established in China, and risk assessment, emergency planning, and emergency exercises were not emphasized. In fact, the national emergency planning system had not been formally established until the State Council issued the National Response Plan for Public Emergencies [CPGPRC 2005] on January 8, 2006. Moreover, the Law of the People's Republic of China on Emergency Response [CPGPRC 2007a], which was issued on August 30, 2007, marked a brand new phase in emergency management throughout China. In the following, we analyze the lessons from this technical disaster for enterprise safety management, safety technical equipment and professional standards, and accident emergency response management [CASST 2006, 2007]. However, we should not forget the overall national situations of emergency management in China at that time; some problems, such as incomplete emergency planning, were not uncommon.

Enterprise's Safety Management

HSE Management System

Before this accident, China's petroleum enterprises had adopted the health, safety, and environment management (HSE) system. The HSE system stressed risk management, especially comprehensive management for serious risk events to avoid the occurrence of serious accidents. However, the 12/23 accident served as an illustrative example of how the petroleum enterprises had not comprehensively implemented the risk management measures during their operation. As for risk source identification, risk assessment, risk control, emergency response management, operation control, education and training, communication, and performance monitoring and measurement, the HSE management was not implemented. The Luojia 16H well-drilling crew established the HSE system and kept the necessary system documents and records. However, during the accident, the HSE system was not fully exerted; the HSE system was not actually combined with the drilling crew's daily operations.

Safety Management System and Supervision

The enterprise's safety management system was incomplete and safety supervision was not sufficient. As observed from the investigation, the safety staff of the drilling crew was changed from full time to part time. The safety supervision responsibility of the drilling supervisor was clearly specified in the drill supervision rules but was not fulfilled. For example, the HSE records were incomplete—gas masks and oxygen respirators were not found in management records and the drilling supervisor gave tacit consent for the removal of the back-pressure valve.

Safety Qualification of the Staff in the Relevant Enterprises

The field workers' violation of the rules was the direct cause of the accident. The continuous incorrect performance of their duty was also the fuse of the blowout accident. The directional well service center engineer did not strictly follow the rules and removed the back-pressure valve; the safety staff of the drilling crew, although knowing that such a decision violated the rules, gave no dissent when commanded to remove the back-pressure valve. The director of the drilling crew, after learning this fact, took no relevant measures to immediately correct the situation, which was the cause of this blowout. After the accident happened, the field rescue commander did not give the ignition order in time, allowing the gas to spread continuously, which greatly increased the losses. Eventually, a series of mal-operations linked together to cause this accident and other serious consequences.

Enterprise's Awareness of Safety Issues during the Exploitation

The petroleum and natural gas mining industry is risky; the safety protection is very important, especially for the exploitation of natural gas with high sulfur content. The safety and environment protection assessment must be performed in advance. As for the Luojia 16H well, there should be no permanent dwellings within 1 km of the natural gas well. However, the nearest peasant household was less than 50 m away from the derrick of the Luojia 16H well, and most of the 2,419 villagers of Xiaoyang and Gaowang of Gaoqiao Town lived within 1 km of the well site.

Safety Education for the Residents around the Well Site

The public safety consciousness and capability are important aspects for reducing the casualties from any serious accident. The well-drilling operation is highly risky. However, the enterprise neglected routine public education; therefore, the public around the well site had no knowledge about the potential health risks and did not know about the available evacuation routes, gathering places, alarm meanings, and emergency rescuing requirement, which all resulted in inactive and inefficient evacuations. Most residents and some village cadres did not even know that there was a possibility of the release of poisonous gas from natural gas exploitation in addition to several other important facts. Residents did not know the seriousness of inhaling H_2S, which resulted in some of the residents not wanting to evacuate during the cold night when the event occurred. In all, approximately 40 individuals who were evacuated from Xiaoyang Village returned to fetch some valuable articles and were poisoned to death because of their lack of information about the effects of the gas.

Safety Equipment and Applicable Standards

Safety Planning

The natural gas from the Luojia 16H well had high H_2S content and a high risk of blowout. Despite these facts, safety planning was badly conducted when choosing the well site. Many important factors (e.g., the distance between the well site location and nearby households, the evacuation safety of residents around the well site, and the emergency escape protocol for the workers) were neglected, which caused the large-scale pollution of H_2S in the residential area when the blowout got out of control. Lack of safety planning caused the following problems for the Luojia 16H well site: no emergency planning zone prescribed before this accident; lack of knowledge about environment and population effects that would be suffered in the event of a blowout accident; a large population living within 500 m of the well site; and no proper evacuation routes for residents around the well site.

Safety Assessment

Lack of risk assessment for a blowout hazard, safety preassessment for project construction, safety acceptance assessment, and periodic operation safety assessment all caused the following problems: no available results of hazard identification and risk analysis; no available blowout risk indexes and classification methods for blowout hazards; no feasibility and applicability assessment on the preventive measures for well kick, blowout, and out of control blowout; and no systematic assessment and analysis for staff-training effects and the staff's abilities to fulfill their responsibilities.

Ignition of the Wellhead in Out of Control Blowout

For a gas well with H_2S under an out of control blowout, the ignition is usually specified as the important means for emergency rescue. The ignition of natural gas containing H_2S may rapidly convert hypertoxic hydrogen sulfide into less toxic sulfur dioxide (SO_2). The burning of natural gas generates a large amount of heat, which may markedly raise the temperature of the mixture of combustibles and air. Consequently, the mixture, under the thermal floating force, moves away from the ground surface, reducing the accumulation of toxic gases on the ground surface and reducing the hazard to human beings.

When this blowout accident happened, there were no regulations on how to ignite the wellhead, who should decide on an action plan, and who should ignite the wellhead in the Chinese petroleum and natural gas industry. The accident emergency response staff had no guidance on rapid ignition of toxic gases, which lead to the delayed action. In addition, the blowout ignition techniques were outdated and there were no effective blowout ignition devices. Therefore, it is not surprising in this disaster that the rescue staff had to use fireworks to ignite the wellhead, which was a great risk.

Safety Technology and Standards for Exploitation of High Sulfur Content Natural Gas

There are no systematic safety standards or codes in the petroleum and natural gas mining industry in China. Currently, the petroleum and natural gas mining safety techniques may be based on general petroleum and natural gas industrial safety standards, and most of the drafters of these standards came from the petroleum and natural gas enterprises, which may limit the usefulness of these standards.

Accident Emergency Response Management

Emergency Resources

The emergency resources are important guarantees for emergency response and rescue work. The 12/23 blowout accident reflected the shortage of emergency resources preparation: air respirators, H_2S concentration monitors, and well-killing materials were not abundant at the well site; the drilling crew had no blowout ignition device; and the emergency response teams did not have enough personal protection equipment and emergency monitoring devices, which seriously affected the emergency response and rescue work, leading to aggravated consequences.

Emergency Plan

Prior to drilling the Luojia 16H well, the Sichuan Petroleum Administration and its subordinates had not made detailed emergency response plans, such as H_2S poisoning prevention measures for nearby residents.

Emergency Exercises

The emergency exercise is a comprehensive check for emergency response capability, and it is helpful in improving and completing the emergency response plans, improving the emergency response coordination, and ensuring efficient emergency rescue actions. Lack of exercise resulted in the failure to identify possible deficiencies in emergency planning and emergency materials preparation and also in the lack of coordination among various emergency response divisions, organizations, and personnel.

Emergency Alarming Capability

No alarm equipment was onsite in advance, which resulted in difficulty alerting the residents quickly after the blowout and, consequently, inefficient evacuation. During the blowout accident, the drilling crew gave the blowout alarm; however, the meaning of the alarm was only understood among drilling workers—the nearby residents had no understanding of it. Moreover, due to the limited coverage of the alarm, the alarm was not loud enough to rouse the residents from sleep. The workers of the drilling crew and other responsible persons had to go door-to-door to notify the residents to leave. Some victims later said: "if the alarm was sent with a tweeter, the casualties may have been much less."

Mass Evacuation

In an emergency response to an accident, evacuation is key to reducing casualties. Therefore, evacuation preparation, prescribed evacuation areas, routes, transportation tools, and temporary sheltering places should be specified in detail, taking into account the number of people, duration, and wind direction, in addition to other factors. The needs of special groups, such as those who are disabled, should also be considered.

During the Luojia 16H well accident, the emergency evacuation involved tens of thousands of residents. Unfortunately, due to lack of safety education from the enterprise, the residents were completely unaware of the sudden disaster, the possible evacuation routes, the hazards of H_2S, and the necessity of self-protection and rescue measures, which made the evacuation inefficient.

Coordination between CNPC Enterprise and Local Government

The CNPC enterprise had no formal coordination channel with local governmental emergency organizations. Its emergency plans had no link with the local government's emergency work, and both the CNPC enterprise and the local government lacked communication and coordination. After the accident, the drilling company did not give a report to the Kai County government, but rather decided to bureaucratically report to Sichuan Petroleum Administration, then to the Chongqing Administration of Work Safety, and then to the Chongqing Municipal Government, who finally notified the Kai County Government. When the Kai County government received the accident report at approximately 11:25 PM, 1.5 hours had elapsed since the blowout. The Kai County Government then notified the governments of Gaoqiao Town, Zhengba Town, Maliu Town, and Tianhe Town to aid in the emergency response. The delayed report from the drilling crew to the local government did not leave enough time for the local government to inform the residents and organize them to evacuate in time.

Accident Response Command

In the different accident response phases, the emergency response command system was not organized in a modular pattern, which caused a lack of flexibility and reduced the efficiency of the whole emergency response system. In the early period of accident emergency response, the local government and the petroleum enterprise had not established an integrated communication and command system, which reduced the efficiency of the whole emergency response system. For example, on the morning of December 24, the petroleum enterprise and the Chongqing Municipal Government established their own command systems that were insufficient for dispatching and using all of the available emergency resources and reduced the emergency response efficiency.

Social and Cultural Consideration on This Disaster

Enterprise Culture and Social Atmosphere

On September 4, 2004, the No. 2 People's Secondary Court of Chongqing sentenced six directly responsible individuals, deeming that the direct cause of the disaster was that no measures were adopted for igniting the blooie pipe, which caused a huge natural gas blowout and spreading gas with a high H_2S concentration. It was determined that Wu Hua, the chief engineer of Chuandong Drilling and Exploring Company and the field chief commander, gave incorrect orders after the blowout became out of control and was thus sentenced to 4 years of fixed-term imprisonment. Wu Hua said that he was afraid that the explosion of the well site might burn the expensive drilling equipment and destroy the well in which millions of Yuan RMB had been invested [Li 2004].

Since the 1960s, the "iron man spirit" of Daqing oil field, which emphasizes the importance of national properties, has always had great influences on China's enterprise culture. Consequently, the petroleum industry has not established a clear position on the priority between human beings and property. It is popular to make the protection of national property a priority and the protection of lives second to it. Although this concept cannot be found in any written manual or document, such a mindset has been conveyed to each staff member. According to this mindset, at the critical moment, Wu Hua cannot make the ignition decision. Seen from this viewpoint, the tragedy was not caused by Wu Hua, but rather by the safety culture of the entire petroleum industry (perhaps even the entire society).

Fortunately, in November of 2003, the Sixteenth Central Committee Conference of the Chinese Communist Party clearly specified the concept of "human-based service" [Central Committee of Communist Party China 2003], which is the belief that people's lives always come first. The conceptual change at the national level will lead to changes in the safety culture for all enterprises.

Responsibility and Accident Investigation

After a technological disaster, the government performs investigations to determine the cause of the accident and, if necessary, the responsible individuals that should be punished. Since the foundation of New China in 1949, the central government has issued five documents on the investigation of accident causes and relevant responsibilities; these are, respectively, *Report and Investigation Rules of Accidents with Staff Casualties* in 1956, *Investigation Procedures (Trial) on Especially Serious Accidents* in 1989, *Reporting and Handling Regulations of Accidents with Enterprise Staff Casualties* in 1991, *Regulations on Administrative Responsibility for Especially Serious Safety Accidents* in 2001, and *Reporting and Investigation Rules of Work Safety Accidents* [CPGPRC 2007b] in 2007. The current *Report and Investigation Rules of Work Safety Accidents* is mainly based on the concept of "four not passed." This comprises the following statements: (1) the investigation should not be terminated if the accident causes are not identified; (2) the investigation should not be terminated if the responsible individuals are not strictly punished; (3) the investigation should not be terminated if all of the responsible individuals are not well educated; and (4) the investigation should not be terminated if preventive measures are not implemented. The second of the four statements concerns the responsibility investigation. The establishment and reinforcement of the responsibility investigation system were originally aimed at making all of the officers and workers pay due attention to work safety; unfortunately, in the practical investigation process, the focus is usually (wrongly) placed on what human factors have contributed to the occurrence of the accident, who could be accused, and how the responsible individuals should be punished. Appropriate attention is not paid to the identification and assessment of the objective causes, and on prevention of similar accidents in the future from a technical viewpoint. During this accident, along with the six individuals from CNPC who were found to be directly responsible by the investigation, Ma Fucai—the chairman of CNPC—also took responsibility for the accident and resigned in April, 2004. Currently, although an expert team is usually established for accident investigations, the expert team may only provide technical evidence to clarify the causes and assign responsibility for the investigated accident. After the accident investigation is completed, the investigation team will usually submit a report to the upper level government, which usually provides some suggestions about accident prevention. However, such suggestions typically only include how to strengthen safety management and education; detailed technical prevention countermeasures are not provided. Fortunately, CASST, CCDC, and similar institutions have already tried to make the subsequent technical field survey with the goal of providing technical suggestions for risk prevention and control of similar accidents. The importance of such technical surveys has been gradually acknowledged by the public, and CASST has made similar technical surveys for three especially serious technical disasters since 2004 [Deng and Jiang 2009].

Several Viewpoints on Reconstruction

Social Economy

Judged from a social economy standpoint, the recovery and reconstruction work after the 12/23 accident was successful. As mentioned before, the huge compensation and relief funds accelerated the development of Gaoqiao Town by at least 10 years.

Social Paradigm

However, the positive economic compensation and relief from the government caused a moral crisis among the local villagers [Song 2004]. The family of one deceased victim was compensated over 140,000 yuan RMB. In the villages of Xiaoyang and Gaowang, many families lost several members. With the compensation for the damaged properties, some families finally received over 1 million yuan RMB in compensation. Since all the villages are located in a poor rural area, per capita annual income at that that time was only about 1,600 yuan RMB and many villagers had never imagined that one day they would have or have the chance to have so much money. As a result, the villagers faced an unprecedented moral crisis. Some villagers that received large amounts of compensation funds began to waste the money and some others that received less compensation tried to get more. Some villagers began gambling; in one case, a young man gambled and lost several hundred thousand yuan RMB, the compensation for the death of his parents and brother, in one night. After this disaster, approximately 30 children were orphaned and their relatives quarreled over their supporting rights. Even worse, one woman murdered her already paralyzed son with rat poison in order to report her son as a victim of the disaster and obtain more compensation funds. It was not uncommon for many villagers to overstate their losses. Although the local governmental departments adopted swift and effective measures to address the social problems, the moral paradigm associated with reconstruction after such catastrophes must be addressed.

Safety Consciousness and Safety Culture

Researchers from Chongqing Technology and Business University and other institutions used a questionnaire to conduct an investigation into the psychological health of victims one month after the disaster [Yuan and Pu 2008]. They found that, before the blowout, 96.4% of villagers of the region knew nothing or just a little about the potential H_2S hazards of a well blowout; after the blowout, although the safety consciousness was slightly improved, approximately 80% villagers still did not have effective knowledge about the hazards of H_2S. According to field investigation results of the CCDC [2007], the rapid economic compensation alleviated the people's psychological pains, which was beneficial to maintaining social stability, but ineffective for improving residents' safety consciousness. Many residents took it for granted that the government should have solved all potential problems and, therefore, there would be no life-threatening risk in the future, so they calmly allowed the CNPC to resume production [Bo 2004; Shen 2005]. To this day, although CNPC has performed the relevant public evacuation exercises, the villagers do not understand the risks. In fact, the villagers are only willing to take part in such activities because they will be paid by CNPC for their participation. Judging from appearances, such public participation does seem to be somewhat useful; for example, a similar but much smaller emergency happened in almost the same location in March of 2006 and more than 10,000 nearby residents were successfully evacuated in time [Xin 2006]. However, any real improvements in public safety consciousness are still a long way off.

Acknowledgments

The research work featured in this chapter was partially supported by the National Science and Technology Infrastructure Program (Project 2007BAK22B05) and the National Natural Science Foundation of China (Project 70833006 and 90924303).

References

Bo, J. 2004. Lives changed by blowout—revisiting Kai County, the place of "12.23" accident after half a year. *Western Time* 21:9.

Central Committee of Communist Party China. 2003. CCCPC decision about several problems of improving the socialism market economic system. Retrieved from http://www.people.com.cn/GB/shizheng/1024/2145119.htm (accessed October 14, 2003).

Central People's Government of the People's Republic of China. 2001. The regulation of State Council about administrative responsibility investigation in especially serious safety accident. Retrieved from http://www.gov.cn/zwgk/2005-06/06/content_4277.htm (accessed April 21, 2001).

Central People's Government of the People's Republic of China. 2005. National response plan for public emergencies. Retrieved from http://www.gov.cn/yjgl/2005-08/07/content_21048.htm (accessed August 7, 2005).

Central People's Government of the People's Republic of China. 2007a. Law of the People's Republic of China on emergency response. Retrieved from http://www.gov.cn/flfg/2007-08/30/content_732593.htm (accessed August 30, 2007).

Central People's Government of the People's Republic of China. 2007b. Report and investigation rules of work safety accident. Retrieved from http://www.gov.cn/zwgk/2007-04/19/content_588577.htm (accessed March 28, 2007).

Chen, X., Zhang, H., Pan, Z., and Jia, Q. 2007. Health impact of serious gas blowout accident on residents in disaster areas of Kaixian county of Chongqing. *Modern Preventive Medicine* 34:1696–1698.

China Academy of Safety Science and Technology. 2006. Study on supervision mechanism of work safety in high risk oil and gas field development. Internal Research Report.

China Academy of Safety Science and Technology. 2007. Study and analysis report on 12.23 blowout accident case of CNPC Sichuan Northeast Gas Division. Internal Research Report.

China Academy of Safety Science and Technology. 2008. Research on safety planning standards for high risk gas field. Internal Research Report.

Chongqing Center of Disease Control. 2007. *Study on comprehensive influences and subsequent effects on the stricken area of 12.23 blowout accident in Kai County Chongqing.* Chongqing: Chongqing Center of Disaster Control.

Deng, Y., and Jiang, C. 2009. Field survey of the mass evacuation during "4.16" accident of Chongqing Tianyuan Chemical Factory. *Journal of Safety Science and Technology* 5(3):30–35.

Li, Y. 2004. Analysis of cultural reasons of Kai County especially serious accident of natural gas blowout. *Journal of Enterprises' Culture* 8–9:97–99.

National People's Congress. 2002. Law of the Peoples Republic of China on work safety. Retrieved from http://www.npc.gov.cn/wxzl/wxzl/2002-07/10/content_297300.htm (accessed June 29, 2002).

Shen, L. 2005. Review of Chongqing Kai County blowout accident: What we get from the loss of 243 lives. Magazine Oriental Outlook (online version) October 27, 2005, 11:15:31.

Song, L. 2004. Problems and countermeasures for citizens' moral construction: Considerations on citizens' moral construction caused by Kai County blowout accident. *Xiamen Communist Party School Journal* 17:46–48.

State Administration of Work Safety. 2008a. *China professional standard on work safety AQ2017-2008, the public risk classification method of natural gas well containing H_2S.* Beijing: Coal Industry Press.

State Administration of Work Safety. 2008b. *China professional standard on work safety AQ2018-2008, the public safety protection distance requirements of natural gas well containing H_2S.* Beijing: Coal Industry Press.

State Administration of Work Safety. 2008c. *China professional standard on work safety AQ2016-2008, ignition time requirements for the wellhead of natural gas well containing H_2S with blowout out of control.* Beijing: Coal Industry Press.

State Council Investigation Team. 2004. Investigation report on 12.23 especially serious blowout accident of CNPC Chuandong Drilling and Exploring Company. Internal Report.

Tang, C. 2004. Investigation of "12.23" especially serious accident of natural gas blowout in Kai County Chongqing. *Journal of Labor Protection* 6:24–26.

Xin, H. 2006. Two different results from two accidents. *China's Disaster Prevention and Control* 4:23.

Yuan, L., and Pu, C. 2008. Investigation of sufferers' mental health after the gas blowout accident on December 23. *Chinese Journal of Health Psychology* 12:321–324.

Zhang, H., Chen, X., Pan, Z., and Jia, Q. 2007a. Analysis of losses and social influence of serious gas blowout in Kai county of Chongqing. *Modern Preventive Medicine* 34:1845–1848.

Zhang, H., Chen, X., Xu, S., and Jia, Q. 2007b. The evaluation of environmental impact of serious gas blowout accident on residents in disaster areas of Kai county of Chongqing. *Modern Preventive Medicine* 34:1700–1701.

V

GLOBAL CHALLENGES AND NEXT STEPS FOR THE TWENTY-FIRST CENTURY

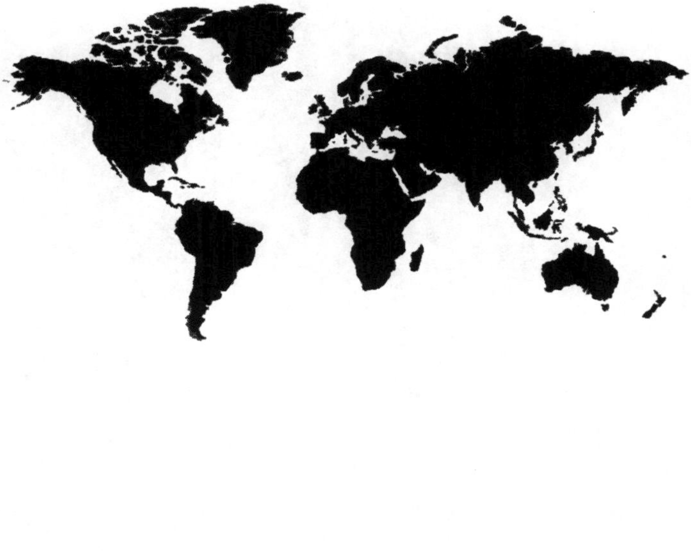

Chapter 16

Collaborative Leadership in Global Health Partnership Strategies of Canadian Research Universities

Anatoly Oleksiyenko

Contents

Background

Global health is the concern of a complex array of organizations, including national governments, international agencies, private foundations, pharmaceuticals, and nongovernmental and academic institutions, among other relevant actors [Cooper et al. 2007; Levin 2004; Merson et al. 2004]. With major infectious killers, such as HIV/AIDS, malaria, and tuberculosis, keeping the world in a state of alarm, scientific and policy communities increasingly confront the urgent need for faster, cheaper, and more accessible solutions [Curran et al. 2005]. This sense of urgency is supplemented by the complex nature of the required responses, which often necessitate better coordinated interinstitutional and international interactions. The recurrence of the diseases also reminds scientists and policymakers that global health challenges cannot be addressed with a one-time solution—they call for sustainable partnership relations across disciplinary, sectoral, and geographic borders.

Establishing effective international cooperation is often hard to achieve given that various political, economic, and cultural factors come into play to deter collaboration [Buse and Walt 2000; McKee et al. 2001]. Given the complexity of institutional networks involved in seeking global health solutions, the roles of some institutions remain vague. As part of traditional academic environments, medical schools experience numerous challenges in determining how global health fits within the traditional framework of health sciences and what makes the research-oriented schools meaningful agents in environments that compel immediate and urgent solutions [Oleksiyenko 2008]. Although the existing literature addresses some aspects of organized research in the medical field [Mallon 2006] and looks at the challenges in global health cooperation [Breman and LeDuc 2001; Buse and Walt 2000; Garrett 2007; McKee et al. 2001], little is known about what determines the level of collaborative capacity at academic institutions.

The shortage of investigations in this field is partially explained by the fact that key mediators of global health solutions do not perceive a clear role and set of responsibilities that can be delegated to academic institutions, which are seen as either dysfunctional and fragmented, "ambiguous and anarchical," or "loosely coupled and loosely coordinated" [Atkinson and Gilleland 2006, pp. 1–2]. Despite the fact that they manage a significant proportion of human resources for healthcare services, as well as provide drug-discovery solutions, most of these schools are latent players rather than leaders in international health decision-making regarding global health program priorities and resource distribution. With growing concern about a low return on research dollars invested in addressing global health challenges, donor communities are increasingly interested in finding organizational models that allow them a measure of control over the process of discovery [Foege et al. 2005; Ganem 2003; Oliver 2004].

Resource contexts evolve as important constructs affecting organizational change. Some researchers argue that adaptation to the environment requires interaction with other actors in the environment. Pfeffer and Salancik [1978] attributed such adaptation to the need for organizational survival, which often depends on the ability of organizations to pursue resources that are critical to their goals. While interacting with other actors in the environment, the organization usually tries to enhance its own autonomy, while at the same time restraining the autonomy of others. Adaptation to the environment depends on how the organization perceives the environment, how it responds to external demands from this environment, and how it tries to influence or manipulate the environment.

Previous research on the organizational strategies of academic institutions reveals significant challenges faced by the university in correlating organizational change with the needs of external environments [Mintzberg and Rose 2003; Oleksiyenko 2008]. Although some researchers provide evidence for small-size and medium-size universities, creating a more responsive, responsible, and relevant mechanism for interdisciplinary, intersectoral, and international research and education [Clark 1998], the evolution of structural flexibility at large research intensive universities, where medical schools are usually housed, is often fraught with numerous challenges [Clark 2001]. The size, scale, scope, and intensity of research have been major factors pulling universities apart. The emphasis on basic research also sustains the traditional paradigm of academic excellence, which is primarily measured in terms of publications and citation rates, collaborative work and partnerships are not regarded as cardinal determinants of excellent research. In the words of some observers: "papers with the largest citation impact do not always come from collaboration . . . interdisciplinary research is often published in journals with a lower impact factor" and "the impacts of interdisciplinary research often take longer to establish" [Rigby and Edler 2005, p. 786].

Although global health is viewed by many researchers as inherently international and collaborative due to the nature of the diseases and problems that the medical profession has evolved to address [Breman and LeDuc 2001; Oleksiyenko 2008], translation of research into practice is hard to achieve and requires effective interfaces of systemic, organizational, and cultural factors that can stimulate the application of scientific findings and interventions to complex epidemiological situations. As an area of applied research and problem-solving, global health often fits uneasily in traditional evaluation frameworks, complicated further by the fact that Canadian medical schools usually do not have special academic departments for global or international health. The interplay of individual, institutional, and policy-level leadership facilitated by the academic environment becomes essential to motivate the institutional mobilization of resources, the development of incentives, and priority-setting in this field [Taylor 2005]. Because researchers often do not have the right incentives and institutional support to engage in major global health efforts and because institutional authorities tend to place priority on the domestic health issues that garner political and economic support and meet the needs of research sponsors, the creation of interdisciplinary niches for global health often remains at the discretion of globally minded scientists [Sá and Oleksiyenko 2008].

As the higher education literature shows, despite the general aspiration for enhanced institutional selectivity and focus in pursuit of collaborative strategies [Taylor 2004], universities experience a significant gap between the top-level aspirational story and the grassroots level empirical story [Oleksiyenko 2008]. As a study of international partnerships at research intensive universities in Canada indicates, institutional planning aimed at organizational adaptation can encounter serious obstacles, dissent, and resistance, rendering strategic plans ineffectual unless they take into consideration that successful partnerships in complex research intensive systems are primarily built at the grassroots level where individual entrepreneurs champion both symmetric and asymmetric collaborative initiatives to advance socially and scientifically significant agendas. The development of institutional frameworks, inconsiderate of the grassroots level strategies, reduces the institutional capacities for adaptation and survival.

Numerous universities, stimulated in various ways by their internal and external stakeholders, keep undertaking organizational efforts to streamline the academic culture in correlation with the contextual changes requiring collaborative work. For example, the February, 2008 issue of *Academic Medicine* describes a number of partnership models that American universities have created to advance institutionalization of global health. The increasingly significant role of partnerships in the area of global health puts emphasis on the capacity of universities to accommodate intrainstitutional collaborative leadership [Beaglehole et al. 2004]. Although the literature discusses various aspects of joint-vision setting, shared values, and team-building to encourage partnership synergies among a range of organizations [Alexander et al. 2003; Lasker and Weiss 2003; Lasker et al. 2001; Umble et al. 2005], little is known about the organizational dynamics of collaboration between the institutional and the grassroots level decision-makers at universities that pursue interdisciplinary, intersectoral, and international synergies. Some researchers point to a need for more research in the field to enrich the analytical and policy-building capacities of schools that aim to remove the organizational impediments to partnerships and to enhance their organizational responsiveness, social responsibility, and professional relevance [Atkinson and Gilleland 2006; Cook et al. 2006; Health Canada 2001].

This study provides some insights into the role of collaborative leadership in the evolution of institutional flexibility and explores the interplay of individual, institutional, and systemic factors that have an impact on institutional capacity for building global health partnership strategies. The proposed findings form a framework for assessing institutional collaborative capacity in global health programs.

Study

The study focuses on the institutional partnership capacities and global health strategies of Canadian research intensive medical schools and is informed by previous research in the area of collaborative strategies at Canadian universities [Oleksiyenko 2008; Sá and Oleksiyenko 2008], which found that the schools exhibited a strong aspiration for international leadership, as well as enhanced collaboration to address global health problems. The study also took into consideration specific contextual attributes of international research in Canada, including scarce funding, indecisive leadership at the level of the superstructure, and heavy commitment by medical schools to the local and national agenda [Jones and Oleksiyenko 2008]. The study explores how institutional flexibility evolves under the influence of individual academic leaders who aspire to successful intrainstitutional and interinstitutional partnerships for global health research and education. Contextually driven principal-agent relations were taken into consideration and the research dependency theoretical framework guided the analytical process [Pfeffer and Salancik 1978].

This chapter is built on analysis of data emerging from 45 semistructured interviews with research center directors, leading scientists, and senior managers from three medical schools in three Canadian provinces. The participants were asked questions related to the role of individual leadership, as well as the influence that funding agencies, institutional plans, and organizational forms play in the overall academic strategy for global health. The data was augmented by content analysis of the schools' annual academic plans, strategic plans, electronic databases, institutional and individual research profiles reflecting a commitment to and research interests in global or international health, crossborder partnerships, collaborative experiences involving international stakeholders, comparative health research projects and publications, and other printed and electronic documentation. Standard qualitative data analysis techniques were used to code and process received data [Miles and Huberman 1994].

Findings

The Canadian medical faculty's response to global health challenges is characterized by a substantial degree of creativity and commitment by individual researchers who, despite institutional incongruities and indecisiveness, take on a leadership role in developing intrainstitutional and interinstitutional collaborations to be active players in global health programming. Institutional strategies make broad claims about international leadership and the importance of addressing global health challenges—such vision statements often belong to the realm of aspiration, which has tenuous connections with practices on the ground. Given the fact that the institutional perspectives are developed within extradepartmental structures, international health initiatives are often eclectic and opportunistic formations of initiatives promoted by individual faculty members in response to funding opportunities propelled by institutional authorities or external donors.

In general, the interplay of institutional and individual leadership is dialectical. On the one hand, global health strategies are driven by a number of forces that reduce the impact of institutional leadership and increase the influence of individual leadership, including a traditional emphasis on individual performance, autonomy, and academic freedom as a platform for successful basic research, decentralization of decision-making within the intricate and extensive academic medical centers, and inadequate diversification of funding sources. On the other hand, there are several forces that push for an enhancement of institutional leadership and somewhat scale down the role of the individual leader. Such forces include the increasing internal and external competition for limited resources, the growing significance of large-scale grants that require effective institutional accommodation and support for individual efforts, the increasing cost of the academic medical centers' infrastructure, and the enhanced role of risk mitigation strategies and ethics compliance, which requires universities to have increasingly efficient legal and regulatory mechanisms. Although these forces often contradict each other and may be viewed as developing competing streams, in the end funding and resource prioritization emerge as key challenges requiring synergetic actions from Canadian medical schools.

The resource contexts in global health research in Canada propel the prevalence or dynamic interchange of one of the following six collaborative leadership structures for global health programs: (1) entrepreneur-based; (2) network-based; (3) team-based; (4) Organized Research Unit (ORU)-based; (5) university-based, and (6) multi-institution-based. The subsequent paragraphs provide brief analysis of each leadership structure.

Entrepreneur-Based Leadership Structure

Many respondents in this study noted that doing global health research in Canada requires a substantial entrepreneurial effort given an environment in which financial support and institutional political endorsement for collaborative projects are hard to secure. With government focused on local and national priorities and without a broad philanthropic network dedicated to global health funding in the country, there is a limited resource basis to support Canadian researchers' intensive collaboration with stakeholders in other countries. Under these circumstances, individual researchers have had to take full responsibility and employ their creative talents to engage governments or industries at home and abroad to support cross-border partnerships. The individual scientists' narratives refer to significant efforts devoted to the conceptualization and implementation of global research initiatives at international agencies, pharmaceutical industry laboratories, nongovernmental organizations, and cultural or religious organizations. Reciprocally, these organizations exert influence on how scientists create bridges for the crossover and interplay of disciplinary, sectoral, and regional expertise.

The entrepreneur-based structure is characterized by a prevalence of charismatic leadership exerted by established scientists pursuing financial autonomy and independent decision-making with regard to their initiatives and partnership choices. These leaders take full responsibility for generating resources and accounting for them directly to their sponsors. In most cases, the choice of partners is driven primarily by trusting personal relationships. The institutional authorities would provide little or no institutional guidance to coordinate or regulate individual initiatives, and neither would a researcher pursue any institutional responsibilities to implement his or her initiative. There is little intrainstitutional cohesion in the entrepreneur-based structure. According to one interviewee, researchers would usually come together to share their success stories; nobody would listen to one another and as a result no shared objectives or initiatives would evolve. Competition for internal as well as external resources is usually fierce.

The overall organizational picture of global health initiatives in such an environment is characterized by a diversity of organizational forms allowing for a variety of leader-driven programs, projects, and initiatives of a small- to medium-range scale and scope. Some argue that the institutional strategy of "observing from a distance" provides a significant amount of organizational flexibility and creativity, allowing for scientific discoveries and individual movement across the geographic, disciplinary, or sectoral borders. Indeed, the entrepreneur is an ultimate decision-maker regarding organizational roles he or she might be playing.

Participants in this study point out that, in cases where universities control and regulate boundary-crossing strategy development but do not provide political, administrative, or financial assistance, researchers often seek a nonuniversity home for their projects. This also tends to be the preferred choice when a researcher feels that it might be better to separate his or her academic and entrepreneurial responsibilities to avoid any conflicts with the traditional performance review systems. Overall, the academic work of such entrepreneurs would be guided by a regular set of institutional research policies and performance criteria, whereas their entrepreneurial engagement would be channeled through and measured by a flexible evaluation framework negotiated with a funding agency. Although initiatives that occur outside of traditional academic settings tend to be innovative and have a significant influence on the processes of drug discovery or treatment delivery, many researchers are cautious about creating too close of an engagement with industrial laboratories. This results from a prevailing notion that the latter tend to restrict freedom of publication and information-sharing. Consequently, many medical entrepreneurs prefer to maintain a distance from industry and some develop their own research enterprises to have more discretion in access to those sectors, but also to retain control over their own discovery and disclosure processes.

Network-Based Leadership Structure

This structure usually consists of a hectic flow of activities supported by individual researchers seeking an exchange of scientific information with their colleagues at home and abroad. The majority of such exchanges occur through disciplinary seminars and conferences, as well as through membership in disease-specific associations. With global health requiring an interdisciplinary approach, new networking arrangements constantly emerge to facilitate interdisciplinary, international, and intersectoral exchanges. The respondents in this study shared numerous stories of boundary-crossing networking spurred on by their personal interest in, and passion for, problem-solving in interdisciplinary or multidisciplinary settings. To galvanize a boundary-crossing discussion, a researcher would often create a separate niche, beginning with a round-table discussion that would be followed by seminars and conferences. Networks can also grow into multistakeholder development projects, which are described elsewhere [Health Canada 2001].

Although medical schools provide spaces for scientific and professional societies, networks, and networking centers, their support is usually nominal, piecemeal, and sporadic. Support to a charismatic faculty member, a "champion," would usually be provided by a charismatic dean or department chair on the basis of a variety of motivators, such as empathy for a noble cause, pressures from institutional stakeholders, recruitment objectives, personal political objectives, alumni relations, or investment plans. In most cases, the nature of such support would be experimental and would be built on the basis of seed funds allocated by the dean's office. The strategy of experiments through seed funds provides limited institutional sustainability. The entrepreneurial pursuit of alternative sources of funding is essential. The continued existence of networks is primarily ensured by personal commitments of equally interested individuals across the sectors (i.e., academic, governmental, nongovernmental, or industrial). Person-to-person relationships are the basis of the network leadership structure.

Team-Based Leadership Structure

Scientific teams are often a result of network activities that stimulate cohesion of ideas and interests among researchers. The pursuit of a specific medium- to large-scale partnership initiative usually requires a collective effort that consists of sharing ideas and resources as well as responsibilities and risks. A competitive environment has proven to be an additional motivator for team-building. An intraorganizational team generally starts as an initiative brought to life by one or two entrepreneurs, who decide to respond to a call for proposals or an invitation from another team or a funding agency to implement a set of specific tasks within specific budgetary and time limits. Given that teams are developed by lead researchers who rely on trusting personal relationships to maximize the scope, scale, or effect of their research, these scientists tend to create a buzz throughout their network, which galvanizes individual responses and triggers the creation of mirror teams at other locations. Frequently, a bilateral or multilateral team is preceded by each group of entrepreneurial scholars building teams inside their institutions that can involve representatives from various departments in the same school, across academic divisions within the same university, or across Canada.

Usually the effectiveness of the team is determined by the cohesion of inputs and the extent of collaboration at the stage of coconceptualization, cowriting, and cosolicitation of political support for project proposals. The longevity of both intraorganizational and interorganizational teams is determined by the cost–benefit ratios of the initiative and the willingness of the team leaders to follow-up on successive steps to solicit and secure funds for subsequent stages of the project. Although most teams prefer to avoid any additional administrative responsibilities, some teams invest their overhead discretionary funds in establishing research centers that report to their home departments or independent nongovernmental organizations located within hospitals or other organized research units working across an academic medical center. Whether an organizational unit evolves or not, team leaders can enhance their chances of securing internal resources and political support by engaging in group pressure.

Although some team members see little value in engaging institutional authorities in their project development and implementation work, most team leaders acting from a medical school base consider institutional support to be crucial because project-related contracts within large-scale initiatives require institutional endorsement and legal guidance. Given the specific contractual reporting and financial accounting within large-scale global health initiatives, which usually require extra attention to results-based management, international transactions, and risk mitigation, medical schools have established separate units or positions to facilitate such interactions. In addition to effective management of contracts and funds, some funding agencies expect that institutions provide in-kind funding through support of administrative salaries and additional facilities. Although the institutional role becomes more pronounced, most medical schools exercise a strategy of engagement with caution to uphold management and financial responsibilities at the level of individual entrepreneurs and their teams, who are ultimately accountable to their funders.

Organized Research Unit-Based Leadership Structure

Organized research units (ORUs) are noted as offering substantial support to the entrepreneurial activities of individual scientists [Mallon 2006]. As networking meetings lead to the creation of teams, more formal institutional arrangements evolve and there is substantial evidence about the growth of research centers, international offices, and nongovernmental organizations [Sá and Oleksiyenko 2008]. Although the key initial decision factor with regard to this evolution is the creation of a more formal meeting venue to coincide with the burgeoning of a global health agenda within the institutional walls, some entrepreneurial faculty members have also noted that moving to a formal status would legitimize their niche, secure some start-up financial support, establish political leverages for resource distribution at the institutional level, and create a distinctive identity in relations with funding agencies.

In some cases, schools establish and support an entrepreneurial unit to generate revenue by responding to calls for proposals through competitive bids and offering custom-designed programs for ministries and health agencies in foreign countries. The latter are triggered by specific professional training demands and rarely by research and advocacy needs. Following revenue streams, some dean's offices placed a significant amount of authority over global health initiatives into the hands of departments responsible for continuing education. Canadian medical schools have not moved as far as some U.S. universities in placing the center of their university-based leadership structure in global health with the President's or Provost's office.

ORU-based leadership is usually characterized by a specific resource context in which university departments facilitate specific research interests of individual faculty members or a team of scientists to pursue problem-oriented research by securing funds from donor agencies. The terms of organized research units are usually dependent on creative talents, fundraising skills, and scientific leadership of their founders and directors. ORUs undertake various collaborative strategies: umbrella (i.e., to create synergies for cross-departmental initiatives); interdisciplinary niche (i.e., to focus on a specific disease or policy problems); or international partnerships (i.e., to engage ORUs in other countries in order to solicit additional financial and intellectual resources). The challenges of leadership relations in ORUs are described elsewhere [Mallon 2006; Sá and Oleksiyenko 2008]. Overall, it is the role of external resource solicitation that determines the dynamics of leadership relations with departments to which ORUs report.

University-Based Leadership Structure

The university-based structure begins to evolve when the number of teams and ORUs engaged in global health grows sufficiently to necessitate coordination to respond to an increasing demand for institutional resources, as well as to reduce internal competition and the related duplication of efforts and costs. Medical schools establish special administrative offices, cross-departmental centers, or to appoint faculty or administrative positions within a dean's office to coordinate resource planning and distribution of seed funds. Deans' offices generally take a proactive approach to contextual scanning, program budgeting, and interdepartmental coordination. Given that resources are usually limited and demands for institutional support are plentiful, the university-based leadership structure would commonly generate entrepreneurial approaches to compete for external resources. In some cases, conflicts and tensions between individual entrepreneurs and institutional entrepreneurial units were observed. To reduce such tensions, some deans' offices or their affiliated coordinating units undertook a portfolio-building strategy. Such strategies are generally market-driven and predisposed to selecting specific countries, scientific areas, and expert capacities that increase the school's competitive advantages. With global health challenges being primarily located in developing countries, where market conditions are rudimentary, revenue-generating university-based leadership structures receive a significant amount of criticism, primarily from a postcolonial school of thought.

In general, the university-based models are usually aspirations rather than effective operational structures. However, some critics argue that the increasing role of global health teams and ORUs within the university will lead over time to the improved position of Canadian research in global competition for research dollars invested by international funding agencies. The increasing influence of major donors in global health, such as the Bill and Melinda Gates Foundation, was noted as raising faculty members' and institutional authorities' awareness of the need to have the requisite capacities and a distinguished record in accommodating complex financial and administrative responsibilities, in addition to delivering excellent research, to be able to compete globally. The other factor that increases the potential growth of university-based leadership structures is the proportion of overhead money available from large scale grants, which allows for significant flexibility for infrastructure development at medical schools. Some also remark that large-scale grants allow for recruiting the best talents from overseas, as well as engaging a larger number of young scholars and students. This is argued to create the critical mass that can influence priority-setting at the highest echelons of the university to make global health research a strategic area of development at the university. The role of deans of medical schools, as well as provosts and presidents of universities, are viewed as particularly important for major decisions to be made.

Multi-Institution-Based Leadership Structure

With an increase in the scope and scale of research requiring engagement by stakeholders from various sectors or constituencies, there is a growing demand for interinstitutional partnerships. The locus of decision-making in an interinstitutional structure is often complicated. Tensions arise when the governance structures are poorly defined. Although funding agencies often play a cardinal role by indicating their preference for the locus of project leadership and logistical responsibilities in a multipartner project, relational balances established by the immediate leaders of the collaborating institutions are not necessarily translated effectively to the second, third, and additional tiers of project responsibilities. The multisectoral and multilevel relations become even more complicated when the projects have to deal with bureaucracies, financial transactions, and clinical studies ethics in the countries where political, economic, and cultural patterns of organizational behavior significantly differ from those in Canada.

Even within the same constituency, the relations are usually complicated. Examples of academic medical centers collaborating among themselves within a successful multi-institutional model are few. One of the collaborative projects in which all three investigated universities were involved was not able to start, despite existing funding support. The major challenges included the lack of trust and the lack of capacity in collaborative leadership at each university; a divergence in the interpretation of healthcare objectives, driven by incongruence of disciplinary strengths within each university; disagreements about the locus of the daily management responsibilities; a complicated governance arrangement for the project; and the complexity of accountability frameworks established by the governmental donor.

Given that medical research is subject to the prevailing framework of traditional performance indicators (e.g., publications, citations, and scientific prizes), which often marginalizes achievements in community-based problem-solving or industry-related product development, most successful large-scale multi-institution initiatives of this type are developed outside of the university. In one case, an extraneous and fully autonomous interinstitutional structure evolved in response to a requirement by a group of funders (a philanthropic research fund, a set of pharmaceutical industries, and governmental institutions located in three different countries) to have a nine-figure research budget managed through a consortium, bringing together three major research universities in their countries to deliver research through a milestone-driven and budget-dependent process. Any excesses in the multiple interpretation of measurement, resulting from a divergence of organizational cultures, were effectively mitigated by clearly defined quantifiable criteria. The leaders within the consortium also noted that they were mitigating challenges related to a shift to a curiosity-driven research framework through enterprise-like structuring and recruitment practices. In community-related projects, the location of multi-institutional relations was often moved to independent organizations, such as nongovernmental organizations and policy institutes. Such arrangements were viewed as creating significant opportunity for administrative efficiency and flexibility. However, further research would be required to explore the correlation between the partnership complexity, organizational flexibility, and the research performance.

Discussion and Conclusion

Individual researchers are key decision-makers influencing how organizational forms and strategies of global health partnerships evolve. However, there has been an increasing demand for universities to step up their commitment to and investment in the field of global health to match the leadership capacities of their individual stakeholders. Institutional capacity is primarily dependent on resource contexts within which discretionary decisions of institutional leaders evolve. The six leadership structures (entrepreneur-based, network-based, team-based, ORU-based, university-based, and multi-institution-based) comprise the most evident outcomes of institutional flexibility resulting from decisions driven by resource contexts. In most cases, an evolution from entrepreneur-based to institution-based leadership can be observed, provided the institution is able to secure and earmark sustainable funds. The development of more complicated structures of leadership, spearheading interinstitutional or multi-institutional partnerships, can occur when the medical school is prepared to prioritize global health and provide institutional support to complex collaboration by establishing institutional offices with significant funds and sustainable administrative and financial support. Within this study, there were few indications of any sustainable modifications taking place in Canada because limited diversification of funding sources supporting global health programs.

The progression to a more sophisticated institutional form of collaboration does not necessarily mean the eradication of less formal leadership structures. The schools studied in this project could be characterized as facilitating a combination of approaches rather than applying any single model exclusively. The leadership structures described above often coexist and compete among each other. Collaborative leadership evolves from dispersed to more cohesive patterns as the medical schools move from entrepreneur-based to university-based structures. Further analysis would be required to understand how the dominance of any particular structure of collaborative leadership or a proportional distribution of various approaches that might constitute a mixed leadership structure is correlated with the quality of global health research, education, and service delivery in a particular school.

Forces that propel individual powers (i.e., traditional performance evaluation, emphasis on basic research, decentralization of decision-making, and entrepreneurialism) and forces that stimulate the enhancement of institutional powers (i.e., internal competition, funding agencies' preference for large-scale grants, expensive medical infrastructure, and risk mitigation) are equally strong and any collaborative synergies require extraordinary leadership capacities at the individual and institutional levels. The medical schools studied indicate that such synergies often occur when an extrinsic authority (i.e., official status and position, capacity of the coordinating office, and discretionary funding) can conflate well with an intrinsic authority (i.e., scientific leadership in the global health area, collegial recognition, strong management skills, and organizational creativity). All these elements can be considered as important measurements for assessing institutional capacities for collaborative strategies.

The locus of leadership in intraorganizational and interorganizational collaborative arrangements is linked to the base of financial responsibility. With multimillion dollar projects becoming a necessity to address global health challenges, the ability to sponsor large-scale projects and to solicit and effectively manage and account for big money become equally important. The locus of leadership that is linked to financial responsibility emerges as an undeniable reality. Few individual entrepreneurs or institutional champions can manage such responsibilities without institutional flexibility in accommodating their internal and external partnerships and arranging new administrative and budgeting solutions to support the entrepreneurs' work. Effective interplay between individual and institutional leadership is crucial when it comes to offering an effective response to global health challenges.

Resource contexts have an impact on how collaborative leadership for global health research is developed. The behavior and funding patterns of governments, research agencies, foundations, and private donors determine the nature of leadership structures. Funding follows the entrepreneurial capacities of individual academic, industry, or government leaders who establish networks, galvanize teams, and arrange institutional frameworks to achieve their specific objectives. International partnerships that are noted as having a significant impact in addressing global health challenges are often a result of institutional flexibility in facilitating leadership capacities combined with the organizational creativity of entrepreneurial faculty members. Overall, institutional flexibility is essential for the evolution of collaborative strategies in global health programming. The malleable intraorganizational arrangements allow for a flexible locus of decision-making, as well as innovative and successful partnerships.

References

Alexander, J. A., Comfort, M. E., Weiner, B. J., and Bogue, R. 2003. Leadership in collaborative community health partnerships. *Nonprofit Management and Leadership* 12:159–175.

Atkinson, T. N., and Gilleland, D. S. 2006. The scope of social responsibility in the university research environment. *Research Management Review* 15:1–8.

Beaglehole, R., Bonita, R., Horton, R., Adams, O., and McKee, M. 2004. Public health in the new era: improving health through collective action. *The Lancet* 363:2084–2086.

Breman, J., and LeDuc, J. 2001. International partnerships in infectious diseases research, training, and control. *Emerging Infectious Diseases* 7:542.

Buse, K., and Walt, G. 2000. Global public-private partnerships: Part II. What are the health issues for global governance? *Bulletin of the World Health Organization* 78:699–709.

Clark, B. R. 1998. *Creating entrepreneurial universities: Organizational pathways of transformation.* Oxford: Pergamon Press for the International Association of Universities.

Clark, B. R. 2001. The entrepreneurial university: New foundations for collegiality, autonomy and achievement. *Higher Education Management* 13:9–24.

Cook, M., Irby, D., Sullivan, W., and Ludmerer, K. 2006. American medical education 100 years after the flexner report. *The New England Journal of Medicine* 355:1339–1344.

Cooper, A., Kirton, J., and Schrecker, T. 2007. *Governing global health: Challenge, response, innovation.* Farnham: Ashgate Publishing Ltd.

Curran, J., Debas, H., Arya, M., Kelley, P., Knobler, S., and Pray, L., eds. 2005. *Scaling up treatment for the global AIDS pandemic: Challenges and opportunities.* Washington, DC: Institute of Medicine, National Academies Press.

Foege, W. H., Daulaire, N., Black, R., and Pearson, C. 2005. *Global health leadership and management.* San Francisco: Jossey-Bass.

Ganem, D. 2003. Bridge building between medicine and basic science: The role of the physician–scientist. Presented at the Institute of Medicine Workshop on Ensuring an Infectious Disease Workforce: Education and Training Needs for the 21st Century. Washington, DC. Institute of Medicine Forum on Microbial Threats.

Garrett, L. 2007. The challenge of global health. *Foreign Affairs* 86:14–38.

Health Canada. 2001. *Social accountability: A view for Canadian medical schools*. Ottawa: Health Canada.

Jones, G. A., and Oleksiyenko, A. 2008. Exploring the inter-relationships among federal, provincial, and institutional policy in the internationalization of Canadian University Research. *Proceedings of the International Workshop on The Internationalization of Higher Education and Higher Education Reforms*. Institute for Higher Education, East China Normal University, Shanghai, PRC, November 4–6, 2008.

Lasker, R., and Weiss, E. 2003. Broadening participation in community problem solving: A multidisciplinary model to support collaborative practice and research. *Journal of Urban Health: Bulletin of the New York Academy of Medicine* 80:14–60.

Lasker, R., Weiss, E., and Miller, R. 2001. Partnership synergy: A practical framework for studying and strengthening the collaborative advantage. *The Milbank Quarterly* 79:179–205.

Levin, R. 2004. *Millions saved: Proven successes in global health*. Washington, DC: Peterson Institute.

Mallon, W. 2006. The benefits and challenges of research centres & institutes in academic medicine: Findings from six universities and their medical schools. *Academic Medicine* 81:502–512.

McKee, M., Garner, P., and Scott, R. 2001. *International cooperation in health*. Oxford, UK: Oxford University Press.

Merson, M., Black, R., and Mills, A. 2004. *International public health: Diseases, programs, systems, and policies*. Sudbury, MA: Jones and Bartlett Publishers.

Miles, M., and Huberman, A. 1994. *Qualitative data analysis: An expanded sourcebook*. Thousand Oaks, CA: Sage Publications.

Mintzberg, H., and Rose, J. 2003. Strategic management upside down: Tracking strategies at McGill University from 1829 to 1980. *Canadian Journal of Administrative Sciences* 20:270–290.

Oleksiyenko, A. 2008. Medicine: International leadership in health and education. *Global portfolios and strategic international partnerships of a major research university*. Ph.D. Dissertation. University of Toronto.

Oliver, A. 2004. Biotechnology entrepreneurial scientists and their collaborations. *Research Policy* 33:583–597.

Pfeffer, J., and Salancik, G. 1978. *The external control of organizations: A resource dependence perspective*. New York: Harper and Row.

Rigby, J., and Edler, J. 2005. Peering inside research networks: Some observations on the effect of the intensity of collaboration on the variability of research quality. *Research Policy* 34:784–794.

Sá, C., and Oleksiyenko, A. 2008. Excellence beyond borders: Organized research units and international collaborations in health sciences. *Proceedings of the 21st Annual Conference of the Consortium of Higher Education Researchers (CHER): Excellence and diversity in higher education. Meanings, goals, and instruments*. September 11–13, 2008, Pavia, Italy.

Taylor, J. 2004. Toward a strategy of internationalization: Lessons and practice from four universities. *Journal of Studies in International Education* 8:149–171.

Taylor, L. K. 2005. Academic development as institutional leadership: An interplay of person, role, strategy, and institution. *International Journal for Academic Development* 10:31–46.

Umble, K., Steffen, D., Porter, J., Miller, D., Hummer-McLaughlin, K., Lowman, M., and Zelt, S. 2005. The National Public Health Leadership Institute: Evaluation of a team-based approach to developing collaborative public health leaders. *American Journal of Public Health* 95:641–644.

Conclusion: Future of Regional Collaboration

Jason D. Rivera and DeMond S. Miller

In this book, contributors discuss various methods employed by different governments, communities, public–private organizations, and nonformal community collaborations seeking to respond to and help communities and regions recover from devastating events. Each chapter reaffirms the notion that disasters often stem from inequity and exclusionary policies that produce adverse impacts on the economy and infrastructure of a country or region in the aftermath of disaster [Rivera and Miller 2010]. This is not to say that government structures are neglecting their fiduciary responsibility of protection; however, these trends point to the consistent inability of many nations, especially in the developing world, to stay in pace with equitable service provision and unplanned population growth [Guo-hua et al. 2009]. Interestingly enough, although much of what is categorized as the Third World is considered to be the most socially vulnerable to disaster events, communities and cultures have evolved within these countries and have been able to bounce back, albeit at varying rates, from catastrophes in the absence of national or international intervention.

> For example, many rural communities in Africa are vulnerable to food shortages because of drought, but they are able to prevent full-blown famine by employing a variety of coping mechanisms that allow them to ride out the hunger until the next harvest. . . . In Zambesia province of Mozambique, the main coping strategy of vulnerable groups to deal with food shortage and drought is to change the diet and reduce food intake. They change the diet from staple crop to fruits, wild fruits, locally grown vegetables, etc. [Medury 2008, p. 348].

This is not to say that these communities could survive indefinitely in the presence of extreme adverse conditions—to do so would require a more profound cultural change; however, the example illustrates the need for communities at all levels to be adaptive to changes in the physical landscape to successfully survive, prevent, and respond to catastrophic events in the future [Miller and Rivera 2008].

When vulnerability to disasters is discussed by politicians and other policy-makers, there tends to be an emphasis on the vulnerability of populations that are socioeconomically less affluent. At the national level, discussion regularly revolves around the vulnerability of marginalized minority groups or other groups within a society that are politically weak in relation to other groups within the society. At the international level, discussions relate to the vulnerability of the Third and Second World nations to cope with extreme natural phenomena, or to rebound from technological disasters. However, what is usually overlooked is the idea that, regardless of the level of income, every nation in the world is more vulnerable than ever to disaster events due to the systematic special and temporal expansion of environmental problems [Beck 1992], as discussed in Chapter 5.

To a certain extent, the most industrialized nations are even more at risk from a disastrous situation occurring due to the invisible and unforeseeable consequences that are created by advanced technological development [Beck 1999; Griner 2002; Jarvis 2007]. According to McGranahan et al. [2007], 10% of the world's population lives within low-elevation coastal zones—areas that are less than 10 m above sea level. On a global scale, this may not seem like a high statistic; however, due to variations in topology across geographic regions and patterns of urban development, susceptibility to flooding varies from nation to nation. For example, in Bangladesh and Vietnam, approximately 46% and 55% of these countries' populations, respectively, live within these high-risk zones [Beatley 2009]. Moreover, when one observes the amount of geographic area at risk to flooding in square miles, the United States has the third-largest amount of land below this 10 m level [McGranahan et al. 2007]. In the United States, 50% of the nation's population resides in coastal areas with some of the highest urban densities, resulting in the physical vulnerability of millions [Crossett et al. 2004]. The notion that vulnerability to disastrous phenomena is a local concern is a moot argument for the twenty-first century, and steps must be taken to further foster regional and international relationships to mitigate physical and economic disaster affects.

Severity of the Problem

Much of humanity's physical vulnerability to natural disasters is significantly influenced by the decisions that individuals make and the various planning and development decisions made by local governments. According to Beatley [2009], both sets of decisions are commonly made without regard for long-term exposure and vulnerability to hazards. For example, in Latin America a great deal of urban expansion has occurred in floodplains, mountain slopes, or in other places ill-suited for settlement that are affected by seasonal weather-related risks. Also, centers of urban development may not have their foundations in hazard-prone areas (i.e., initial city centers during the founding of a city may have been strategically situated in a physically safe zone); however, changes in regional planning, demographic shifts, politics, and economic factors have forced internal and sometimes international population shifts that increase low-income immigrant populations into urban centers. According to Hardoy and Pandiella [2009], city authorities in Santa Fe, Argentina, have recognized that in the past 50 years there have not been any official urban land policies created and that people settle where and how they can, prioritizing proximity to work places or social networks. By prioritizing informal settlement around daily social interactions, people are increasing their physical and social vulnerability to disasters.

> In informal settlements, houses are usually built incrementally over a number of years, with materials of diverse origin and quality, and not always following accepted techniques. These houses rarely comply with official safety standards and there are no controls in place. Most buildings are used intensively—with high levels of overcrowding and a mix of living and working spaces. This is often combined with a lack of maintenance and with environmental conditions (e.g., humidity from proximity to river edges and coastal areas) that cause rapid deterioration. Houses are not as solid or as insulated as they should be and are often built on inadequate foundations (many on landfill or unstable land) [Hardoy and Pandiella 2009, p. 211].

In addition to Latin America, there are many places throughout the world, especially in developing nations, plagued with the practice of informal settlement. Bangladesh, which is globally known for its flooding disaster issues [UN/ISDR 2007], continually faces redevelopment concerns associated with informal settlement. As Marshall et al. [2005] explained, disasters tend to reveal the relationships that individuals and communities have with the local biophysical environment for good or bad.

Frequently, human society is faced with disasters that are caused by metrological phenomena directly connected with the global climate; however, less common are disasters, such as earthquakes and volcanic eruptions (see Chapters 11 and 14), that tend to occur independently of changes in global temperatures. Moreover, disasters that are not influenced by natural phenomena also impact urban communities, nations, and even whole geographic regions. Many of these disasters take the form of technological failures (see Chapter 15), but human and socially induced disasters such as wars and other military actions also contribute to the devastation of communities, which have severe impacts on the social vulnerability of populations, thereby making them more vulnerable to natural phenomena.

In reference to devastating natural phenomena, the raw number of people affected and the economic impact of natural disasters has significantly increased

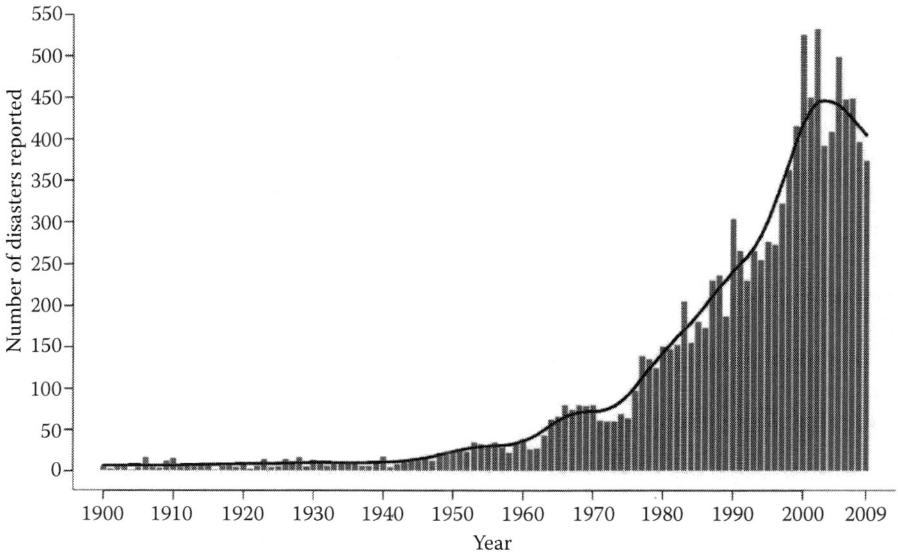

Figure A.1 Natural disasters reported, 1900–2009. (Reproduced from EM–DAT: The OFDA/CRED International Disaster Database, http://www.emdat.be, Université Catholique de Louvain, Brussels, Belgium. With permission.)

over the past century. As is shown in Figure A.1, there has been an observable increase in the number of disasters that have occurred globally between the beginning and the end of the twentieth century. Subsequently, when one observes the number of specific types of disasters that have affected people throughout the world, there are similar increases in the number reported (Figure A.2).

When observing the number of recorded natural disasters, it is interesting to note the significant increases in the number of reported disasters in reference to floods and storms.* These disasters, which may have occurred regularly throughout time, have increased or have been reported more frequently in recent times due to increased urban development in areas with a higher physical vulnerability to catastrophic natural disasters, bringing society into direct contact with these occurrences. However, this explanation excludes the impacts of global climate change, which is also a significant contributing factor to the increase of extreme weather events, such as storms, floods, drought, and possibly epidemics.

* According to EM–DAT: The OFDA/CRED International Disaster Database [2010], in 2009 alone, there were 150 floods and 87 storms reported to occur throughout the world. The increase in the number of these events is illustrated in the number of similar reported events that occurred 30 years earlier in 1979, 34 and 31, respectively.

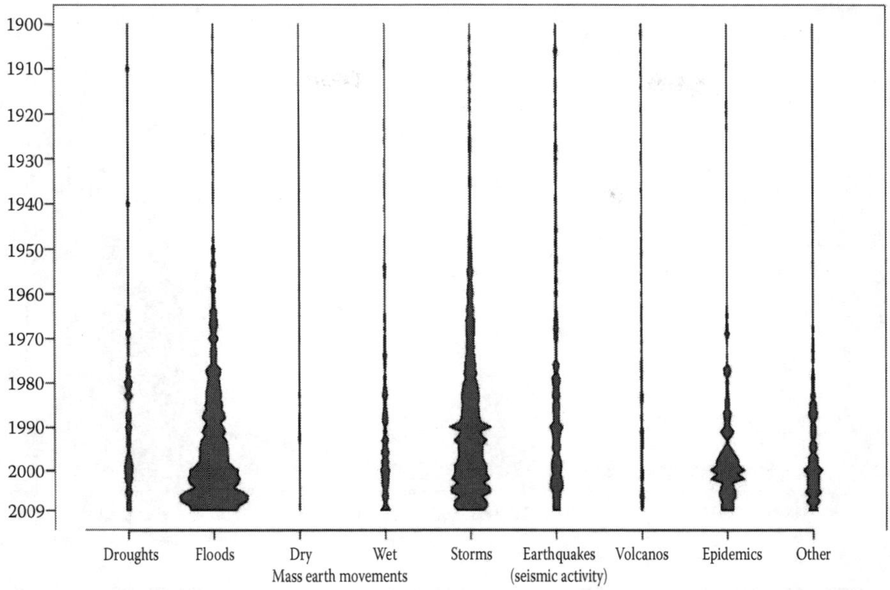

Figure A.2 Number of natural disasters reported, 1900–2009. (Reproduced from EM–DAT: The OFDA/CRED International Disaster Database, http://www.emdat.be, Université Catholique de Louvain, Brussels, Belgium. With permission.)

Observations of technological disasters show similarly disturbing trends on a global scale.* As shown in Figure A.3, the number of technological disasters that have been reported dramatically increased in the last 40 years of the twentieth century. This can not only be attributed to the increased use of new and aging technologies, but also to variances in vulnerability among different populations that have been studied within the field of environmental justice.

Moreover, the frequency of events in different regions of the world seems to illustrate that a variety of places have a tendency toward the occurrence of technological disasters (Figure A.4). Although the specific reasons for the higher rates of technological disasters is beyond the scope of this volume, it is important to understand that the advent of technological disasters is irrespective of a county or region's level of development. The frequency of events may be linked to regions where unregulated rapid industrial growth and environmental policies are weak in the face of multinational profit-generating corporations; however, research on what

* According to EM–DAT: The OFDA/CRED International Disaster Database [2010], in 2009, there were 229 recorded technological disasters throughout the world. The increase in the number of these events is illustrated in the number of similar reported events that occurred thirty years earlier in 1979.

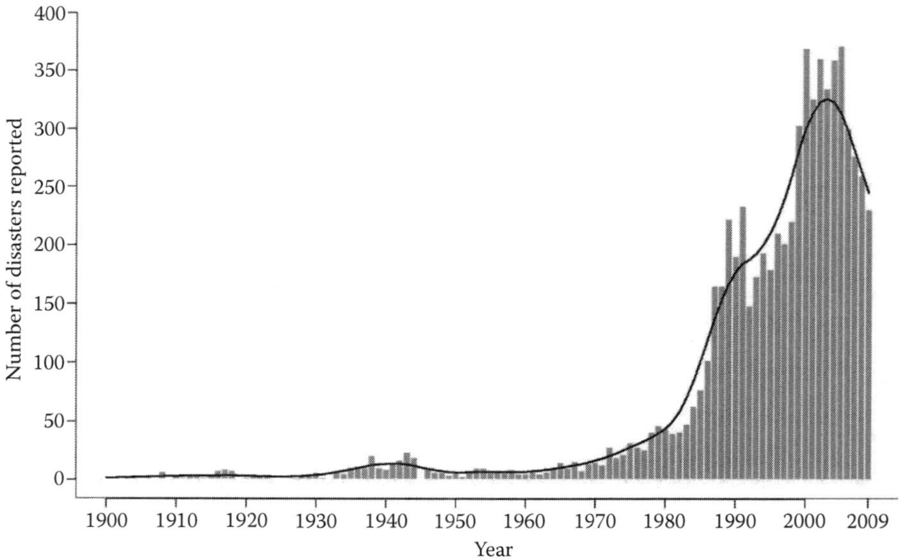

Figure A.3 Technological disasters reported, 1900–2009. (Reproduced from EM-DAT: The OFDA/CRED International Disaster Database, http://www.emdat.be, Université Catholique de Louvain, Brussels, Belgium. With permission.)

are the leading causes of technological disasters in these regions needs to be more adequately examined.

Although Figures A.1 to A.4 highlight the pattern and frequency with which disasters occur throughout the world, these illustrations do not depict the impacts of economic and social development. Oftentimes, events occur that are not officially construed as disasters. When matters of national improvement and benefit, such as the construction of a dam, requires the displacement of whole communities, occurrences similar to these can be equally as devastating to traditional cultures and local communities as severe natural or technological events [Heming et al. 2002; see also Jacob 2010]. Additionally, human-induced disasters, such as military incursions and other socially constructed events that can often occur over the course of several years and have profound and detrimental cultural, socioeconomic, and political consequences for affected groups, occur on a continual basis throughout the world. These events, which are sometimes overlooked as disasters because they do not tend to be naturally induced,* have had equally, if not more,

* The authors concede that although sometimes military actions may result from pressures placed upon political systems as a result of natural disasters (i.e., the need to gain access to food or water supplies), actual acts of violence are human choices and the disasters that occur as a byproduct of these decisions are therefore human-induced.

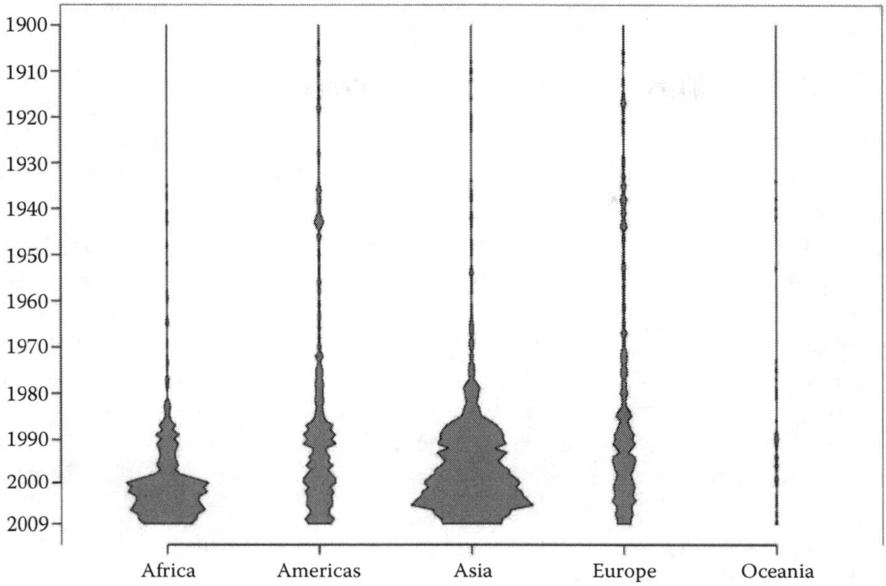

Figure A.4 Number of technological disasters reported, 1900–2009. (Reproduced from EM–DAT: The OFDA/CRED International Disaster Database, http://www.emdat.be, Université Catholique de Louvain, Brussels, Belgium. With permission.)

devastating effects on regions' economies, internal infrastructure, and, sometimes, the survivability of entire human communities. To guard against the detrimental effects of disasters, more collaboration must occur between different nations and, where possible, throughout different regions of the world.

Transnational-Regional Disaster Management

According to Miller and Rivera [2008], an understanding of the physical landscape that communities and nations inhabit is paramount in reducing a population's level of vulnerability to disasters. By understanding the physical environmental dynamics of the location in which people live, which encompasses a place's interactional past and interactional potential [Milligan 1998], the general public can begin to internalize its own specific responsibilities when it comes to mitigating, responding to, and recovering from disasters. The process of understanding these dynamics is an unconscious one, which inevitably occurs as an individual, community, or entire nation resides in a specific location. Over time, these social groups experience the workings of the surrounding environment and contribute the natural phenomena, such as natural disasters, to memory [Casey 1987], sometimes learning from it and adapting development patterns and sometimes ignoring it. What is sometimes lost in collective memory is the occurrence of past traumatic experiences, such as disasters, that have the potential to either be forgotten or reinterpreted [Katovich and Couch 1992; Katovich and Hintz 1997; Milligan 1998; Zerubavel 1996; Miller and Rivera 2008].

One of the largest obstacles that impinge governments from proactively mitigating natural disasters—and sometimes technological ones as well—on an international scale is the notion that to acknowledge that a society is vulnerable is to look weak in the metaphorical eyes of the rest of the world. This notion is supported by the public's view that their vulnerability to disasters is the responsibility of the government and that disaster preparedness should not be a priority in their daily lives [Pelling 2003]. Moreover, most national leaders have illustrated a decided preference for dealing with our national and international vulnerabilities behind closed doors [Flynn 2007].

Along these lines, Swartz and Campbell [2010] and Kaufmann [2004] maintain that the government uses practices of secrecy and monopoly on intelligence resources that distort the public's perception by publicizing favorable analyses and neglecting to publicize contrary information. By adhering to practices such as these, the public does not pressure government authorities to mitigate potential disaster events because, for the most part, they are kept ignorant of the eventuality of catastrophic occurrences. This is not to say that governments should continually emphasize the potentiality of destruction to the point where it demoralizes the population; however, populations should be better educated on the eventuality of disasters so that they are more inclined to support disaster mitigation projects when they are proposed before their respective government officials.

Educating entire populations in reference to any subject can be difficult. Various venues to educating the public have been proposed and implemented, such as posting educational documents and facts on government Web sites, introducing disaster studies into higher education curriculum [Rivera and Miller 2009], and even integrating disaster education into children's books. According to Dunlap [2008], formal and informal education at the elementary to university levels has continued to integrate an ecological world view in people since the 1970s. Although these methods do have the potential to succeed in educating a broader public on their vulnerability to disasters, there are gaps as to who can access some of these materials. In reference to integrating disaster studies and factual knowledge onto Internet Web sites and into college curriculum, these methods only reach specific demographic groups that are either able to afford a computer, and subsequently Internet service, and those individuals with the socioeconomic standing to attend institutions of higher education. In addition, placing regionally specific disaster-related material in children's literature requires both individuals and schools to actively search for these materials if they want to expose the youth to these subjects.

A more practical and cost-effective approach to educating the public that reaches an audience that is comprised of every socioeconomic standing, in addition to communities that are relatively isolated from the rest of the nation, let alone the world, is being employed in Afghanistan. In Afghanistan, Tearfund and the BBC World Service Trust began a program in 2006 that integrates disaster risk reduction messages into the story lines of the radio show "New Home, New Life" (NHNL). Because 80% of Afghans have radio sets in their home and between 60% and 68% of them listen to NHNL already, the BBC believed it was a good method of reaching a large portion of the population. Stories for the program are developed from real-life community disaster reduction experiences and packaged to the public in an entertaining way that meets the needs of the people of the region.

> One community recalled when forests of tamarind trees surrounding their village had been cleared so that the trees could be sold, and this resulted in heavy destruction caused by flood. After hearing on NHNL, a few years ago, that destroying forests could lead to destructive flood impact on farms and houses, the community established a local council and decided to fine anyone who cuts trees. The forest trees have now been growing for 4–5 years and the community is happy with the benefits from them: no flood impact on houses, less dust, good pastures for animals, etc. [UN/ISDR 2007, p. 4].

According to the UN/ISDR [2007], the potential for replicating or augmenting these programs to reach other nations or regions of the world is promising. One obstacle to educating people has been the isolation of rural communities. If methods similar to those being used in Afghanistan are augmented to engage the public through a region's more popular media outlets, governments at all levels will be in better positions to interact with disaster survivors. In theory, after learning more, citizens will be more educated on what to do in times of crisis. Moreover, through public education, local disaster mitigation efforts will lead to a decrease in the potential for certain human tragedies and loss of life to occur.

To advance serious political discussion regarding disaster mitigation, response, and recovery at the regional, national, and international levels, the public's knowledge and understanding of both the environment and the risks associated with developing in certain areas must be broadened. As Dunlap [2008] argued, society is in the midst of a paradigm war in which both sides are attempting to give significantly different interpretations of ecological realities. In Inglehart's [1995] opinion, this metaphorical war has resulted in society shifting its values and goals toward quality of life as opposed to economic growth and consumption. Changes in society's social, political, and economic values have direct impacts on the transformation of lifestyles, fertility rates, and other issues that have significant impacts on the priorities people give to environmental issues. In the case of population and demographic shifts, these social changes have the potential to impact urban development patterns that directly influence humanity's exposure to disaster situations.

Historically, society has been in favor of environmental protection and disaster risk-reducing policies; however, the public seems to alter this position in the face of economic hardships. Developing public policy that incurs massive economic costs, often associated with disaster mitigation and resilience techniques, poses a political risk in democratic societies. In Cuba,* which has an authoritative government structure, disaster management policies are more easily developed and implemented. This is not to say that whole-scale disaster policy development and implementation can only occur in authoritative government structures; however, for it to occur more widely within democratic governments, additional social change and political will that places more value on beneficial interactions between humanity and the natural environment are required.

In an effort to stimulate progressive social change that proactively pursues efforts to mitigate or deal with global climate changes regional cooperation must be at the forefront of the conversation. Just as with the formation of regional governments or cooperatives at the local level, construction of regional cooperation or disaster management agreements have the potential of promoting economic development, reducing inequality, and addressing social, economic, and environmental externalities [Basolo 2003; Lowery 2000, 2001]. In specific reference to disaster management, issues of national sovereignty preclude, in many situations, responses to occurrences that overlap different national jurisdictions [Downs 1994; Katz 2000; Warner and Hefetz 2002] or borders. Therefore, according to Feiock [2007], governments are unable to cope with spillover problems that result from the choices, or maybe the lack thereof, that are made in neighboring states or nations. Although Feiock [2007] referred to issues associated with smaller local governmental units, the same can be said of the risks posed to neighboring nations when disaster effects spill over into other countries, such as the creation of migratory refugee populations or the depletion/destruction of valuable resources, which can have severe economic, social, and political outcomes for all countries within the affected region.

* According to the United Nations' Economic Commission for Latin America and the Caribbean (UN/ECLAC) [2010], Cuba employs the best practices to date in all of Latin America, in addition to other countries within the western hemisphere, in reference to disaster preparedness, resilience building, and climate change adaption. Although the UN/ECLAC does not specifically state why this country is more efficient in implementing their whole-scale disaster risk reduction programs, it can be theorized that the country's political structure allows for a top-down model of decision-making to occur that places policies in motion while other governments would resort to democratic debate.

Where Do We Go From Here: from Regional Risk to Regional Resiliency

Systemic risks* have the potential to economically, socially, and politically destabilize developed countries just as much as developing ones. Although this has been the case for many years, current climate patterns in conjunction with human development and the increased frequency of natural disasters makes these risks ever more important to address. One issue that tends to confront organizations or governments when attempting to set up international and regional mechanisms for disaster management is the way in which different states, nations, or even populations view potential risks. At the micro-level, specific populations of people may not perceive natural or technological disasters as high-risk scenarios because their day-to-day lives bring them in contact with superseding risks to their physical safety that warrant more attention than situations that may not ever occur [Nathan 2008]. On a macro-level, Sarewitz [2004] argues that the emphasis on scientific studies and reports hinders environmental decision-making and risk assessment because stakeholders (i.e., the general public) have a difficult time understanding the information. This results in fewer stakeholders participating in decision-making processes, reducing the favorability of political decisions made by representatives, especially in circumstances that require large monetary investment. To overcome this situation, Stirling [2002] recommends that more stakeholders be brought into the decision-making process. By adhering to these explanations, national governments or appointed national committees should be brought together within a geographic region to discuss the risks that each and all of them have the potential of experiencing. While meeting under this premise, these representative groups can develop cooperative agreements that address issues of national sovereignty that have frustrated international humanitarian efforts in the past [Lai et al. 2009].

Government relationships that are based on issue-specific circumstances usually have the tendency of disintegrating over time if the similar inclusionary circumstances do not present themselves to the stakeholders [Rivera et al. 2008]. Therefore, regional governmental bodies should be specifically designed to address inter jurisdictional matters of disaster management so that all parties are included in the processes of mitigation, response, and subsequent recovery from catastrophic experiences. According to Strand [2004], governance at the regional level has three potential advantages: involvement of a small number of actors, provision of opportune linkages, and enhancement of connections between national and global governance systems. By reducing the number of actors (or in the case of disaster management the number of international organizations or national governments) that participates in a disaster management situation, the prospects

* According to Deere-Birkbeck [2009, p. 1177], systemic risks are "risks embedded in a larger web of social and economic consequences, [that] are not confined to national borders and defy management by a single sector."

for cooperation and efficiency increase [Axelrod and Keohane 1986; Betsill 2007; Oye 1986]. Regional governmental bodies also provide the potential for developing linkages between disaster management issues and other shared interests among the member states more easily than single-issue treaties [Betsill 2007]. According to Betsill [2007], regional governmental bodies have the ability to address multiple issues simultaneously that encompass a wider range of concerns. Finally, regional organizations have the ability to develop global and national governing systems by enhancing the capacity of national governments to implement global commitments and standardize national activities and positions [Hoballah 2004; Lian and Robinson 2002], which theoretically would make responding to disaster situations more efficient.

The importance of making disaster management more efficient cannot be overstated. Current international relation trends emphasize the importance of a nation's stability to help avoid international security threats. According to Carment et al. [2007], good governance is related to security, not only at the local level but at the state, regional, and global levels as well. Economic and social threats, including poverty, disease, environmental degradation, internal conflicts, weapons of mass destruction proliferation, terrorism, and transnational organized crime have the potential of degrading not only local and state political and social relationships, but also economic, social, and political international relationships [United Nations 2004]. In the aftermath of disasters that occur over international boundaries, even humanitarian responses that provide resources to government agencies in the name of strengthening institutions have the ability to create resentment and armed conflict within the disaster-stricken area or region, jeopardizing peace and the recovery process [Call 2008a, 2008b]. Moreover, the development of these situations contributes to the formation of mass corrosive communities in a region that leads to its destabilization.

Picou et al. [2004, p. 1500; Marshall et al. 2003] stated that there are three significant factors that lead to the development of corrosive communities: the mental and physical health of the victims [Arata et al. 2000; Baum and Fleming 1993; Freudenburg and Jones 1991; Green 1996; Picou and Gill 1996, 2000], perceptions of governmental or organizational failure or recreancy [Couch 1996; Freudenburg 1993, 1997, 2000; Marshall 2004; Marshall et al. 2003], and protracted litigation [Gill and Picou 1991; Marshall et al. 2004; Picou 1996a, 1996b; Picou and Rosebrook 1993]. When disasters occur across international borders but are confined to a geographic region that has subsequent political, economic, or social ramifications, most of these characteristics have the potential of developing. Further contributing to the issue is the bias through which humanitarian aid and recovery operations are conducted in nations throughout the world when Western nations are bearing the charge. According to Levin and Dollar [2005] and Carment et al. [2007], there are darling and orphan nations throughout the world. Darling nations, which are strategically important, receive more aid than one would expect; orphan nations receive comparably smaller amounts of aid due to the fact that they are strategically insignificant. This practice creates animosity among nations that receive and do not receive aid based on their supposed allegiance to Western principals or security.

The picking and choosing game that occurs in reference to the amount of aid granted to different disaster-stricken nations, in combination with the protracted international legal or organizational discussions that occur with regard to allocating the aid, and the short-term and long-term physical and emotional affects felt by victims, can create international corrosive communities that have the potential of taking one disaster and turning it into many subsequent human-induced disasters, which affects not only the nations that are involved but the global community as well. By turning to a regionalist governance or organizational approach to disaster management, we can shift the focus from "risk societies" to "risk populations" [Marshall 1999]. In this way, the globalized human community in which we all operate today can shift from looking at disasters as incidents that destabilize a specific town, state, or nation to disasters that affect or destabilize entire international regions and, to some extent, the political, social, economic, and cultural landscapes of the entire human community.

References

Arata, C. M., Picou, J. S., Johnson, G. D., and McNally, T. S. 2000. Coping with technological disaster: An application of the conservation of resources model to the *Exxon Valdez* oil spill. *Journal of Traumatic Stress* 13:23–39.

Axelrod, R., and Keohane, R. O. 1986. Achieving cooperation under anarchy: Strategies and institutions. In *Cooperation under anarchy*, ed. Kenneth A. Oye, 226–254. Princeton, NJ: Princeton University Press.

Baum, A., and Fleming, I. 1993. Implications of psychological research on stress and technological accidents. *American Psychologist* 48:665–672.

Basolo, V. 2003. U.S. regionalism and rationality. *Urban Studies* 40:447–462.

Beatley, T. 2009. *Planning for costal resilience: Best practices for calamitous times*. Washington, DC: Island Press.

Beck, U. 1992. *Risk society: Towards a new modernity*. London, UK: Sage.

Beck, U. 1999. *World risk society*. Cambridge, UK: Polity Press.

Betsill, M. M. 2007. Regional governance of global climate change: The North American Commission for Environmental Cooperation. *Global Environmental Politics* 7(2): 11–27.

Call, C. T. 2008a. The fallacy of the 'failed state.' *Third World Quarterly* 29:1491–1507.

Call, C. T. 2008b. Conclusion. In *Building states to build peace*, eds. C. T. Call and V. Wyeth. Boulder, CO: Lynne Rienner.

Carment, D., Gazo, J. J., and Prest, S. 2007. Risk assessment and state failure. *Global Society* 21(1):47–69.

Casey, E. 1987. *Remembering: A phenomenological study*. Bloomington, IN: Indiana University Press.

Couch, S. R. 1996. Environmental contamination, community transformation and the Centralia mine fire. In *The long road to recovery: Community response to industrial disaster*, ed. J. K. Mitchell, 60–84. New York: United Nations University Press.

Crossett, K. M., Culliton, T. J., Wiley, P. C., and Goodspeed, T. R. 2004. *Population trends along the coastal United States: 1980–2008*. Washington, DC: NOAA National Ocean Service. Retrieved from http://oceanservice.noaa.gov/programs/mb/pdfs/coastal_pop_trends_complete.pdf (accessed July 28, 2010).

Deere-Birkbeck, C. 2009. Global governance in the context of climate change: The challenges of increasingly complex risk parameters. *International Affairs* 85:1173–1194.

Downs, A. 1994. *New visions of metropolitan America*. Washington, DC: The Brookings Institute.

Dunlap, R. 2008. The Hew Environmental Paradigm Scale: From marginality to worldwide use. *The Journal of Environmental Education* 40:3–18.

EM-DAT: The OFDA/CRED International Disaster Database. 2010. Université Catholique de Louvain. Retrieved from http://www.emdat.be/advanced-search (accessed August 25, 2010).

Feiock, R. C. 2007. Rational choice and regional governance. *Journal of Urban Affairs* 29:47–63.

Flynn, S. 2007. *The edge of disaster: Rebuilding a resilient nation*. New York: Random House.

Freudenburg, W. R. 1993. Risk and recreancy: Weber, the division of labor, and the rationality of risk perceptions. *Social Forces* 71:909–932.

Freudenburg, W. R. 1997. Contamination, corrosion and the social order: An overview. *Current Sociology* 45:19–40.

Freudenburg, W. R. 2000. The 'risk society' reconsidered: Recreancy, the division of labor, and risk to the social fabric. In *Risk in the modern age: Social theory, science and environmental decision-making*, ed. M. J. Cohen, 107–120. New York, NY: Palgrave.

Freudenburg, W. R., and Jones, T. J. 1991. Attitudes and stress in the presence of technological risk: A test of the Supreme Court hypothesis. *Social Forces* 69:1143–1168.

Gill, D. A., and Picou, J. S. 1991. The social psychological impacts of technological accident: Collective stress and perceived health risks. *Journal of Hazardous Materials* 27:77–89.

Green, B. L. 1996. Traumatic stress and disaster: Mental health effects and factors influencing adaptation. In *International review of psychiatry*, eds. F. L. Mark and C. C. Nadelson, 177–210. Arlington, VA: American Psychiatric Press.

Griner, S. 2002. Living in a world risk society: A reply to Mikkel V. Rasmussen. *Millennium: Journal of International Studies* 31:149–160.

Guo-hua, C., Tao, L., and Hua-wen, Z. 2009. Study on the methodology for evaluating urban and regional disaster carrying capacity and its application. *Safety Science* 47:50–58.

Hardoy, J., and Pandiella, G. 2009. Urban poverty and vulnerability to climate change in Latin America. *Environment and Urbanization* 21:203–224.

Hoballah, A. 2004. *Stakes of environmental governance at regional level: The Mediterranean case*. Paris, France: Institut du Developpement Durable et des Relations Internationals.

Heming, L., Waley, P., and Rees, P. 2002. Reservoir resettlement in China: Past experience and the Three Gorges Dam. *Geographical Journal* 167:195–212.

Inglehart, R. 1995. Public support for environmental protection: Objective problems and subjective values in 43 societies. *PS: Political Science and Politics* 28:57–72.

Jacob, M. 2010. Ethnography, memory, and culture: Healing the soul wound of technological disaster. In *How ethnically marginalized Americans cope with catastrophic disasters: Studies in suffering and resiliency*, eds. J. D. Rivera and D. S. Miller, 37–49. Lewiston, NY: The Edwin Mellen Press.

Jarvis, D. S. L. 2007. Risk, globalization and the state: A critical appraisal of Ulrich Beck and the World Risk Society thesis. *Global Society* 21(1):23–46.

Katovich, M. A., and Couch, C. J. 1992. The nature of social pasts and their use as foundations for situated action. *Symbolic Interaction* 15:25–47.

Katovich, M. A., and Hintz, R. A., Jr. 1997. Responding to a traumatic event: Restoring shared pasts within a small community. *Symbolic Interaction* 20:275–290.

Katz, B. 2000. *Reflections on regionalism*. Washington, DC: The Brookings Institute.

Kaufmann, C. 2004. Threat inflation and the failure of the marketplace of ideas: The selling of the Iraq war. *International Security* 29:5–48.

Lai, A. Y., He, J. A., Tan, T. B., and Phua, K. H. 2009. A proposed ASEAN Disaster Response, Training and Logistic Centre enhancing regional governance in disaster management. *Transit Studies Review* 16:299–315.

Levin, V., and Dollar, D. 2005. The forgotten states: Aid volumes and volatility in difficult partnerships countries. Retrieved from http://siteresources.worldbank.org/INTLICUS/Resources/TheForgottenStates.pdf (accessed August 19, 2010).

Lian, K. K., and Robinson, N. A. 2002. Regional environmental governance: Examining the Association of Southeast Asian Nations (ASEAN) Model. In *Global environmental governance: Options and opportunities*, New Haven, CT: Yale School of Forestry and Environmental Studies.

Lowery, D. 2000. A transaction costs model of metropolitan governance: Allocation vs. redistribution in urban America. *Journal of Political Administration Research and Theory* 10:49–78.

Lowery, D. 2001. Metropolitan governance structures from a neoprogressive perspective. *Swiss Political Science Review* 7:130–136.

Marshall, B. K. 1999. Globalisation, environmental degradation and Ulrich Beck's risk society. *Environmental Values* 8:253–275.

Marshall, B. K. 2004. Gender, race, and perceived environmental risk: The 'White Male Effect' in Cancer Alley, LA. *Sociological Spectrum* 24:453–478.

Marshall, B. K., Picou, J. S., and Bevc, C. A. 2005. Ecological disaster as contextual transformation: Environmental values in a renewable resource community. *Environment and Behavior* 37:706–728.

Marshall, B. K., Picou, J. S., and Gill, D. 2003. Terrorism as disaster: Selected commonalities and long-term recovery for 9/11 survivors. *Research in Social Problems and Public Policy* 11:73–96.

Marshall, B. K., Picou, J. S., and Schlichtmann, J. 2004. Technological disasters, litigation stress and the use of alternative dispute resolution mechanisms. *Law & Policy* 26:289–307.

McGranahan, G., Balk, D., and Anderson, B. 2007. The rising tide: Assessing the risks of climate change and human settlements in low elevation coastal zones. *Environment and Urbanization* 19:17–37.

Medury, U. 2008. Toward disaster resilient communities: A new approach for South Asia and Africa. In *Disaster management handbook,* ed. Jack Pinkowski, 337–354. Boca Raton, FL: Auerbach Publications.

Miller, D. S., and Rivera, J. D. 2008. *Hurricane Katrina and the redefinition of landscape*. New York, NY: Rowman & Littlefield Publishers, Inc.

Milligan, M. J. 1998. Interactional past and potential: The social construction of place attachment. *Symbolic Interaction* 21(1):1–33.

Nathan, F. 2008. Risk perception, risk managements and vulnerability in the hill slopes in the city of La Paz, Bolivia. A preliminary statement. *Disasters* 32(3):337–357.

Oye, K. A. 1986. Explaining cooperation under anarchy: Hypotheses and strategies. In *Cooperation under anarchy*, 1–24. Princeton, NJ: Princeton University Press.

Picou, J. S. 1996a. Compelled disclosure of scholarly research: Some comments on 'high stakes' litigation. *Law and Contemporary Problems* 3:149–157.

Picou, J. S. 1996b. Toxins in the environment, damage to the community: Sociology and the Toxic Tort. In *Witnessing for sociology: Sociologists in court*, eds. Pamela Jenkins and Steve Kroll-Smith, 210–223. Westport, CT: Praeger Publishers.

Picou, J. S., and Gill, D. A. 1996. The *Exxon Valdez* oil spill and chronic psychological stress. In *Proceedings of the Exxon Valdez Oil Spill Symposium*, eds. Stanley D. Rice, Robert B. Spies, Douglas A. Wolfe, and Bruce A. Wright, 879–893. American Fisheries Society Symposium.

Picou, J. S., and Gill, D. A. 2000. The *Exxon Valdez* disaster as localized environmental catastrophe: Dissimilarities to risk society theory. In *Risk in the modern age: Social theory, science, and environmental decision-making*, 143–170. New York: Palgrave.

Picou, J. S., Marshall, B. K., and Gill, D. A. 2004. Disaster, litigation, and the corrosive community. *Social Forces* 82(4):1497–1526.

Picou, J. S., and Rosebrook, D. R. 1993. Technological accident, community class action litigation and scientific damage assessment: A case study of court-ordered research. *Sociological Spectrum* 13:117–138.

Pelling, M. 2003. *The vulnerability of cities: Natural disasters and social resilience*. Sterling, VA: Earthscan Publications Ltd.

Rivera, J. D., and Miller, D. S. (2010). From recovery to resilience: Long-lasting social change and disaster mitigation. In *Community disaster recovery and resilience: Exploring global opportunities and challenges*, eds. D. S. Miller and J. D. Rivera, 563–572. Boca Raton, FL: Auerbach Publications.

Rivera, J. D., and Miller, D. S. 2009. Environmental education with a focus of risk. *Journal of Applied Security Research: Prevention and Response in Asset Protection, Terrorism and Violence* 1:60–67.

Rivera, J. D., Miller, D. S., and Wright, D. D. 2008. The future effectiveness of racial-political coalitions in American politics. *Journal of Public Management and Social Policy* 14:5–17.

Sarewitz, D. 2004. How science makes environmental controversies worse. *Environmental Science and Policy* 7:385–403.

Stirling, A. 2002. Making better environmental decisions: An alternative to risk assessment. *Environmental Values* 11:518–521.

Strand, J. R. 2004. The case for regional environmental organizations. In *Emerging forces in environmental governance,* eds. N. Kanie and P. M. Hass. Tokyo, Japan: United Nations University Press.

Swartz, O., and Campell, K. 2010. Creative democracy and diversity in the struggle for minority group empowerment and survival. In *How ethnically marginalized Americans cope with catastrophic disasters: Studies in suffering and resiliency*, eds. J. D. Rivera and D. S. Miller, 337–357. Lewiston, NY: The Edwin Mellen Press.

United Nations. 2004. *A more secure world: Our shared responsibility*. Report prepared for the United Nations' High Level Panel on Threats, Challenges, and Change. Retrieved from http://www.un.org/secureworld/ (accessed August 19, 2010).

UN/ECLAC. 2010. *Analysis of extreme events in the Caribbean 1990–2008*. Port of Spain, Trinidad and Tobago: United Nations. Retrieved from http://www.eclac.cl/publicaciones/xml/3/39533/LCARL.254.pdf (accessed August 5, 2010).

UN/ISDR. 2007. *Building disaster resilient communities: Good practices and lessons learned.* Geneva, Switzerland: United Nations. Retrieved from http://www.unisdr.org/eng/about_isdr/isdr-publications/06-ngos-good-practices/ngos-good-practices.pdf (accessed July 28, 2010).

Warner, M. E., & Hefetz, A. 2002. The uneven distribution of market solutions for public goods. *Journal of Urban Affairs* 24:445–459.

Zerubavel, E. 1996. Social memories: Steps to a sociology of the past. *Qualitative Sociology* 19:283–299.

Afterword

This is an important book that addresses the most serious problem facing humankind in the twenty-first century. It is apparent that both natural and technological disasters are increasing at an alarming rate. Over the last 40 years, this global trend has also tragically revealed that human organizational responses to massive catastrophes are woefully inadequate. *Comparative Emergency Management: Examining Global and Regional Responses to Disasters*, edited by DeMond S. Miller and Jason D. Rivera, moves the reader to a new level of discussion of society's obvious limitation for responding to catastrophe. I will briefly discuss problems that loom on the horizon that entail a sober and possibly frightening consideration of our continuing challenges posed by future catastrophic events.

The mindset of prevention in disaster response oftentimes involves interagency practice in anticipation of future catastrophe. The fact that in July, 2004 a mock drill for Hurricane Pam was conducted in New Orleans, Louisiana reveals the limitations of this practice. This scenario was played out in the context of 120 mph winds, the destruction of 500,000 homes, massive rainfall, and involved local, state, federal, and volunteer organizations and agencies. The action plan included accounting for shelter, search and rescue, medical impacts, educational disruption, evacuation, and removal of debris estimates. Ron Castleman, FEMA Regional Director stated: "We made great progress this week in our preparedness efforts" [FEMA 2004]. Such progress was completely absent when Hurricane Katrina slammed into New Orleans, killing over 1,800 people, flooding 80% of the city, and threatening helpless survivors for over a week. Simply put, except for the illegal action taken by the Coast Guard, Cajun Flotilla, and brave residents, there was no organized emergency response. The Hurricane Pam exercise also pales as a meaningless exercise given Katrina's aftermath.

This failure of organized emergency response has been repeated countless times for both natural and technological disasters. A review of response effectiveness to the *Exxon Valdez* oil spill (1989), Hurricane Andrew (1992), the 9/11 Terrorist attacks (2001), the Indian Ocean Tsunami (2004), the Haitian earthquake (2010), and the BP oil spill (2010), among many other catastrophes, validates this claim. Ineffective emergency response creates a social context for the emergence of a series of secondary disasters that prolong community disruption, create family instability and cause serious mental health problems. This "new normal" is not normal and as disaster researchers and emergency response managers we need to enhance timely community recovery and the restoration of supportive social structures, culture, and quality of life for disaster survivors. Everyone would agree to this humanistic contention.

The fact that we need to rethink our orientation to catastrophic disasters is clearly related to the shifting paradigm of how we view the risk of such events in the twenty-first century. The sociologist Ulrich Beck [1999] has expressed this issue as follows:

> The concept of risk reverses the relationship of past, present and future. The past loses its power to determine the present. Its place as the cause of present-day experience and action is taken by the future, that is to say, something non-existent, constructed and fictitious. We are discussing and arguing about something which is <u>not</u> the case, but <u>could</u> happen if we continue to steer the same course as we have been [p. 137].

Certainly emergency management systems learn from both failures and successes. Responding to catastrophe is the science of defining vulnerability, building resiliency, and organizing responses. The excellent chapters in this volume make important positive strides toward this end. We do learn from our achievements and mistakes. But I now ask a sobering question in the context of Beck's shift of our timeframe for the risk of future catastrophe. Maybe we will be unable to prevent and respond to catastrophic events that will occur in the twenty-first century. This is a real possibility. I will briefly discuss one challenge—the possibility of extreme global heating and rapid climate change. There are many others.

Despite the existence of a variety of skeptics that deny climate change and/or global warming, I am thoroughly convinced that this important issue has become politicized to the degree that climate science has been seriously compromised. Global heating, according to the controversial scientist/ecologist James Lovelock, is beyond the control of human intervention. The basis for this claim implies that the scientific predictions of the Intergovernmental Panel on Climate Change (IPCC) have been compromised by politics and the alternative energy movement (wind, solar, biofuels) has been compromised by financial gain and government subsidies. Using sea-level rise as the best available empirical measure of our planet's current pattern of heat absorption, Lovelock [2009, pp. 40–45] convincingly argues that the IPCC's simulated climate model predictions are highly inaccurate. Indeed, over the last decade sea-level rise is occurring 1.6 times faster than predictions issued by IPCC's models and the decline of floating ice in the Arctic ocean is currently at the level predicted for the year 2025 [Lovelock 2009, pp. 41–43].

In short, the hubris of climate science is noted as follows:

> I find it extraordinary that, given the depth of our ignorance, scientists are willing to put their names to predictions of climates up to fifty years

> from now and let them become the basis of policy. Surely they are not predictions, just speculations to assuage the fear of the dark clouds that look on the climate horizon [Lovelock 2009, p. 61].

The consequences for the human community of this emerging bogus climate science are beyond catastrophe. As we conduct business as usual significant social upheaval will occur from this permanent catastrophe. Natural disasters will continue to become more frequent, volatile, and destructive to the point that technological systems will collapse and massive migrations of people will become a global lifestyle. Space limitations preclude detailing Lovelock's Gaia theoretical platform for these predictions and his vision of purposive responses to this permanent state of emergency. Nonetheless, by employing worst case thinking, emergency management should begin to restructure the models of the past and entertain visions of the future [Clarke 2006]. Lovelock's controversial predictions are not without critics and his survival options include nuclear power and a poorly informed sociological view of human behavior. Despite these limitations, the prelude to "starvation, competition for space and resources, and war" involves massive flooding, drought, and elevated instances of powerful storms [Lovelock 2009, p. 31]. All of the above provides the context for twenty-first century emergency management.

One must now ask, "What are our alternatives?" The gloom and doom vision I have all too briefly etched above can be dismissed as fiction and an oversimplified interpretation of James Lovelock's theoretical and empirical accounting of climate change as catastrophe. Such dismissal ignores elaborating alternatives for emergency management and global regional collaboration. Globalization and individualization are two sides of the same coin [Beck 1999]. The critical contribution of the present volume elaborates emerging regional partnerships, networks, cooperatives, and other collaborative systems for responding to disasters. These activities are important and critical for our future. However, if we view globalization not only from above, but from below, insights for effectively responding to the possibility of a global climate catastrophe can be gleaned and the imagination of social change can be used to protect human kind.

If one defines globalization as the integration of regional economics, societies, and cultures through complex systems of communications and emerging systems of interdependence, we document globalization from above [Beck 1999]. However, the concept of individualization reframes our perspective for understanding the structural potential of people for responding to catastrophe in the twenty-first century. As Beck notes, individualization "does not mean individualism" nor "individuation — how to become a unique person" [1999, p. 9]. It means that:

> . . . people are invited to constitute themselves as individuals: to plan, understand, design themselves as individuals and, should they fail, to blame themselves. Individualization thus implies, paradoxically, a collective lifestyle [Beck 1999, p. 9].

The implication of this theoretical insight for emergency management is that responses to megacatastrophes, such as rapid and irreversible climate change, massive toxic contamination, and destructive natural disasters, requires continued education and the empowerment of people who exist in harm's way. Indeed, global and regional collective emergency management response should also be constructed and implemented at the level of individual citizens, thus making every citizen prepared and informed. "First responder" victims must be transformed to active promoters of community recovery and, through collective knowledge and resources, they become active agents that rebuild cultural and social capital [Picou 2000, 2009]. This objective should be integrated into all phases of emergency management. This becomes an imperative for the twenty-first century.

In summary, I have attempted to take the reader on a brief sojourn that involves thinking outside of the box with regard to the nature of an ultimate climate catastrophe and a globalization response from the bottom. The challenges posed by future disasters are real and daunting. We must respond with innovative, collective, and individual agency through institutional leadership. We cannot be limited by past and present procedures and time-honored organizational structures. Indeed, individualization is a potential mitigating solution to modern disaster threats. The future will soon become the present, and our ability to effectively respond relates to both regional cooperation and individual empowerment.

J. Steven Picou
Department of Sociology
The University of South Alabama
Mobile, Alabama

References

Beck, U. 1999. *World risk society*. Cambridge, UK: Polity Press.

Clarke, L. 2006. *Worst cases: Terror and catastrophe in the popular imagination*. Chicago: University of Chicago Press.

FEMA. 2004. Hurricane Pam exercise concludes. Federal Emergency Management Agency News Release. www.fema.gov/news/newsreleases.fema?id=13051. Retrieved September 15, 2010.

Lovelock, J. 2009. *The vanishing face of Gaia*. New York: Basic Books.

Picou, J. S. 2000. The "talking circle" as sociological practice. *Sociological Practice: A Journal of Clinical and Applied Sociology* 2(2):77–97.

Picou, J. S. 2009. Disaster recovery as translational applied sociology: Transforming chronic community distress. *Humboldt Journal of Social Relations* 32(1): 123–157.

Index

Page numbers followed by f and t indicate figures and tables, respectively.

D

E

F

G

O

P

T

9781439804919